Alternative Housebuilding

Alternative Housebuilding

Mike McClintock

Drawings by Richard J. Meyer

Sterling Publishing Co., Inc.   New York

For Inez B. McClintock

Library of Congress Cataloging-in-Publication Data

McClintock, Michael, 1945–
   Alternative housebuilding / by Mike McClintock ; drawings by
Richard J. Meyer.
      p.   cm.
   Reprint. Originally published: New York : Grolier Book Clubs,
1984.
   Bibliography: p.
   Includes index.
   ISBN 0-8069-6995-4
   1. House construction.   2. Building, Wooden.   3. Masonry.
I.   Title.
TH4812.M2374   1989
690′.837—dc19   88-38140
                                                                    CIP

   5 7 10 8 6

Published in 1989 by Sterling Publishing Co., Inc.
387 Park Avenue South, New York, N.Y. 10016
Originally published in hardcover by Grolier Book
Clubs, Inc., copyright © 1984 by Michael McClintock
Distributed in Canada by Sterling Publishing
℅ Canadian Manda Group, P.O. Box 920, Station U
Toronto, Ontario, Canada M8Z 5P9
Distributed in Great Britain and Europe by Cassell PLC
Artillery House, Artillery Row, London SW1P 1RT, England
Distributed in Australia by Capricorn Ltd.
P.O. Box 665, Lane Cove, NSW 2066
Manufactured in the United States of America
All rights reserved

Sterling ISBN 0-8069-6995-4 Paper

Produced by Soderstrom Publishing Group
Book design by Pam Forde Graphics

If one advances confidently in the direction of his dreams,

and endeavors to live the life which he has imagined,

he will meet with a success unexpected in common hours.

Henry David Thoreau, *Walden*

Photo of Thoreau house timber-frame replica, courtesy of Roland Wells Robbins.

CONTENTS

METRIC SYSTEM

UNIT	ABBREVIATION		APPROXIMATE U.S. EQUIVALENT		
Length					
		Number of Metres			
myriametre	mym	10,000	—————— 6.2 miles		
kilometre	km	1000	0.62 mile		
hectometre	hm	100	109.36 yards		
dekametre	dam	10	32.81 feet		
metre	m	1	39.37 inches		
decimetre	dm	0.1	3.94 inches		
centimetre	cm	0.01	0.39 inch		
millimetre	mm	0.001	0.04 inch		
Area					
		Number of Square Metres			
square kilometre	sq km *or* km²	1,000,000	0.3861 square miles		
hectare	ha	10,000	2.47 acres		
are	a	100	119.60 square yards		
centare	ca	1	10.76 square feet		
square centimetre	sq cm *or* cm²	0.0001	0.155 square inch		
Volume					
		Number of Cubic Metres			
dekastere	das	10	13.10 cubic yards		
stere	s	1	1.31 cubic yards		
decistere	ds	0.10	3.53 cubic feet		
cubic centimetre	cu cm *or* cm³ *also* cc	0.000001	0.061 cubic inch		
Capacity					
		Number of Litres	*Cubic*	*Dry*	*Liquid*
kilolitre	kl	1000	1.31 cubic yards		
hectolitre	hl	100	3.53 cubic feet	2.84 bushels	
dekalitre	dal	10	0.35 cubic foot	1.14 pecks	2.64 gallons
litre	l	1	61.02 cubic inches	0.908 quart	1.057 quarts
decilitre	dl	0.10	6.1 cubic inches	0.18 pint	0.21 pint
centilitre	cl	0.01	0.6 cubic inch		0.338 fluidounce
millilitre	ml	0.001	0.06 cubic inch		0.27 fluidram
Mass and Weight					
		Number of Grams			
metric ton	MT *or* t	1,000,000	1.1 tons		
quintal	q	100,000	220.46 pounds		
kilogram	kg	1,000	2.2046 pounds		
hectogram	hg	100	3.527 ounces		
dekagram	dag	10	0.353.ounce		
gram	g *or* gm	1	0.035 ounce		
decigram	dg	0.10	1.543 grains		
centigram	cg	0.01	0.154 grain		
milligram	mg	0.001	0.015 grain		

WHY ALTERNATIVE?

Why is there so much interest in alternative building systems? Aside from the particulars about energy efficiency and low-cost methods for owner-builders, I'm looking at four beautiful reasons as I write this.

I get to work at home most of the time, in an office attached to the back of the house, which sits at the foot of a fairly steep hill. We are the first terrace of a small ravine split by a brook at the bottom.

The office windows face the hillside, and now, with a purple-pink sunset ending a subdued, gray day, four deer are picking their way across the slope, nibbling at new growth on the ground and stretching up for leaves.

Not that I live in the wilderness. This is suburbia—well, pretty rural suburbia, considering I can get to New York City in an hour if I push it. If you have to operate on a shoestring and produce equity with your sweat to glance up from work and watch something like this, it's worth it. Two of the deer are only 15 feet away so I can see their expressions, a sniff, ears a little darker around the edges on one.

If you have to bargain with a building inspector or mortgage banker, do it. If it takes away from your vacation time to attend one of the owner-builder schools, do it. If you must pass up the sports or stock pages, the thriller, the romantic novel, whatever, because there are so many alternative building books to read, do it.

Write, call, get brochures, make sketches and scale models. Make contact with professionals in the field. A lot of them I've talked with are really wonderful people, accessible, willing to talk, to share information. Yes, they make a dollar or two, but money is only part of the picture. These alternative building fields have attracted many people, particularly one-man and small design-plus-construction firms, who really enjoy what they do for a living. It shows. You'll enjoy them as I have.

If you work without professional help you will undoubtedly make some mistakes. Expect them. Learn from them. And press on. Alternative housebuilding won't stump you with high-tech engineering trickery. Alternative materials make sense. They will be familiar and comfortable from the start, even though you may need some time to use them with great skill.

The deer grazing on the wooded slope are beautiful, an elegant reward for going a little out of the way, and going to some trouble to make a place to live. But what stays with me, aside from the vast array of alternative houses on upcoming pages, and their handcrafted details, is the process of designing and building.

You don't often hear from new homeowners about the wonderful experiences of building and buying their split-level, high-ranch, split-ranch development house. It's usually described as a battle: squabbling with the builder, juggling closing dates and escrow accounts, confronting last-minute surprise costs, despairing over the unfinished items on the checklist, discovering all the new work that needs immediate home repair and improvement. The more you hear, the more it sounds like an uncomfortable, often harrowing experience, worth the trip only because in the end you do wind up in a new home.

Many of the alternatives to minimal, trouble-prone housebuilding are not new ideas. That's reassuring. Structural concrete, forms, weather-resistant lime mortar, rammed-earth walls, and many other ideas have histories of trial and error and improvement that span centuries. The alternatives to worry about are the recent and expedient changes such as aluminum wiring, urea-formaldehyde foam insulation, and underbuilt "engineered" frames—all efforts to build it cheap, not necessarily build it better.

Most alternative owner-builders and buyers I talk to have rearranged those priorities. To them, craftsmanship really counts. And they have powerful positive memories of building their home, even when they work with a professional contractor. They have snapshots and stories. Everywhere they look—at the biggest saddle notch in the corner, at the luck-striped stone above the mantel—there is a scene, a summer day, friends who helped, or the private moment of concentration and strength when the 6 × 10 that might have been set by crane got the last boost from their shoulders.

After a while, you do take it for granted. It is just a rock wall, just a log rafter, just a house with a kitchen and a bathroom—at least most of the time. It's just a house until that rafter grabs your eyes as you sit back in the chair, and then you again see the V-wing of geese that made you stop work on the rafter to look up a year or more ago.

Too softhearted and mushy-headed? Hardly. If you're satisfied with match-stick homebuilding, and ready to outshark the home repair and improvement sharks you'll be dealing with, fine. Good luck. But what a difference, what a treat, to take the trouble, build something personal and something special, and enjoy the trip as much as the destination.

MIKE McCLINTOCK

LOG HOUSES

Photo by Don Smith.

Somewhere in the virgin woods of British Columbia or the New England north country, somewhere there is a stand of tall, straight trees waiting for you. Sealed beneath the bark are carefully scribed joints winding gracefully with the grain to seal out the weather, and notched corners, and purlins, and even a one-of-a-kind outrigger rafter with a raccoon's face carved into the end. It's all there, and more, waiting to get out. There is immense structural strength, natural insulation, security, and durability in those trees, and something else—something special that most people don't associate with construction. Serenity.

To pull all this out of the timber you may mix your sweat with the sap and strain your muscles against the grainy wood fibers. You may struggle, but like the trappers and the Pilgrims and the pioneers who used the woods before you, you can succeed. From the moment you start to build, and as long as you use your log home, you will have a kinship with every lean and resourceful pioneer family that hewed a home out of the forest, with every wily and free-spirited trapper who snored peacefully inside his log walls while a blizzard howled outside.

And one night, when dark winter clouds scatter the moonlight and the only noise is the fire popping in your wood stove, when icicles reach from the eaves down into the snowdrifts piled high against your log walls, the courage and determination of all the log

Photo by Neil Soderstrom.

DEVELOPMENT, DESIGN, AND COMPONENTS

builders before you, and the discoveries of all the log builders to follow, will run together in the heartwood of your log home. One night you'll feel it, because when you work with logs, you work with nature and with history; you work with your mind and your hands in the past, the present, and the future.

EARLY LOG HOMES

At most local lumberyards you can order redwood from California, longleaf pine from Georgia, and practically any other building material no matter where it is found or processed. Our transportation capability has many benefits—eating reasonably fresh Florida grapefruit even though you live in Chicago, for instance. But it has tended to make our architecture homogeneous, so that you may see nearly identical houses in suburban developments on the East Coast, the West Coast, and anywhere in between.

Before there were freight cars and semitrailers, the character of building systems and housing styles was inexorably tied to the character of building materials at or near the building site. That's one reason igloos

are made of ice and snow. There are not a lot of options on the barren and frozen tundra. But in central and eastern Europe, where log building flourished, there was an abundant supply of tall, straight conifers in dense evergreen forests.

Unlike the stone structures of antiquity that still attract tourists, the earliest log homes have long since returned to the soil. The very first ones may have been built in Russia. But it is certain that the system of laying horizontal logs on top of each other and interlocking the corners (with many variations from region to region) was used widely in the forested areas of Europe, particularly Scandinavia.

Log homes make so much sense when you think about it. Half-timber homes (massive timber frames with spaces between timbers filled by some combination of mud, masonry, and straw binder) were common in western Europe and, for a while, in the American Colonies. But the Scandinavians and, later, many of the colonists learned that solid wood walls offered more protection in harsh climates. And, just as important, they learned that solid wood provided better insulating value than half-timber walls. That's right: Solid wood (8 to 12 inches of it) is a fairly good insulator.

This is Andrew Jackson's birthplace. (Reprinted courtesy of Dover Publications.)

As you approach this house, built by Pacific Log Homes, from the water, the sweeping roof line gives the impression of a bird about to take off. The large covered porch, with a railing but without full-height posts, is protected with roof decking on two extra rafter pairs, tied to the extended stack of wall logs.

Log construction, an import from Europe that traveled in the minds of early settlers, is also rooted deeply in American history. This is one of the cabins occupied by Lincoln. (Reprinted courtesy of Dover Publications.)

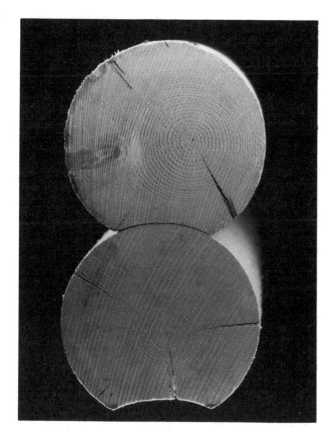

Cope construction is used by a few log-home companies. This corner detail on a Rocky Mountain Log home is even and uniform and machine-made, even though it looks more like a handcrafted job.

Scandinavian log homes added a few unique wrinkles to the system, like large-diameter wall logs hewn into oval shapes to minimize the disparity in thickness between the depth of the log and the depth of the joint. But most of the structural components (and some are pretty sophisticated) we recognize in log homes today can be found in records of construction that are centuries old. To deal with heavy snow loads, for instance, an extra-long log was set at the top of each eave wall so it would cantilever well beyond the projecting corner notches and support additional rafters and substantial roof overhangs (15 to 20 feet on some European chalets and barns) at the gable ends. This was commonsense architecture, a design that enabled a farmer to get at the barn doors without shoveling away 6 feet of snow.

The real beauty of typical early Scandinavian log structures was the chinkless construction. Chinking, like mortar between bricks, is a system used to fill up the irregular spaces between horizontal wall logs, ensuring a complete seal on the cabin shell against wind and water. Take a look at any 10 trees in the forest and it becomes clear why chinking is normally necessary. Trees, like people, come in different shapes and sizes, with peculiar bumps and curves that make standardized design impossible. And this means that builders of chinkless log walls had to be carpenters and sculptors, literally carving the irregularities of the first log into the bottom edge of the second log, at the same time allowing for the depth of half-notches where logs met at the corners. Chinkless construction is fine cabinetwork on a massive scale, accomplished on early log homes in Scandinavia with a concave, saucer-shaped cut (often called cope construction) along the bottom of each log, with the outside edges of the saucer cut scribed to the contours of the log below. The finished product is a stunning combination of high-quality woodwork and wood in a nearly natural state displaying all its unprocessed beauty and warmth.

Another unique feature was developed by Swiss log builders. Although the Swiss were among the first to employ full-diameter log construction, this was relegated to outbuildings, while the post-and-beam, half-timber system common to western Europe was used for residences. But the Swiss refined this system (again, to get more insulation than wattle and daub, the half-timber filler, could provide) and used square-hewn and sawn timbers. The squared timbers also had a small, saucer-shaped channel filled with moss or a similar natural insulator to make an energy-efficient wall.

Both basic log systems, the full-diameter and flat-sawn types, were brought to America in the minds and skilled hands of settlers. Flat-hewn walls were (and still are) built predominantly in the southern United States. Full-log walls with chinkless construction were constructed first in Swedish settlements in Delaware in 1638. Yet many of the first structures of the colonists were true emergency shelters cut into embankments or roughly framed thatched or sod-covered wigwams built over dirt-floor dugouts.

The textbook illustrations of cozy family scenes with everyone in front of the crackling fire, glass-globed oil lamps illuminating the friendly, slightly cluttered cabin interior, do not paint an accurate picture. Before interior chimneys became common in the 1600s, log cabins were smoky, grimy places. And before glass was imported from England (starting about 1640) for extremely small windows, log cabins were dark and dingy. Life in these cabins was not colorfully homespun. It was claustrophobic, dark, damp, and dirty.

DESIGN PRINCIPLES

Today's log buildings are anything but emergency shelters or suitable only for barns and outbuildings. Log structures have been built to serve as stores, restaurants, golf course clubhouses, even airplane hangars. Log homes now have attached greenhouses, skylights, bay windows, sliding glass doors, all the light you want. The *Log Home Guide*, a major periodical in this field, has even reported on a log home with an indoor swimming pool. Have log homes made it into the mainstream of code-approved residential construction? Are they complete with plenty of closet space and all the amenities? Well, how about this modest little cabin: 2,581 square feet of handcrafted space; enter under a massive, peaked overhang into a 22 × 11-foot foyer; to the right, a kitchen, bath, laundry, and a 32 × 14½-foot dining room; straight ahead a sunken 22-foot-square living room; to the left, a 22 × 20 master bedroom with bath and walk-in closet, opening to a 22 × 18 covered atrium; plus bedrooms and a huge workroom on the second floor.

This Pacific Log home has a massive, cantilevered entry overhang and spacious interior rooms.

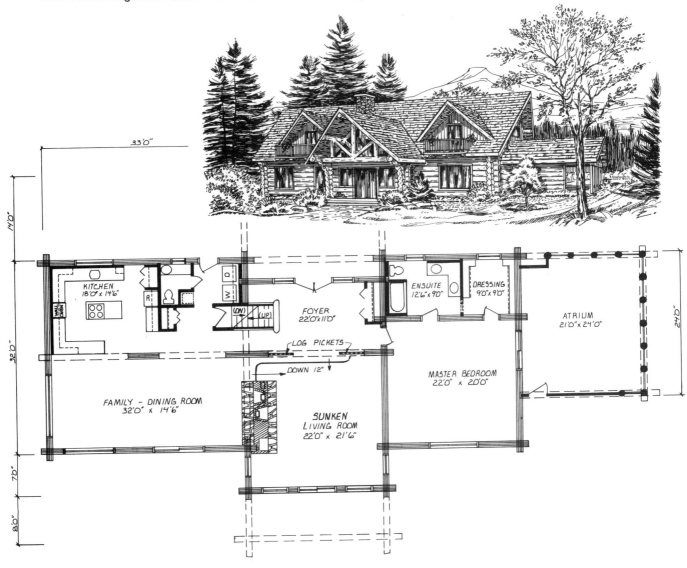

Early log cabins made the most out of what amounted to no more than one small room. One of the materials not available in the forest—glass—made even small windows a luxury. Today's building inspectors no longer approve of slanted poles as chimney supports. (Reprinted courtesy of Dover Publications.)

The log home you're thinking about may not be extravagant, but the possibility is there. Wood is the most workable building material. That's important whether you decide to go to a log-building school and try a handcrafted chinkless cabin from scratch, or to purchase a precut log home that comes on a truck in a lot of big pieces with an assembly manual. Workable wood logs tolerate minor mistakes and changes. Logs can be hewn or sawn, squared on the interior face only or on all faces, cope-cut for chinkless seams, machined at the mill to absolutely uniform dimensions with double tongue-and-groove joints, joined in half-notch or dovetail corners. The design possibilities are almost endless, in pine, oak, pressure-treated timbers guaranteed for 25 years and more.

The countless choices of building materials and construction details are matched by the choices in overall design. There are literally hundreds of companies offering complete plan books that include everything from basic one-room cabins to spacious and elaborate floor plans pushing 4,000 square feet. Many of these firms allow modifications on these plans as well. And, of course, there are handcrafters, individuals and small firms who can custom-build just about any log structure you dream up.

But if the job is to be done from scratch, the design must take into account the site—in particular, the trees on the site. All the normal considerations of site selection still apply: access, availability of water and

utilities, solar orientation, drainage, and so forth. To build a log house from scratch, the site must have the trees you need in a location that is accessible to the site. That can mean accessible by horse and tow harness, by snowmobile and tow chain, by one-man log wagon, even by a system of ropes, pulleys, and winches. But before we get to those details, there are two topics to investigate: basic layouts—and that includes finding the best practical and social solutions to the site, your needs, and your budget—and basic log engineering, so you can understand how to make that layout hang together and stand up.

Even though there are thousands of floor plans available, I urge you not to follow blindly any stock plan, even if it seems just about perfect. With workable logs you can get just the right ceiling height, roof pitch, window placement, and more, because logs are a perfect medium for personal design. You don't have to worry about modular framing and 16-inch centers. Don't panic because you're not an architect. The pioneers weren't architects, and they came up with some very sensible design principles that can still serve well as guidelines.

Remember the obvious: Logs are long and straight. They are big enough and strong enough to make a wall, hold up a roof, and keep out the weather. Logs are just about the opposite of piecework and conventional stick-built homes, where trees are taken apart at the mill and then rearranged at the site to form walls and

Log homes need not be rustic little cabins in the woods with kerosene lamps and drafty rooms and tiny windows. This modest site and design completed by Rocky Mountain Log Homes is an eye opener. Photo below: Inside its cathedral ceilings and unique staircase form a sweeping roof line to surround the water's-edge porch.

partitions and ledgers and nailers and firestops and wind braces and endless other individual parts. Designs that take advantage of log length and strength make a lot more sense than designs calling for irregular walls with a ton of partitions and corners.

Each log is labor-intensive. But each log goes a long way to completing the home. That's one reason why the typical pioneer log cabin was a straight-edged rectangle, about 16 by 20 feet, built on a stone foundation or a simple mudsill, using round or square-hewn logs, often with chinked seams to minimize time-consuming scribing work.

The typical cabin had a chimney at one of the short, gable-end walls and a doorway centering one of the longer eave walls. Nothing too exotic or sophisticated? Let's look further. The roof was sharply sloped (like an 8–in–12 pitch). That's not practical for construction because an 8–in–12 roof is not walkable. You need roof jacks, shingle holders, flexible muscles, and at least one helper. But the steep pitch is the right choice for construction efficiency, so that rafters can be modest in size while supporting a roof requiring a heavy snow load design. The steep pitch is also a good choice for minimizing maintenance. Materials that shed the elements efficiently tend to last longer than materials that hold them. The severe pitch also provides enough room for a full sleeping loft. Among many benefits, that means the precious 320 square feet of floor space is not used inefficiently for beds (probably the largest

On the second story of this Rocky Mountain Log home, round-log gable ends are complemented by massive square-cut rafters, with knee walls faced with knotty pine in board-and-batten herring bone pattern.

pieces of furniture in the house), which occupy a lot of space and are needed only at night. The loft also has social importance, providing at least a little privacy. (Some things haven't changed much. Mom and Dad can still use a little time to themselves with the kids out of the way and off to bed.) The minimal-depth rafters (the steeper the slope, the less timber needed to carry the load) pay extra dividends by maximizing headroom in the loft.

You can find more, but this short look should tell you that these cabin components form a pretty tight design circle. And that is the way I analyze building design. I have the notion that elegant designs standing alone, isolated from their environment, show self-indulgence on the part of the designer. They may be beautiful, but in a way, so what? Think of building design using a circle to represent the whole structure. I want one little arc of the circle (the roof design, for instance) to hook up with and help another part of the circle—in fact, as many parts as possible should work together. This kind of structural and social cross connection (really design integration) makes the circle, and the home, stronger in every way and the home a place where you might be content for a lifetime.

This utterly basic cabin has more design precepts to offer you: Locate the chimney on a short gable wall. Why? Fireplace or wood stove heat is radiated heat, which will warm the largest volume of interior air and be absorbed by the greatest surface area of wood when it emanates from the head of the rectangle—the gable

end. Also, this means an exterior flue has structural support on the gable wall along most of its height. That's where the roof peaks, after all. In the case of an interior flue, the gable-end location allows more exhaust heat to be recaptured in the masonry and re-radiated into the living space. Now put the beds at the center line of the sleeping loft, and you get a location with the most warmth and the most headroom. Also, 2 feet above the ridge line is almost always the best bet for proper draw in the fireplace. When you look closely, the utterly simple cabin design is unquestionably efficient.

Now, how does the thing stand up? Before tackling the tables and formulas in the upcoming pages, be advised that the basic log cabin rectangle (or series of overlapping rectangles) is, by design, exceptionally strong. And unless you have a lot of general construction experience or at least a modest amount of experience building with logs, it is unwise to try something exotic like an octagon or a hybrid design of vertical poles and horizontal logs.

Right-angle, interlocked corners do take time to build, but they are almost impossible to rip apart. In fact, basic-box, solid-log construction is so sound structurally that log construction is free of many engineering concerns common to stick-built homes. Lateral thrust, for instance, acts as long, low-slope rafters push out on long exterior stud walls. Stud walls have great compressive strength (you can put a ton of bricks on top of the double plate) but little lateral strength,

The airy back porch of this Rocky Mountain Log home is supported by concrete piers set below frostline. Here such posts could support a second-story deck or a projecting room.

so that windows and casings, doors and jambs, can be pushed out of plumb.

One engineering factor of log design requires special attention, and that is *settling*. Logs do some settling among themselves, but that action may be moderate and determined more by log species and moisture content—a question of shrinkage. The settling that can cause real trouble is overall settling from the great weight of the solid log walls. This weight varies according to wood species as well as moisture content. In western regions, you might be likely to use white cedar, western hemlock, Douglas fir, or redwood. In eastern regions, you might use poplar, pine, or red oak (yes, red oak is murder to handle when dry, but it's not bad to cut and hew when green). Some southern pines weigh 50 pounds per cubic foot when green, while western cedars may be only a bit more than half as heavy. So if you are building on soft clay, a light species of wood is in order. And even though a southern pine might lose 8 pounds per cubic foot drying from green to nominal 19-percent moisture content, you can't plan against settling by figuring what your cabin will weigh when it dries out. Be pessimistic with structural design. Figure the worst, not the best, case.

The great weight of logs is not a problem for the other logs in the wall, not even the sill log taking all the weight at the bottom of the pile. Suppose you are using Douglas fir. Construction-grade timber of that species weighs about 35 pounds per cubic foot. So a 20-foot-long, 12-inch-deep, 6-inch-wide hewn log (10 cubic feet total) weighs about 350 pounds. You need about 15 of them (plus chinking) for an 8-foot-high wall. That's 5,250 pounds of wood. But the fir sill has a crushing strength (technically, compression perpendicular to grain, notated $F_c \perp$ on the National Design Specification and other tables) of 400 pounds per square inch. A little multiplication shows that the surface of only 1 linear foot of that sill log (12×6 inches = 72 square inches, each inch with a resistance of 400 pounds) can carry up to 28,800 pounds. No problem at the sill, even if it rests on a series of vertical wooden piers with the grain running parallel to the compressive force. Resistance of compression parallel to the grain (notated F_c on tables) is about four times greater than perpendicular resistance that would occur in all species of wood.

But what will all that weight (including people, furniture, and a roof covered with snow) do to the footing system and the ground below it? The key is to use a footing system designed to carry roughly 150 percent of the total building load, to build it roughly 20 percent lower than the depth called for by frost-penetration tables, and to be certain of the bearing-pressure capacity of the subsoil that is carrying the footing. You'll get all the nitty-gritty help you need for figuring loads as we progress. Just keep in mind that a little extra care (like getting a few SPTs, or standard penetration test core holes) and careful planning up front will ensure that the corner joints and door casings you work on will stay where you put them.

DESIGN DECISIONS

If you decide on some form of precut log home (ranging from raw but limbed trees delivered to the site to prenotched frame and skin with windows, doors, and trim), many design options are decided for you by the logger or manufacturer. The following list of possibilities can help you decide a few crucial questions:

1. Do you want a precut or handcrafted home?
2. If you want a precut home, what basic type makes the most sense for you?
3. If you want to get involved in the log-building process, where and how should you join in?

Not everyone has a year or more to design and build a log home using authentic, even antique tools and techniques. (That year or more is not a few hours each weekend; that's practically full time.) But you can at least put some of your talents and enthusiasm to work,

whether you decide to start from scratch, logging your own trees, or to get involved at the design end and let a contractor put the pieces together for you.

Logging vs. log buying. Cutting your own logs requires expertise few people have. Logging accidents, although not necessarily more frequent than in other trades, tend to be severe. If in doubt about your skills in the forest, think about working through your county agricultural agent or USDA Forest Service representative to locate a reputable log broker. Several firms can deliver fresh-cut, limbed trees to your site, so you can start almost from scratch.

Seasoned or unseasoned logs. Air-drying (seasoning) for at least one year, preferably three or four depending on species and conditions, removes all

Because of the demand from handcrafters, both professionals and owner-builders, Appalachian Log Structures started to sell CCA salt-treated flat-hewn (as shown) and full-round logs. Here logs are preassembled to assure fits before shipment to a customer.

but about 20 percent of the moisture in wood. Seasoned logs (less all that water) are much lighter, which means they are easier to work. Also, they are done shrinking, which is nominal by log length but 4-5 percent by log width—enough to open corner joints and to bow door frames. Seems as though unseasoned logs are a bad choice, except that these fresh trees are much easier to cut, hew, and notch. Also, even though they are heavier, they are ready when you are, and you can accommodate the shrinkage with a system of grooves and splines around window and door frames.

Round or hewn logs. Flat-hewn logs are considerably lighter than full-round logs and are generally easier to put in place. Their right-angle corners are easier for most people to notch. (Working with curvilinear shapes like logs is a unique experience for most builders and do-it-yourselfers alike.) Also, flat-hewn walls are considered more finished looking; round logs, more rustic. If you off-center hewing lines, you can rescue a bowed log by cutting a straight section out of a curved tree. But pros and cons don't seem to matter much in this case. Several builders and manufacturers tell me that most people love the hewn look and hate the full-log look, or vice versa. People want the style they like regardless of cost or time on the job. A few companies, such as Appalachian Log Structures in Ripley, West Virginia, mill square-edged logs, then give them a surface hand-hewing for appearance's sake—a sensible, time-saving option for many owner-builders.

Chinked or chinkless seams. Chinkless construction easily adds 20 percent to labor time—much more if you're learning on the job. It's painstaking work; V-notching the bottom of each log (albeit with a chainsaw), rolling the log in place for scribing, rolling it back off for shaving, rolling it back in place to scribe notches, off again, and so on. Also, chinkless seams can always be improved with a little more time and attention. You rarely get a near-perfect fit both inside and out, which means constant compromise, a tough, ongoing emotional and physical battle as you build—excruciating if you're a perfectionist. The result is stunning, but chinked walls are not necessarily second best. Carefully insulated chinking has a rugged, utilitarian beauty of its own.

Open space or room floor plan. Be cautious about chopping up precious log cabin space with partitioned rooms. This can be done without jeopardy in kit and precut plans, which are sometimes no more than log-faced development houses to begin with. But in handcrafted cabins, conventional, stick-built partitions can just kill the character of log facings along

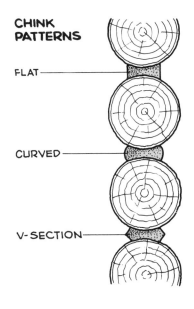

CHINK PATTERNS

FLAT

CURVED

V-SECTION

Chinked seams between logs, above, means an additional step in construction. But the chinking can absorb many irregularities of the logs and your layout. The chinkless seams below require high-class, if large-scale, carpentry joints.

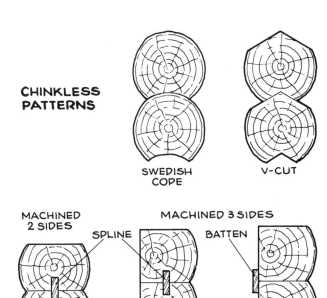

CHINKLESS PATTERNS

SWEDISH COPE

V-CUT

MACHINED 2 SIDES

SPLINE

MACHINED 3 SIDES

BATTEN

exterior walls. Log partitions look great, but they use up a lot of floor space; they're heavy, requiring bigger piers and footings just like the exterior walls. A lot of full-log partition walls will slow down progress on exterior wall construction dramatically as you stop to cut extra sets of notches for interlocked T-joints. A log home should not be valued like a city apartment—the more rooms, the better.

All-in-one or unit-built plan.

All-in-one means, for example, that you add a shed-roof extension off the back of the cabin to create storage space or a shop, plus a porch off the front, plus an extension for an airlock entry—everything accessible from inside the cabin. Unit building means, for example, that your cabin is one independent 16 × 20-foot building; your barn for storage is another; so is your 4 × 6-foot privy—more like a farm with outbuildings. The great advantage of unit building is that you get practice, time to develop skills on the 6 × 6-foot wellhead enclosure, then on the barn, which you might use for temporary shelter while working on the cabin. Better to butcher the notches on a 6-foot log for your shed than to miscut a 20-foot log for your cabin. One nice compromise: two rectangular cabins under a common roof, creating a protected breezeway, entryway, storage area between them.

Continuous or pier foundation.

Log buildings need to hug the ground. They look incongruous, even downright silly, up on piers and poles. Masonry piers will suffice. Treated poles will be quicker and easier yet. But you will have to come back with some kind of skirt construction later on to keep even an insulated floor from the cold and damp (and maybe from a few animals as well). Continuous foundations certainly require more time and material, with endless batch mixing on remote sites not accessible to a ready-mix truck. If you have the time, local stone and mortar over a concrete footing is excellent. But go for a continuous reinforced concrete pour (yes, even with all the formwork) before opting for raw block walls. Blocks and logs are oil and water to the eye; beyond that, they are disparate structural systems—one is long, solid, and continuous, sitting on the other, which is short, piecemeal, and mortar-jointed.

Full cellar or crawl space.

An insulated crawl space (possibly with a small root cellar) is usually the most sensible option, particularly on remote sites. Before you bring the idea of the conventional, development-home, full cellar with you into the woods, think, first, of the endless digging and undiscovered boulders and, second, of where the heck you are going to put all that dirt. Excavating an 8-foot cellar for a small 16 × 20-foot cabin produces enough fill to lay 4 inches of dirt over an entire baseball diamond. If you have to move the dirt by hand, 3 cubic feet of earth per large, contractor's-type wheelbarrow translates into 853 backbreaking trips from cellar to dirt pile.

Most turnkey log homes and log-home kits are designed to sit on a conventional full-perimeter foundation.

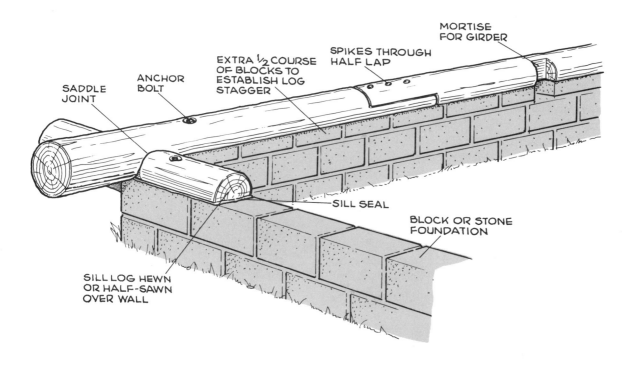

Steeply pitched gable ends, under construction on this Pacific Log home (floor plan shown earlier), and massive cantilevers make possible a spacious second floor. Small cantilevers built into the wall support the entry roof.

Single or double story. You will always save money (and some time) by building up before building out. (No new foundations and no cold corners at the end of the extension that need another chimney and wood stove.) It might even make sense to go to a full second story instead of a full basement. The floor space up there will be warmer in winter and easier to ventilate in summer. Plus the view is better. Dead loads increase with two stories, which means beefed-up footings. No problems there, but long, two-story walls do need collars (horizontal ties to prevent bowing).

Collar beams on rafters help, but second-floor joists interlocked with exterior walls provide the stiffness you really need. And you might think of cantilevering the joists past the exterior wall to support a free-floating balcony. Two-story cabins, particularly those with modest floor areas, do "sit up" a bit, looking more like Swiss chalets than ground-hugging homesteads. Also, they need extended eave and gable-end overhangs (from 2-3 feet, for instance) to give reasonable weather protection to first-floor openings and help extend the life of the walls.

Pier footings simplify excavation, and notching sill logs is not that difficult or time-consuming. But to keep your log home from "floating," some skirting is needed.

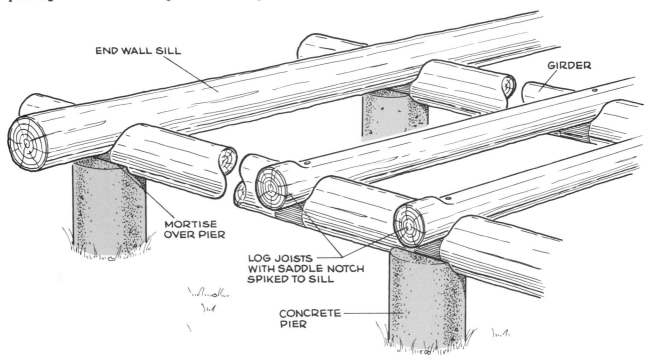

END WALL SILL

GIRDER

MORTISE OVER PIER

LOG JOISTS WITH SADDLE NOTCH SPIKED TO SILL

CONCRETE PIER

Openings. Even allowing for differences in taste, sliding glass doors (patio doors), so efficient for access, light, and ventilation, generally look bizarre in log homes. To a lesser degree this is true also of large, single-light casements, particularly when they are installed in multiples with fixed and operator frames joined together. Operator skylights, in the same category of modern-looking glass, can be tucked away, at least slightly out of sight, on the roof. They can cause trouble in shake roofs but offer an efficient, time-saving alternative to dormers. Six-over-six-light or nine-over-nine-light double-hungs look great; so does nine- or even huge twelve-light barn sash, top-hinged for an awning swing. Whatever glazing you use, try to keep openings 3-4 feet away from critical corners.

Authentic or efficient construction. Some will argue that the "or" should be changed to "and," that these two characteristics are completely compatible. I sympathize with the thought, but after many years of custom homebuilding and furniture-making and more, I am not willing to discount all modern tools and technology just because some of them are corner-cutting and geared to making homes less costly at the expense of durability. I'm not a purist. I don't think you have to drop $100 or more for an authentic, antique broad axe in order to hew logs. You can, and that's fine. But you also can hew them with a brand-new axe or even buy milled beams. Don't feel belittled or short-changed somehow because you're not building a log home exactly the way pioneers did. After all, they didn't decide against chainsaws in favor of axes. They had no choice; they had to use the tools and technology of the time. You have a choice. You can log and hand-hew heart redwood for sills. That's fine. You can also order special penta- or salt-sealed logs guaranteed for 25 years. That's fine, too.

These basic decisions alone give you a lot to think about. Obviously, I've included some of my personal opinions. But there is room for your opinions, too. Consider the issues, but don't make any basic design decisions yet. Read on. Send for plan books. Make sketches. Build cardboard scale models. Remember that these basic choices affect the amount of material, time, and money needed throughout the building process—as well as your ultimate satisfaction.

PREPARATIONS FOR CONSTRUCTION

SITE SELECTION AND ORIENTATION

Housebuilding is a complicated process. You know that. But housebuilding is not an exact science where questions have black-and-white answers, where you are all right or all wrong. Sometimes you can afford to make a mistake. But there is a catch. Sometimes a seemingly innocent mistake can have powerful consequences. And I'm not talking about a structural miscalculation, where the first foot of snow on the roof makes everything collapse. Believe it or not, that kind of mistake is difficult to make.

Suppose you locate large kitchen windows at mid-wall, for symmetry, and because you want the kitchen table in the middle of the room between wood stove and china cabinet. But once moved in, you find the wood stove too hot at that range, so you have to move the table and lose that ideal view of the lake during meals. Too bad, but no disaster. But make a mistake orienting those 3 × 6-foot double-hung windows away from morning sunlight, and you may need an extra cord of wood in the kitchen stove every winter, year after year. You can decide to forgo the solar benefits for a prime view, for example. But take some time with orientation so you know what's at stake.

A good site offers a good view, plenty of light, good ventilation, adequate drainage, and reasonable access, not necessarily in that order. A bad site is socked in by terrain and vegetation to block light and air transmission; it's in the path of drainage runoffs or, worse yet, is itself a drainage-collection point; it's lower than the adjacent landscape, which limits the view; and it's not readily accessible (the dirt road seems to turn to mud when the humidity increases).

Of course, there are exceptions. But a higher, more open site on a natural rise with a stand of evergreens to the north acting as a windbreak and with deciduous trees to the south, offering summer shade but full winter sun, is almost always easier to take care of and a more pleasant place to live than a damp, dark, water-logged depression between surrounding hills.

Orienting your building on the site is a personal, commonsense decision based on three major considerations: aesthetics, engineering limitations, and weather efficiency. Even though you may be well along the design path (you know you need about 750 square feet, hewn log walls, a steep-slope roof), resist the temptation to finalize cabin plans until you have finalized site plans. Don't try to force one plan down the throat of a site where it just doesn't fit in. Keep your mind open. Let the site dictate the final plan as well.

Aesthetics. This includes the view of the sunrise from the porch, the view of the sunset out the living room windows—every possibility from every part of the home. And that may have you twisting the floor plan back and forth in your mind. It also includes the view of the home from the site—where the building looks best. This is somewhat less important unless you spend more time outside than inside. These considerations are purely personal. What looks good to you is exactly what you should do, period.

Engineering limitations. This consideration includes picking the most structurally sensible location for the building, out of the way of water, which, along with fire, ranks as the worst enemy of wood. You want solid ground for the foundation, although with enough technology and enough money you can build anywhere. In fact, one of the restaurants at the Walt Disney Epcot Center in Florida sits just over a sinkhole on welded pylons driven 80 feet down into the muck. Bedrock, at the other end of the spectrum, is as solid as you can get (most of Manhattan's skyscrapers sit on it) but often requires irregular footing formwork stepped up and down with the rock—a real pain. Local soil engineers, structural engineers, architects, or building contractors may be able to help you determine what kind of subsoil you have to work with. And you may be able to get reliable soil analysis from an existing survey, from records of building and highway departments in local townships, even from detailed, regional geological maps produced by the U.S. Geological Survey National Center, Reston, VA 22092.

The point is, different subsoils have different bearing capacities; i.e., your 200,000-pound home will rest on some but sink through others. Soft clays and silts, for example, will bear about 2,500 pounds per square foot; compacted gravel-sand mixtures, about 10,000 pounds per square foot; sound rock, about 10 times that. Evenly graded gravel or gravel-sand mixtures with little grit (technically called fines, since they will pass through a ³/₁₆-inch-square mesh sieve) are excellent for foundation support, having good drainage characteristics and only marginal frost action. Gravelly mixtures with increased silt or soft clay also perform well, although with reduced drainage capacity. Sand-silt and sand-clay mixtures are adequate. The real culprits are soft silts and both organic and inorganic clays. They offer poor drainage, moderate to heavy frost action, and only marginal bearing capacity. Highly organic bases like peat are only one step up from quicksand.

Strength is one issue; the other prime concern is area drainage, because it can change the shape and strength of all that stuff holding up your foundation. Remember that tree roots and topsoil soft enough to poke a stick into will hold groundwater. But ledgerock near the surface, inorganic clays, hardpans, and claypans (some nearly as hard as rock) act more as impervious, underground aqueducts.

Weather efficiency.

Solar orientation, the issue you hear so much about, is only part of the picture. If you must play the solar game, the simplest way to do it is to orient glazing in homes in northern latitudes to absorb as much winter sun as possible; in southern latitudes glazing in homes should be away from summer sun. The most complicated way to do it is to use the sun charts in a reference like Ramsey and Sleeper's *Architectural Graphics*, for instance, to compute seasonal sun bearing and altitude angles for your site. You can track down figures for a site in Florida, for example, as follows: on February 1, at 2:00 P.M., the sun bears 45° SSW at an angle of 45°. At a site in Washington on the same day at the same time, the sun bears 32° SSW at an angle of 28°. It gets very specific. But with a little work, data from solar tables can be translated into practical plans. The Florida cabin, for instance, might be most efficient with the bulk of the glazing facing south, exposed to moderate temperatures on a clear January day generating some 1,600 BTUs per square foot of wall area. Turn the glazing north, and it is exposed to only 250 BTUs per square foot on the same winter day. The difference is big enough to alter fuel requirements substantially. You can spend days crunching solar numbers. You can also visit the site, talk to neighbors, open your eyes, and take some pictures.

Controlling wind is less technical. If there are sea-sonal prevailing winds (not true for all sites), leave evergreens or other trees (or plant new ones) to reduce winter wind-chill factors. The cold without the wind is enough to deal with. Dense, interlocked tree stands can effectively shelter a ground area extending away from the trunks equal to four times the height of the trees.

You can't control the amount of rainwater you get, but you can plan to get rid of it downhill (even a slight gradient will do) from the house. If you have no natural rise to make a well-drained site, think about a large dry well. And if you use a backhoe on the job, think about it while the backhoe is still there. If you get stuck siting the house in a natural depression that collects groundwater, the answer is an area drain. This is simply a trench filled with gravel or rock, functioning like a moat around part of your property. Only a few feet deep and arced like a shield against the water, an area drain captures groundwater on the way in, carries it sideways, and releases it out of harm's way.

One short note on site safety. I suggest you take your plot plan sketches to the local fire department (volunteer or professional) and ask a simple question: If I build a log home here and it catches fire somehow, do you have a shot at saving it, or will it be impossible? Maybe if you site downhill, a bit closer to the road, they'll be able to reach you. But uphill, across the little footbridge, you'll be on your own. Other major considerations being equal, this issue or one like it—say, access for an LP gas delivery truck—may tip the balance.

However, of the three prime concerns—aesthetics, engineering limitations, and weather efficiency—aesthetics usually decides matters at the site. And why not? As long as you and your home are reasonably safe, maybe what you see everyday should be more important than a few thousand pounds of bearing capacity, more important than degree days and sun angles.

PRELIMINARY PLANS

At some point, you have to stop adding pieces to the puzzle and start putting everything together. This is it. If you decide on some form of precut cabin, here are a few points not to miss.

1. Find out just how much "customizing" is allowed; who turns your special need for a larger kitchen into working plans, blueprints, and specifications; and how all this alters the price.
2. Make sure you have access to a set of blueprints before you turn over all the money to the builder. Then make sure your local building department is happy with the plans. Company literature will undoubtedly say "code-approved," but it pays to

check plans and specs, just in case they violate some peculiar local code convention. Be advised that you are more likely to get hassled the closer you get to densely populated metropolitan areas, where log homes are uncommon.

3. Find out whom you are dealing with and who is responsible. The manufacturer may warranty the logs but not the installation. That may be covered by the manufacturer's local distributor, an affiliated installer, or a completely independent contractor. When responsibility for the finished product is diversified (spread around among manufacturer, distributor, shipper, foundation contractor, frame erector, mechanical subcontractor, and others), each party can blame mistakes on another, while you get caught in the middle.

4. If you buy a precut log home and decide to assemble it yourself, get a copy of the manufacturer's assembly manual first. Some are thorough. Some boil down the relatively complicated process to a disarmingly simple 15 pages or so, leaving out all kinds of little details. All the manuals make some assumptions about your knowledge of tool use and basic construction techniques.

Looking like a giant jigsaw puzzle, the "Log Layout Sheet" is from Timber Log Homes. Manufacturer precutting, including gable-end angles, gives you a complete frame without using a saw. When batter boards are clamped and nailed (only after establishing vertical elevations and right-angle corners), a ruler and a level at the foundation string lines will produce accurate corner stakes.

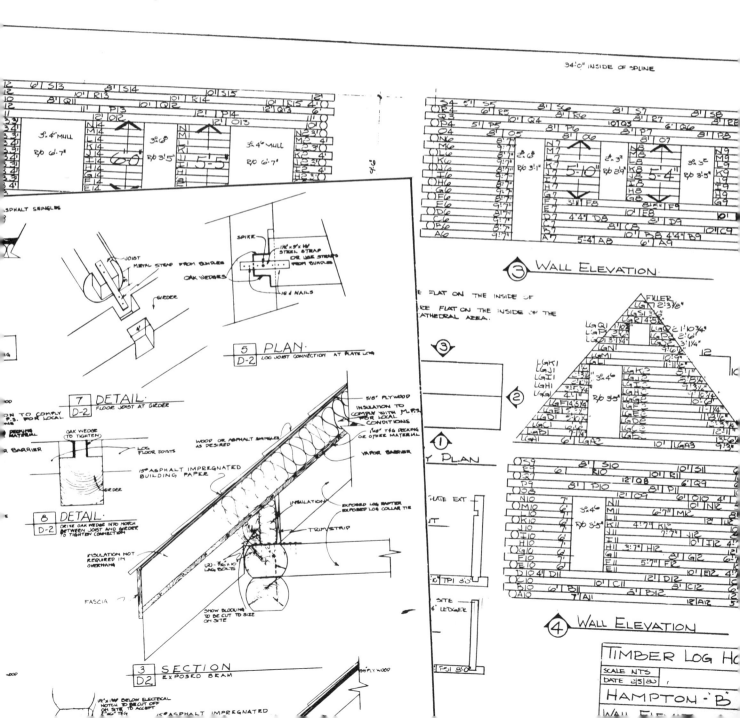

Completely precut kit homes like this one from Real Log Homes (through their regional dealer, Berkshire Hills Vermont Log Homes of Poughquag, New York), arrive at the site on flatbed trucks. At right, dealer Clint Bruggemann supervises offloading, which many firms leave to the owner. Below, numbered logs are sorted and stacked off the ground. Working one piece at a time, a big burly owner-builder could assemble a kit house solo. But the job is faster and safer when you have a few helpers. (Photos on these pages by Neil Soderstrom.)

If you decide on a handcrafted cabin, you won't have the help of blueprints and material lists and assembly manuals. And to get a building permit (you'll need one almost anywhere you build), you have to provide a set of plans for the inspectors in the building department. But in many cases you do not have to turn in professional blueprints prepared by an architect or structural engineer. Many building departments (particularly in more rural areas) will accept a neat, home-drawn version, as long as it includes the following: a plot plan locating and orienting the building on the site and showing survey vectors with prominent terrain features (taken from a survey); a floor plan; and a framing plan (usually shown as a section view), on which the size and length of typical framing members are detailed. Neatness helps. But most inspectors are more concerned with what you know than how well you draw. If your plans show nail and bolt specs, typical joints, roof slope, concrete-reinforcement details, and more, you may get by on your own. If your plans show 4×4-inch joists for a 16-foot floor span, the inspector may conclude (and rightly so) that your home-drawn plans cannot be saved with a few suggestions and adjustments. Then he may insist on an architect's or engineer's seal.

JOB SEQUENCING

When I was building stick-frame houses, I started many workdays at the lumberyard. As the job progressed, it was my custom to take a tour of the project at the end of each day, notebook in hand, deciding on the work for the next day. In the morning, it was a simple matter to pick up those few extra 2×4s for firestops and other bits and pieces as needed. This is a luxury that cuts waste and prevents overordering, a luxury you rarely have on a handcrafted log home. A strict schedule can be as valuable as a thorough materials list. You do not need a schedule of completion dates for different parts of the job, but you do need to organize each stage of work so that one flows logically and efficiently into another.

PLANNING.   The obvious first step is to begin collecting ideas from plan books, company catalogs, architectural history books, and your own imagination.

DESIGN.   As hard on the brain as the physical work will be on the back, this is a time of synthesizing the ideas, using site visits, sketches, and models to blend aesthetics, engineering, budget—all the requirements and considerations.

PLANS AND SPECS.   This is the straightforward process of representing your decisions on paper, using the language of architecture and construction. If you are very thorough, you can uncover flaws and construction bottlenecks on paper, where they can be changed with an eraser. The plans and specs let you talk intelligently to building departments and banks.

SITE PREPARATION.   In the final stage before construction begins, you make way for the job, providing access, preliminary utilities, and power. On some sites this means full septic systems, water lines, and electrical service. On some sites, it may mean only a shallow dug well and cans of fuel for the chainsaw.

At top, lap joints are tightened with a come-along before spiking. Above, seasoned wall logs are short enough to place without mechanical help but require splice joints. Below, a good crew can close-in a gambrel kit like this one in about a week.

LOG ORDER AND STORAGE.   Turnkey companies will bring everything on a truck or two, off-load, and erect, all in a few days. But many handcrafters start behind the eight ball when the trucker pops the chains and deposits a tangled logjam 200 yards from the foundation. Plan to minimize log handling. Lay a series of sleepers with 4-foot centers to keep the log pile off the ground. Use peaveys and crowbars to get an inch or two of airspace between the logs. Most builders of hand-hewn log cabins debark by early spring, leaving the bark until then for water protection (even at the risk of some insect damage). Don't wrap the log pile in plastic or tarps. That makes a wonderfully hot and humid environment for insects and fungus. A breezy old barn would be an ideal storage space. If you're out in the open, a fly tarp for water protection, with plenty of ventilation along the sides and ends of the pile and plenty of airspace between the logs, will do pretty well.

CONSTRUCTION

FOOTINGS AND FOUNDATIONS

This crucial part of the job requires close attention in two areas: proper layout and adequate strength. Basic rectangular layout is a subject covered in many how-to books. But here are a few tips and procedures that have worked for me out in the field, where textbook methods can get stuck in the mud, literally.

1. Make a rough layout first, measured with long tapes, squared up roughly by measure and by eye, marking the corners.
2. Extend your layout lines about 4 feet past the corners, away from the active work area, and use the extensions as two legs of a 4-foot-square box.
3. At every corner of the box except the corner where the house will be, take a 6-foot 2×4, cut at one end to a sharp 60° point, and drive it about 3 feet into the soil, keeping it as plumb as possible.
4. Brace each 2×4 in two directions with 1×2s nailed between the vertical 2×4 stakes and small ground stakes.

This gives you very strong and solid supports on which you can set horizontal batter boards. These can be made of 1×4s, for example, clamped to the stakes at first, so you can make all kinds of adjustments for square and level without pulling and redriving nails. You should spend considerable time on the layout. And with these fortress-type batter board stakes, you won't have to start from scratch when some turkey throws a shovel of dirt in the wrong direction. Build batter boards well off the foundation line, and make them very strong. Make all adjustments on horizontals without nailing. Nail only once, backing up the 2×4 stake with the head of a sledge to minimize movement from nailing.

There are two good ways to check square: overall, with diagonals, which should be equal for any rectangle, and at each corner, with the properties of a 3–4–5 triangle, where two legs of the corner projecting from the 90° corner are 3 and 4 respectively (it doesn't matter which is which), while the line connecting the ends of the two legs (the hypotenuse of the triangle) is 5. Any convenient multiple of 3–4–5 will work—6–8–10, 9–12–15; the longer the legs, the more accurate the triangulation.

You may have strings and lines and tapes all over the place, but don't nail yet. I've always tried to kill two birds with one batter board; i.e., to adjust both square and level before nailing the horizontals. There are two good ways to check level. If you have, or can borrow, a really accurate builder's dumpy level (on tripod), the job is fast and easy. Make a story pole (or use a surveyor's measuring rod), and have a helper hold it at each corner while you sight through the crosshairs. Using the dumpy level, you can mark minuscule changes in elevation, so, for example, a pencil line on each horizontal is exactly 3 feet from the top of a future concrete footing. (Who cares if the trench is a little out of level as long as the top of the footing is level, right?)

Without a dumpy level, I always used one of the more basic construction tools—a water level. Mason's levels aren't long enough, and line levels are a joke, bending and wobbling with each breath of wind. But clear plastic, ¼-inch tubing filled with water is absolutely accurate. (Most water-level kits include a red dye for easy visibility.) You and your helper will spend

When batter boards are clamped and nailed (only after establishing vertical elevations and right angle corners), a ruler and a level at the foundation string lines will produce accurate corner stakes.

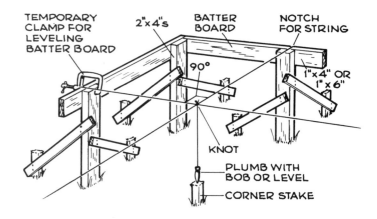

TEMPORARY CLAMP FOR LEVELING BATTER BOARD

2"x4"s

BATTER BOARD

NOTCH FOR STRING

90°

1"x4" OR 1"x6"

KNOT

PLUMB WITH BOB OR LEVEL

CORNER STAKE

some time yelling back and forth across the excavation ("Up a little—no, down a hair"), but it does work. And it is satisfying to cash in on a simple principle of physics: Water always seeks its own level. If the tubing is not kinked and twisted and filled with air bubbles, the water line at one end will settle exactly level with the water level at the other end—even if the ends are hundreds of feet apart. Adjust the batter board horizontals to the level water lines, cross-check, recheck, then nail.

LOADS

It is unwise to go by conventional design-load tables when building with logs. Dead loads (literally the combined weight of all the materials in the structure) are relatively standardized in stick-built construction, even allowing for variations such as paneling as opposed to wallboard or plaster. This doesn't apply to logs. Some are light (just over 20 pounds per cubic foot), and some are heavy (just over 50 pounds per cubic foot), and that's dried out to 12-percent moisture content.

If you use your common sense and check local codes, you could probably design a footing and foundation system to the following guidelines:

Location. In a roughly level trench with an undisturbed or compacted soil base, placed so the logs will center on the footing.

Excavated depth. Equal to the depth called for by frost-penetration tables, tempered by local conditions as approved by the local building inspector.

Concrete width. Equal to roughly twice the wall thickness; i.e., 12 inches across for logs 6 inches in diameter, 18 inches across for logs 9 inches in diameter.

Concrete depth. Equal to the approximate log diameter, or average diameter in the wall, but at least 8 inches.

Concrete mix. From a certified ready-mix company or mixed to a proportion of approximately 1 part cement, 2½ parts sand, 3½ parts aggregate, and no more than 7 gallons of water per bag of cement.

Reinforcement. Use two ⅜-inch rebar lines on commercial hangers (or brick or rock supports), roughly 2 inches in from each side of the form or earthwall trench, 2 inches off the ground, with standard 20-foot lengths overlapped 18 inches, 36 inches around corners, and secured with three wire ties on straight runs. Assuming you have a uniform mix, good aggregate distribution, and protection from freeze-thaw cycles, these rough guidelines should get you past most building departments. Of course, you can alter concrete mix ratios, even add special chemicals, to overcome all kinds of special conditions and to achieve all kinds of special characteristics. Most important, though, are the strength changes you can get by jiggling water-cement ratios. The table below shows how less water and a stiffer mix makes stronger footings.

EFFECT OF WATER RATIO ON CONCRETE STRENGTH	
Compressive strength (psi)	Gallons of water per bag mix
2,500	7.3
3,000	6.6
3,500	5.8
4,000	5.0
4,500	4.3

Disarmingly simple tools are often the best. For only a few dollars, and a little time to make sure you've got all the air bubbles out of the line, a clear-plastic water level produces unfailingly accurate readings.

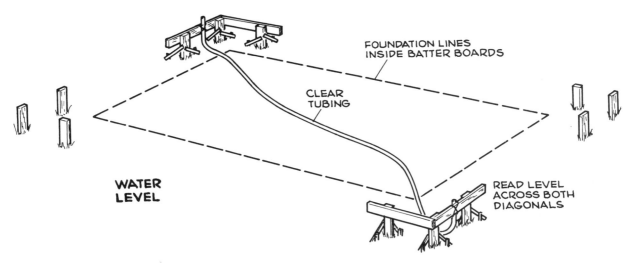

FOUNDATION LINES INSIDE BATTER BOARDS

CLEAR TUBING

WATER LEVEL

READ LEVEL ACROSS BOTH DIAGONALS

Instead of sending you off in search of books about concrete, I want to make sure you get these additional and essential points.

1. *Don't make soup.* The more water you add and the easier it is to mix, the more the concrete turns to soup. Resist the natural temptation to add water as you go to make the mix workable. Maintain high compressive strength, even if it means sore shoulders from mixing.

2. *Allow for curing time.* A general-purpose mix (Type I cement) gets stronger and stronger as it cures and as the migrating cement filaments in the mix bind the components together.

STRENGTH INCREASES WITH CURING AGE

Days of cure	Compressive strength (psi)
3	2,100
7	3,200
28	4,500
90	5,000
365	5,600

You can see on the "Curing Age" table that the first 28 days are important. The concrete continues to gain strength, but you could build a fortress of 18-inch-diameter logs on a system with 4,500-psi strength with no problems. Wet curing (burlap soaked with water regularly) or film curing (plastic sheeting or proprietary sealers sprayed or rolled onto the concrete surface) prevent premature drying, which stops the binding process. Try to plan the job so the concrete has at least 7 days' cure without a construction load. Better yet, spend 28 days on log preparation, hewing, well digging, and other projects before starting to load up the foundation. The first 3 days of curing are crucial. *Lack of curing can cut 28-day strength in half.*

3. *Beware of the rebar Catch-22.* Unless you can maintain the roughly 2-inch margin with rebars, don't use them. Rebars at the very edge of the pour will weaken the footing and cause premature deterioration. And don't rely on Rube Goldberg balancing contraptions to keep the rebars in place during a ready-mix pour. Concrete comes out of the chute with enough torque to knock down forms unless they are well braced. And try to pour from directly above the trench. Bouncing the pour off sidewalls of forms or trenches tends to segregate the mix (make the aggregate bunch up).

Reinforcing slabs and foundations is crucial, as long as the rebar and wire are in, and not at the edge of, the pour. Manufactured wire hangers like the one shown at right prevent the reinforcement from moving as the concrete is poured.

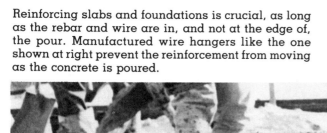

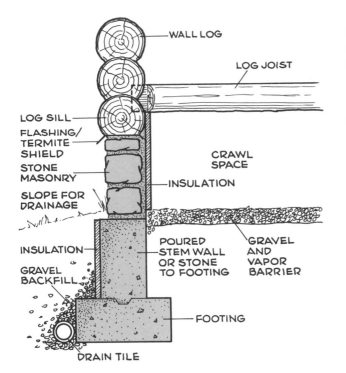

WALL LOG

LOG JOIST

LOG SILL

FLASHING/
TERMITE
SHIELD

STONE
MASONRY

SLOPE FOR
DRAINAGE

CRAWL
SPACE

INSULATION

INSULATION

GRAVEL
BACKFILL

POURED
STEM WALL
OR STONE
TO FOOTING

GRAVEL
AND
VAPOR
BARRIER

FOOTING

DRAIN TILE

This foundation system combines the traditional stone-and-log sill above grade with a modern, insulated, concrete footing and foundation below grade.

4. *Pay for trenching instead of forms.* If soil conditions and your excavation elevations permit, you can save time, money, and material (and aggravation) by paying for quality backhoe work. A sharp-cut 18- or 24-inch bucket can leave a straight trench of uniform width, with plumb walls and a relatively flat trench floor. You can undermine the sidewalls at the bottom of the trench a few inches on each side, line the trench with plastic (particularly if the soil is very dry and would pull water from the mix), set rebars, including vertical rebars extending to serve as sill anchors, then pour carefully, in stages called lifts. If you want concrete with a finished appearance projecting above grade (about a foot is a minimally healthy separation be-

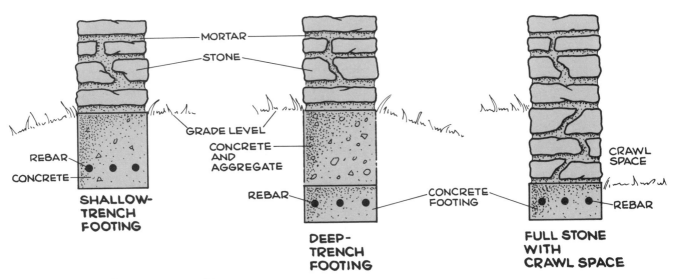

MORTAR

STONE

GRADE LEVEL

CONCRETE
AND
AGGREGATE

REBAR

CONCRETE

SHALLOW-
TRENCH
FOOTING

REBAR

DEEP-
TRENCH
FOOTING

CONCRETE
FOOTING

CRAWL
SPACE

REBAR

FULL STONE
WITH
CRAWL SPACE

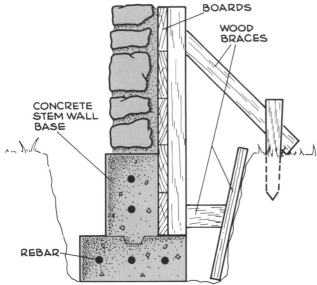

BOARDS

WOOD
BRACES

CONCRETE
STEM WALL
BASE

REBAR

FACE STONE
WITH CRAWL SPACE

STONE FOUNDATION SYSTEM

Depending on the depth required by frost penetration in your area, you can build (a) a shallow concrete footing with stone foundation above grade, (b) a deep trench concrete foundation wall above the footing to grade height, (c) a concrete footing and cellar stone foundation wall facing an excavated crawl space, or (d) a stem wall with facing stone (shown with forms in place).

tween soil and sill), staked plywood forms can be used on this top portion only. Two-side, full-depth forms consume time and money at an incredibly fast rate.

It is worth knowing at least this much about concrete, because that first piece of construction holds up everything that follows. It is also important to note that you can grunt-and-groan large, flat-faced rocks into a rectangular layout, hew a 4- to 6-inch bearing surface on the bottom of sill logs, notch the corners, lay them in place so you get log bearing on rock every few feet, then fill in the gaps with mortar and stone. If the soil is stable and the rocks are big, you can often do without a backhoe and without ready-mix—just like the pioneers.

It's not that difficult, though, to keep track of building loads and make a more precise footing calculation. There are two kinds of loads: dead loads (the weight of materials) and live loads (the weight of people and things in the house, plus environmental loads, even if they are intermittent, from snow and wind, even earthquakes).

Architectural reference books such as *Architectural Graphic Standards* and the *Building Construction Handbook* contain weights for all kinds of building components, although they generally do not include logs. In any case, these basics should help:

Concrete. At roughly 125 pounds per cubic foot, a typical 8-inch-deep by 24-inch-wide footing would run about 165 pounds per linear foot (running foot of perimeter foundation); an 8-inch-wide by 36-inch-high poured foundation would run about 250 pounds per linear foot.

Logs. The table on the next page gives you an idea of how log loads vary depending on wood species. Typical weights given are in pounds per cubic foot for thoroughly air-dried logs at 12-percent moisture content (MC). But, obviously, not all aspen weighs 26 pounds per cubic foot and not every log in the pile is at 12-percent MC. But if you design with a safety factor, idealized tables like this one will work. (And that's one very good reason for safety factors.)

You have to do a little math, depending on the size and shape of the timbers you use. For hewn beams, volume is simple: base times height times timber length. For round logs, the formula for volume is: $V = \pi r^2 h$, where π is 3.142, r is the radius (one-half the diameter) of the log, and h is the log length. Here's an example.

Using 9-inch-diameter logs 20 feet long:

$$
\begin{aligned}
V &= 3.142 \times 4.5'' \times 20' \\
&= 3.142 \times 20.25'' \times 240'' \\
&\quad \text{(radius squared, feet converted to inches)} \\
&= 15{,}270 \text{ cubic inches} \\
&= 8.837 \text{ cubic feet} \\
&\quad \text{(divide by 1,728,} \\
&\quad \text{the cubic inches in one cubic foot)}
\end{aligned}
$$

At 8.84 cubic feet per log, Engelmann spruce (at 23 pounds per cubic foot) weighs about 203 pounds, while white oak (at 47 pounds per cubic foot) weighs about 415 pounds.

Once you make these basic calculations, simply add up the number of logs in the wall, multiply by the weight per log, and you have just done a wall-load estimate as well as a licensed structural engineer—really.

The next step is to include major timbers like floor girders, then add in finishing components. As a rough

Dimensional timbers backed up with stakes provide the simplest forms for masonry above grade. With a clean cut trench, this may be all you need. More elaborate forms made of planking or plywood must be braced strongly to support concrete before it sets.

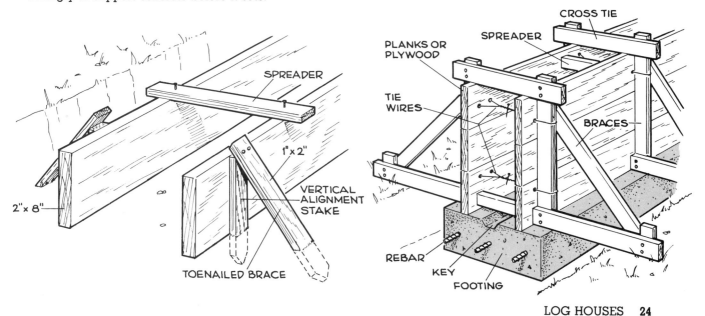

LOG SPECIES CHARACTERISTICS (AT 12% MC AIR-DRIED)			
Species	Weight (lb./ cu. ft.)	Decay resistance	Shrinkage
Aspen	26	low	low
Basswood	26	low	high
Cedar, eastern red	33	high	minimal
Cedar, western red	23	high	minimal
Cypress	32	high	low
Fir, Douglas, coast	34	moderate	moderate
Hemlock	28	low	low
Oak, white	47[1]	high	moderate
Pine, longleaf	41[2]	moderate	low
Pine, ponderosa	28	moderate	low
Pine, white	25	moderate	low
Poplar, yellow	28	moderate	moderate
Redwood	28	high	low
Spruce, Engelmann	23	moderate	moderate
Spruce, Sitka	28	moderate	moderate
Tamarack	36	moderate	low

[1] Exceptionally heavy at 63 pounds per cubic foot when green.

[2] Heavy at 55 pounds per cubic foot when green.

For most people, two-handed control on a drawknife works better than peeling with an axe. A large drawknife like this makes short work of peeling. The larger the blade, the more energy you can devote to control and the less energy you need to peel the bark. (Photo by Don Smith.)

guide you might use about 7.5 pounds per square foot of floor space for joists and flooring, about 15 pounds per square foot for rafters and shakes.

Now for the live loads: 40 pounds per square foot of floor space is a safe live load for people and possessions, excluding, of course, one-of-a-kind items like the 400-gallon fish tank in the kitchen. Special items need special, beefed-up construction right underneath them. Environmental loads are slightly more complicated. You can arbitrarily add in another 40 pounds per square foot as an environmental safety factor, or you can use these proven ratios:

1. Increase load design by 15 percent for nominal, two-month load duration, as for snow. Consult local snow-load tables for details about your site.

2. Increase load design by 25 percent and up to 33 percent for severe wind load and earthquake resistance.

Any structure exposed to special environmental factors (extreme fire hazard, excessive wind loads) must be treated as a special case, and should be analyzed by a professional.

Before you leave the concrete and get to the good stuff (the logs), think about adding perimeter insulation (one or more inches of rigid styrene-type board) to keep the cellar or crawl space temperate; adding foundation vents under the sill line to evacuate moisture that would otherwise be absorbed by the joists and flooring; adding plastic sheeting, tar paper, or other vapor barrier directly over the ground (for a crawl space) to minimize the effects of ground moisture to begin with; adding a termite shield (aluminum is okay) protruding at least an inch, angled downward, from each side of the foundation, and adding a sill sealer (hand-cut 3½-inch batts of fiberglass work fine) to prevent air infiltration at the critical energy seam between concrete or stone foundation and wooden sill.

PEELING

This is the perfect job to occupy your time while the concrete cures. You'll get different opinions on this one: Debark in the spring, just after felling, when sap is up, and the bark on many species pops off the tree with only a little help; or leave on the bark until you're ready to build. I suggest the latter, for weather protection, to minimize checking and discoloration before the logs are placed, and for protection in transit as you haul logs from the stack to the site.

Most builders start debarking by slicing off a 4- or 5-inch swath of bark along the log length with an axe. Short, controlled strokes, swung tangentially to the log, minimize gouging. One swath gives you two lines of entry with a barking spud, the largest, longest, heaviest chisel you can locate, or even a tough garden spade sharpened up with a mill bastard file. This last one will make you look a few bricks short of a load to passersby, but it does work.

Working away from the swath in both directions, use the spud firmly and under control, almost like lifting a Band-Aid without ripping the hairs on the arm underneath it. You may get the bark in large sections or little strips. No matter. If you gather what will wind up being an enormous pile of bark, then protect it from the weather, it will dry out and make great, if bulky, kindling.

Debark up to and around raised knots, burls, and limb stubs, then cut these deformations (bark included) flush to the log with a chainsaw. If you poke and push and the bark won't budge, hit it with the axe.

If this leaves you with a scraggly looking log with some raw wood, some bark, and some blotches of bark skin, shave the tree with a drawknife. This is tedious work, but some people prize the slightly mottled, softly striped effect produced by the drawknife.

SILL LOGS

Each wall log you place gives you practice and experience. At this rate, the logs at the top of the wall would be the easiest to set if it didn't become increasingly difficult to get them up to the top of the wall. This self-regulating feature keeps the job moving along

Specialty logs like these from Timber Log Homes are highly engineered to provide weather seals with a system of caulking and splines. Positioning on the sill is critical in order for the system to work.

at a consistent pace—once you get past the sill log.

This first course of logs is worth extra care because it lies at the bottom of the pile, because the sill is an important seam to control for energy efficiency, and because this seam is the transition point from masonry to wood. These two building materials have dramatically different properties and react differently to construction stresses and weather conditions.

First, for buyers of prefab cabins, prepare the sill (or the box-beam frame or foundation beneath it) according to manufacturer's specs. In fact, many kits come with a specially cut sill log that may not work on every foundation. If you want to do something a little special—with a fieldstone facing, for instance—check with the prefab company first.

For handcrafters, try to select the largest, straightest, soundest logs from the pile to serve as sills. If you are hewing flat-faced timbers, consider using oversize logs, scoring and chopping off most of the sapwood, which is weak, on the outer portions of the log to leave weather- and insect-resistant heartwood. Avoid logs with significant taper. (Trees do tend to be thicker at the bottom and thinner at the top, after all.) Also avoid trees with significant crowns, a bowing over the length of the log. Slightly crowned (bowed) logs are okay if they are set crown up. This is even desirable on sills clear-spanning independent piers, as the crown acts as a built-in counterforce to deflection. It's an old-fashioned, proven counterpart to the modern innovation of prestressed concrete beams.

Sills for pier construction require the least amount of preparation because they bear only at intervals. Sills for full-perimeter foundations are prepared using the same techniques. (See illustrated instructions beginning on the next page.)

The same channeling and chiseling procedure shown for the sill-log pocket works for hewing sills on full foundations and for hewing logs on four sides. But with long logs you should space the chainsaw cuts 3-4 inches apart, saving a lot of time, and clean away the channeled wood with an axe. Rescoring the faces and finish hewing create a flat-surfaced, elegant-looking timber.

I would not tell you which brand or size chainsaw to use. Veteran log builders swear by different brands for different reasons. But do avoid buying one of the "toy" chainsaws, the tree-pruning types too often pushed on homeowners as tree-felling saws. I can tell you that among log builders two names pop up on a regular basis: Jonsereds and Stihl. I have used the monster Stihl 031 regularly, and now use the 048, which I find efficiently designed for real work in real conditions, and incredibly reliable. Also, I have used Skil electric chainsaws on several commercial pole-building jobs with a lot of heavy-duty production cutting and, of course, power at the site.

HOW TO CUT
A SILL POCKET

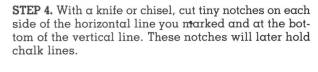

The following steps show how to cut a pocket (notch) like this where a single-sill log crosses a heavy timber beam. It's better that the fit be loose rather than too tight. (Photo sequence by the author.)

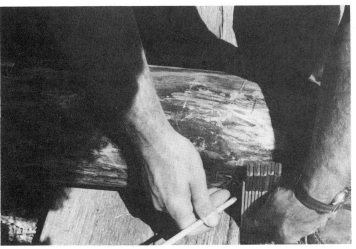

STEP 1. Place the sill log crown-up on the beam and mark the beam edges on both sides of the log.

STEP 2. After bracing the log temporarily, mark the centerline verticals (plumb lines) on each end.

STEP 3. Mark a horizontal line about an inch from the bottom, showing the depth of cut for the eventual pocket. Here the 1-inch depth will result in a flat bearing surface of 5 inches or more, yet won't be so deep that it would weaken the log.

STEP 4. With a knife or chisel, cut tiny notches on each side of the horizontal line you marked and at the bottom of the vertical line. These notches will later hold chalk lines.

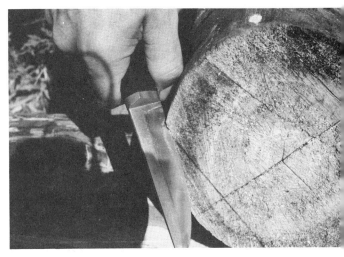

SILL LOG NOTES

Carefully marked verticals and horizontals, connected with chalk lines, help you visualize a single plane through the log cylinder. Without chalk lines, deformation on the log surface—even wood tone and grain changes—can trick your eye into cutting bearing pockets that are not in the same plane, tilting in different directions. Later, the chalk lines will wear away or wash away.

STEP 5. Roll the sill log bottomside up to allow you to snap chalk lines set in the tiny notches at each end of the log.

STEP 9. Hit the kerf edges with a hammer to break out the remaining ridges.

STEP 6. To secure the log for cutting of the pocket, brace it with scrap lumber as shown, or with wedges, or with metal log dogs. Sturdy bracing lets you perform upcoming cutting steps with vigor and safety.

STEP 10. At this point, your cutting depth shows how accurately you followed the chalk lines. Accuracy here minimizes cleanup chores and helps provide an even bearing surface.

If you notch too deeply or not deeply enough, so the bearing pockets are in the same plane but slightly out of level with each other, you can make up the discrepancy by shimming. Don't shim with wood, as many books suggest, particularly wood with longitudinal grain open to deterioration and lacking strength against compression from the log wall. Instead, use slices of slate.

Also, remember to allow for cantilevers (say, for a porch) extending beyond perimeter piers. If more than one log is required for a long, straight run, try to use logs with roughly equal diameters, and cut each one long enough to lap the pier completely. Don't cut them to butt each other at the center of the pier. On full foundations, allow for overlapping-splice joints at intervals that let you use log lengths efficiently.

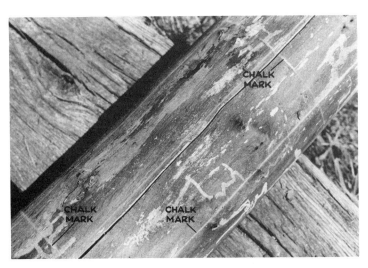

STEP 7. The snapped chalk lines give three parallel lines the length of the log. The two outside chalk lines and the two across-the-log score marks (parallel with beam edges) show the outline of the pocket to be cut.

STEP 8. The many chainsaw kerfs here minimize handwork in the next steps. (Note: On log and timber work you can cut directly on your marks; whereas in finished carpentry it is customary to leave the mark.)

STEP 11. A sharp chisel, guided with hand pressure, should be enough to clean broken ridges. (I've also used an old paring slick 2½ inches across the blade.)

STEP 12. After sealing the freshly exposed grain, roll the log pocket into place on the beam, locking the sill log in place. The pocket I've made is fairly flat and shallow enough not to reduce the sill's strength.

SILL LAP NOTCHES

To make bearing pockets on the sample 8-inch log shown on preceding pages, I suggested a horizontal 1 inch up from the bottom of the vertical. To make a half-lap splice joint, work down from the top of the vertical. If the two logs to be spliced are not equal across the diameters, it's better to have the discrepancy show up temporarily, at the bottom of the splice, in the area that will be removed for the bearing pocket in any case. Here are the steps of the sill sequence, modified to give you accurate half-laps.

1. Mark centerline verticals on each log face, as shown on preceding pages for the single-sill log.
2. Measuring down from the top of the centerline vertical on the larger log (assuming the two are not equal), mark the horizontal line right through the center (midpoint) of the vertical.
3. Snap chalk lines running back 12 inches (the size of the pier) from each end of the horizontal.
4. Make repetitive chainsaw kerfs from the 12-inch mark out to the end of the log, and chisel out the channeled wood on the top half of the log. Alternatively, if you're really good and really careful with a chainsaw, you can make a 12-inch rip cut following the horizontal, with a crosscut at the 12-inch mark to make the top half of the splice fall away in one piece.
5. Measuring down from the top of the vertical on the smaller log, mark the horizontal equal to the radius (one half) of the larger log, even if this point is slightly off-center on the smaller log.
6. Use kerf cuts and chiseling or cross and rip cuts to clear out the bottom half of the smaller log. These cuts will give you a 12-inch lap and will keep the top of the sill, the part you build on, level from one log to the other.
7. Seal the endgrain and spike the lap joint, or drill for alignment with an extended rebar or anchor bolt with countersink for washer and nut. Take care not to spike or bolt in the line of a mortise that will accommodate a girder or floor joist.

At a lap of sill logs over a beam, allow each log to bear fully on the beam rather than butting them at the center of the pier. Final notch marks on the bottom lap should be made as shown previously for the single-sill log.

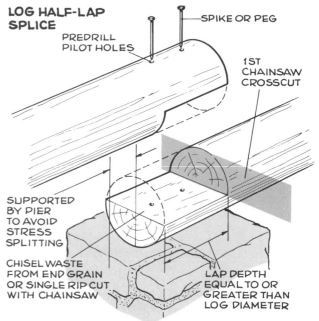

LOG HALF-LAP SPLICE
SPIKE OR PEG
PREDRILL PILOT HOLES
1ST CHAINSAW CROSSCUT
SUPPORTED BY PIER TO AVOID STRESS SPLITTING
CHISEL WASTE FROM END GRAIN OR SINGLE RIP CUT WITH CHAINSAW
LAP DEPTH EQUAL TO OR GREATER THAN LOG DIAMETER

Lap or scarf joints also can be made in heavy, square-hewn timbers. But if possible, try to find large, straight (or slightly crowned) single logs to use as girders. Bolts through half-laps may be a case of overkill. If the logs are approximately the same diameter, a few spikes will hold them in position.

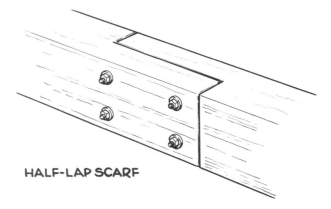

HALF-LAP SCARF

Purposeful log stagger must be carefully planned. If you were to use the big sucker in the middle of this wood pile and one of the smaller-diameter logs in the same course, you would get a stagger that could drastically upset the saddle notches.

With backfilling completed and the house completed, the skimpy-looking half-log sill along the short wall may not be so noticeable.

Below: With a concrete foundation supporting exterior wall weight, sill logs can be raised above grade on a log post. Stone and mortar fill between.

SILL CORNER LAP

With pier construction, it is easy to begin the log stagger where the center points of the logs on one wall line up with the seams between logs on the adjoining wall. For a typical rectangular cabin, sill logs and an interior girder, if required, may be set at the same level only if half logs or some other system is used to initiate the stagger.

Over piers it makes sense to install sills on the long side of the rectangle first, complete with bearing pocket notches. Then, all floor joists, and the joists at the outsides of the joist run, which serve as sills along the short walls, can be saddle-notched to rest on the long wall sills that bear directly on the piers. This procedure establishes log stagger, but there are drawbacks. You can imagine the amount of work required to saddle the second course of logs over every one of the floor joists, locking them into the wall system. This extra saddle-notching is worth the effort if you are after cantilevered

This drawing, adapted from the Beaver Log Homes construction manual, shows a full-log sill. In this case, a belt is recessed into the foundation, and joists are suspended from hangers.

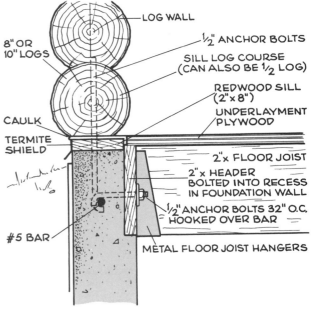

LOG WALL

8" OR 10" LOGS

½" ANCHOR BOLTS

SILL LOG COURSE (CAN ALSO BE ½ LOG)

REDWOOD SILL (2"x 8")

UNDERLAYMENT PLYWOOD

CAULK

TERMITE SHIELD

2"x FLOOR JOIST

2"x HEADER BOLTED INTO RECESS IN FOUNDATION WALL

½" ANCHOR BOLTS 32" O.C. HOOKED OVER BAR

#5 BAR

METAL FLOOR JOIST HANGERS

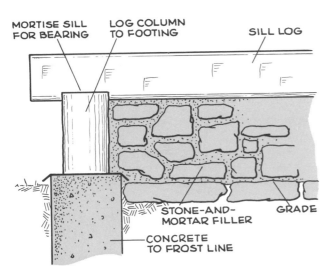

MORTISE SILL FOR BEARING

LOG COLUMN TO FOOTING

SILL LOG

STONE-AND-MORTAR FILLER

GRADE

CONCRETE TO FROST LINE

joists—to support a porch or balcony, for instance. And this is a very nice touch on second-story loft joists, cantilevering them out a few feet for a balcony outside glass doors to the bedroom.

Another approach is to establish the stagger on perimeter walls and stick to it, without any extra saddle notches, making progress at the rate of roughly one full course per day on a modest-sized cabin. (That's true for a reasonably competent handcrafter.) With this process, you treat the floor-joist system as an independent unit that can be hooked onto the log walls in a number of ways.

1. Joist logs can be flattened on top, for bearing against flooring material, and tenoned at each end to sit in mortises in the sill log or, for that matter, in mortises in the third or fourth log in the wall to create an unexcavated crawl space. Chainsaw plunge cuts (absolutely the most dangerous, kick-back-producing type of cut) into the sill can define the limits of the mortise, which is then cleaned out with mallet and chisel. Tenons, tapered in from the edges of the log, should be left at least half as wide and half as thick as the joist diameter.
2. A ledger to support floor joists can be cut into an oversize sill log.
3. Square pocket mortises, cut about halfway through the sill log, can be cut to accept milled, dimensional joists.
4. If you don't mind straying from handcraft methods and materials, the inside face of the sill log can be hewn or milled flat to accept a standard, dimensional timber belt (a 2 × 10, for instance, spiked and glued to the log face), from which conventional joists and joist hangers can be suspended. You might feel better about this efficient but somewhat

contradictory method knowing that it can be buried, out of sight, under the roughest, most rustic floorboards.

Another sill option is to use full-diameter logs in one direction and half logs (sawn in half along the full

Plunge cuts with a chainsaw are the most dangerous cuts to make. Instead of making repetitive kerfs the way you would on a sill-bearing pocket, consider using the saw only to outline the mortise, then doing all the cleaning work with hammer and chisel.

If you use joists 16 inches on center, working with endless, individual mortise pockets is needlessly time-consuming. A full ledger, milled, hewn, or cut with a chainsaw, allows for fast dimensional flooring.

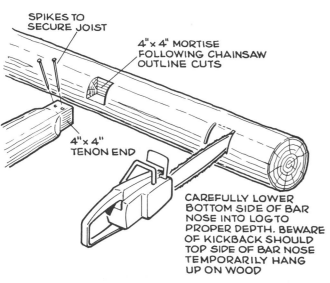

SPIKES TO SECURE JOIST

4" x 4" MORTISE FOLLOWING CHAINSAW OUTLINE CUTS

4" x 4" TENON END

CAREFULLY LOWER BOTTOM SIDE OF BAR NOSE INTO LOG TO PROPER DEPTH. BEWARE OF KICKBACK SHOULD TOP SIDE OF BAR NOSE TEMPORARILY HANG UP ON WOOD

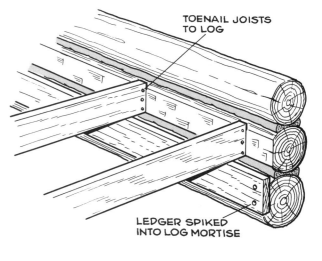

TOENAIL JOISTS TO LOG

LEDGER SPIKED INTO LOG MORTISE

Corner notches do not adhere to formulas and mathematical tables. As you can see in the photo above and on the previous page,there are big notches cut most of the way through medium-size logs and small notches cut only partially into huge logs. (Photos by Pacific Log Homes.)

length, leaving a half-moon section at each end) in the opposite direction. With this method, the saddle notch at each end of the full-diameter sills must be cut a tad deep to accommodate the loss of diameter due to hewing a bearing surface. You can use this method on a single-level full foundation, get full log to masonry bearing on all sills, and establish a log stagger at the same time. With all the advantages I still find the appearance of only half a log at the bottom of the pile a little disconcerting.

With square-hewn sills, pockets for individual joists can be milled, or cut (most of the way) with a circular saw. The flat interior surface also can be used to support a conventional ledger or beam hangers.

Although a conventional floor frame like this one does not exactly compliment the appearance of your log walls, it will be buried under the finished floor. With minimal trimming, a flat bearing surface on the sill log can hold a dimensional timber belt.

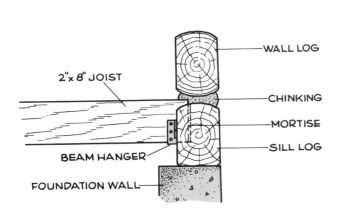

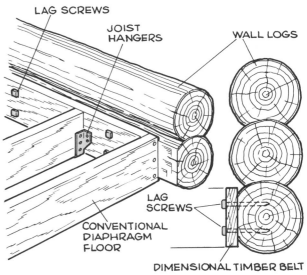

CORNER NOTCH DESIGN

Many of the sill and joist systems will bring you to the fundamental skill of log building: fitting two cylinders together, tightly and securely, so that two stacks of cylinders can be meshed together at right angles. There are easily 25 ways to accomplish this and structurally interlock log-wall corners. Several corner systems are purely curvilinear, and I think it takes a special kind of person to understand and execute these rounded-notch systems. If you continue to have trouble with the saddle notch as you read and look further, please don't take my last comment as a criticism. It's only natural that with linear building materials, linear lay-outs, linear houses and furnishings, linear distance measurement, and even a linear number system, most of us wind up being a lot more at ease, and a lot more competent, with linear carpentry—cutting straight lines and square mortises. I think this natural affinity for linear systems of all kinds is at work in the decisions of many owner-builders to use flat-hewn logs with square edges and square corners. With these materials, there are enough variations of lap joints, proportional rabbets, dovetails, even bizarre-looking compound dovetails, to keep every one of your linear-thinking brain cells in action. Here is a preview of the possibilities for flat-hewn and full-log notching systems.

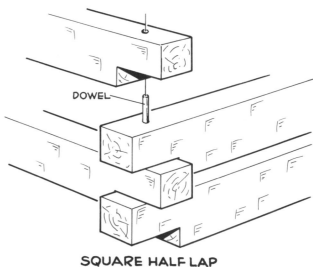

SQUARE HALF LAP

With square lumber, lapped timber can be cut to project beyond the corner or trimmed flush to make a finished, hard edge more like a piece of furniture.

Half-laps. Probably the easiest corner to cut, a half-lap meshes hewn logs but does not structurally interlock the timbers. Without spikes or dowels to lock the laps together, you could literally kick this type of corner apart.

Lock joints take lap joints a step farther. First, they allow the logs or timbers at the corner to nestle together. Second, they capture each log in the joint system so that it cannot slide loose along the horizontal or pull loose along the perpendicular.

Lock notches. These should remind you of Lincoln Logs from your childhood. Lock notches produce an exceptionally strong corner and an elegant, overlapped design. Each notch need be no deeper than a quarter of the log depth, leaving half the log intact.

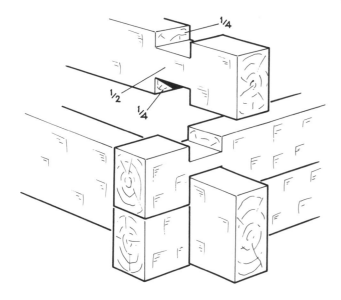

LOCKED LAP IN HEWN LOG

LOCKED LAP IN ROUND LOG

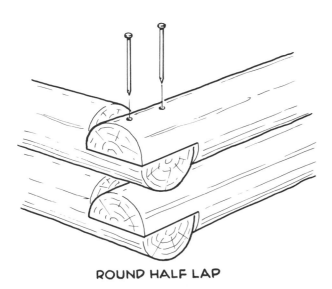

ROUND HALF LAP

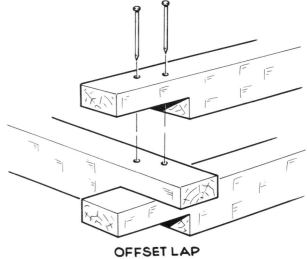

OFFSET LAP

Half-laps, the easiest corners to cut, can be pinned with dowels or spiked. What works well and looks good in flat-hewn logs, however, does create some peculiar shapes and joints in round-log walls.

Rabbeted joints. (Half rabbets are the same as half-laps.) Rabbeted joints can be cut to vary the monotonous appearance of half-lap joints; two-thirds to one-third, for instance, with a small lap section showing on one log, a large lap showing on the next. This system requires pinning as well.

V-notches. These use log weight against angled notches to keep corners interlocked. This system produces a lighter, more delicate overlap but requires a lot of measuring, cutting, and cleaning. You have to love the look to spend the extra time.

Although V-notches can be cut in logs, they are more successful in square timbers. After all, when the joint itself is made of straight lines, you will have more trouble carving them into round logs.

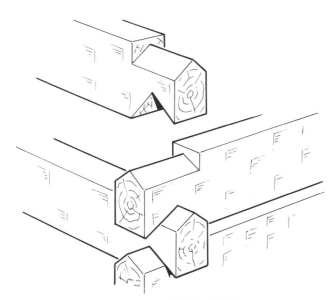

V-NOTCH IN HEWN LOG

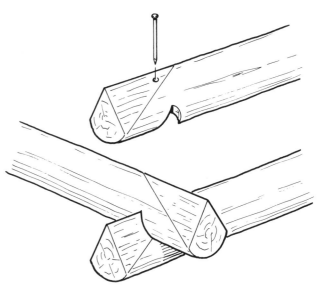

V-NOTCH IN ROUND LOG

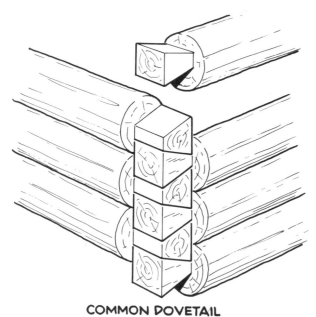

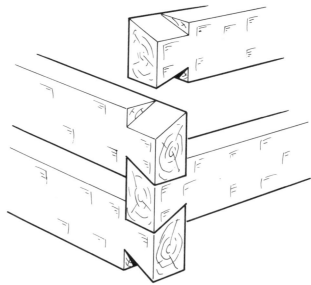

COMMON DOVETAIL

COMPOUND DOVETAIL

Dovetails, which you see on dresser-drawer corners, make an excellent notch if, and only if, you use a compound cut. The common dovetails shown on the round log corner look nice but do not keep the logs from pulling out of the joint. Only a spike does that. With compound cuts, each timber is held in horizontally and, by the angled cuts, along the perpendicular.

Compound dovetails. These corners transplant fine furniture construction on a grand scale. Dovetails are used on dresser drawers, for example, to keep corners that take a lot of pushing and pulling from separating. Good dovetail joints are stronger than

the wood itself. No spikes or dowels are needed. Compound dovetails take a lot of time and careful handwork. They do lock each timber into the wall from two directions, and that's good. But unlike drawer dovetails cut with a router and steel dovetail template in flat, straight hardwood, many large-timber dovetails have to be rescribed and retrimmed in place, up on the wall. Slightly bowed 6×12s often do not line up with mathematical perfection. Note that a common dovetail (without compound angle cuts on the notch faces) does not lock up the corner in two directions, and so is really not worth all the effort.

Double notches. Like shallow saddle notches cut with an alternating bevel on the top and bottom of each log, double notches appear in the finished wall as conventional saddle notches. Theoretically, you get more holding power with the beveled saddles. But this system can drive you crazy because it attempts to transplant a linear design (compound dovetails) onto curvilinear materials (full-diameter logs). Somehow, they just don't go together.

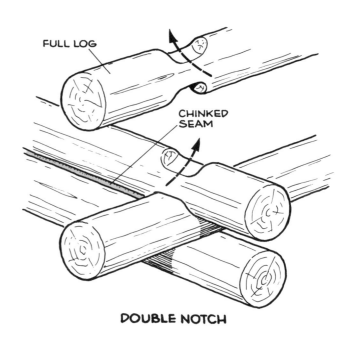

DOUBLE NOTCH

The double-notch modification of the simple Lincoln Log notch uses a beveled seat to help hold the corner together. It requires a lot of cutting (really sculpting). This complex notch becomes nearly impossible when you want fully seated logs without chinked seams.

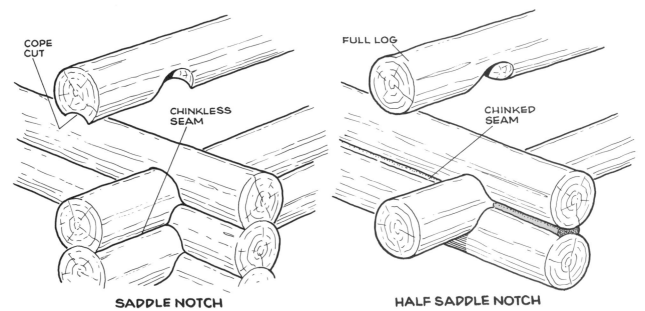

SADDLE NOTCH

HALF SADDLE NOTCH

Saddle notches. These notches are curvilinear, used for joining full-diameter logs—the classic, highly efficient, exceptionally strong log cabin corner system. They work for chinked or unchinked walls and shed water well. Cut halfway through the log but fitted snugly around the full log below, 2- or 3-foot overlaps can be made, enough to support large cantilevered logs to carry mammoth overhangs and porch roofs. No spikes or dowels are needed. In fact, spiking may hang up logs as they settle.

Saddle notches, in all shapes and sizes depending on the shape and size of the logs, can be cut for full seating (one log bears on the other) or for partial seating (space between logs is chinked).

Locked saddle notches. A special case, saddle notches provide exceptional holding power in all directions. They are ideal for girders and leave a lot of the log intact, hidden inside the saddle, as a key that fits into a slot cut into the log below. This key-and-slot design makes locked saddle notches ideal for large-scale, leading-edge cantilevers. Plus, I've included it because the system is intriguing, a structurally efficient delight. Can't you imagine someone rebuilding your cabin a hundred or so years from now, uncovering one of these beauties, and smiling?

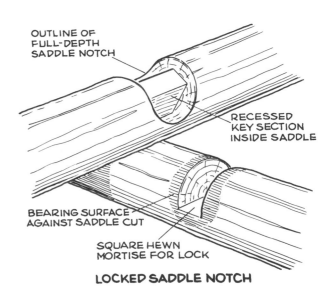

LOCKED SADDLE NOTCH

Truly a piece of sculpture, this time-consuming notch might well be reserved for special conditions where extreme security is needed.

The handcrafted log home below, built by Krissa
Johnson of Adirondack Log Building (shown right),
uses logs of both large and small diameter. The difference is dramatic on the second-story corners. Saddle notches, as shown being cleaned by Johnson with
a chainsaw, are the most efficient notches for homes
with great variety in log size.

CUTTING SADDLE NOTCHES

Before you take a crack at this classic, bear in mind that layout is two thirds of the job. The cutting is straightforward, using the familiar method of making multiple kerfs with a chainsaw about an inch or less apart, but this time cutting each one to a slightly different depth. Cleaning out the saddle notch with mallet and chisel is another familiar and straightforward job.

If you make a mistake, and you will, it will most likely be the result of scribe lines in the wrong place. The answer, not surprisingly, is practice, practice, practice. Here is a close look at the layout and cutting sequence.

First, let's agree on terminology. The bottom log on which no cuts are made, I'll call the sill log. The log on top of it, the one getting the saddle cut, I'll call the notch log. A scriber, the tool used to transfer the contour of the sill log onto the notch log, is a heavy-duty version of the compass you used in high-school geometry. One end can carry a pencil, or you can simply mark the log with scratches from the pinpoints at the end of the compass.

Also, you should make a few design decisions about the wall log to be notched before starting up the chainsaw. If you plan notch depth according to the radius of the log at its tip, where it rests on the sill, you may be in for trouble, even if you cut a perfect notch, because the log may not "lie down." The radius across the tip may increase slightly as you move back down the log to the thick end, called the butt end, of the tree. Also, the radius may change markedly at irregularities in the log—bumps and burls that hit the sill log first, preventing the rest of the log and the saddle notch from seating.

If you have a stack of irregular logs with a lot of taper end to end (tip to butt), and a lot of bowing, consider the benefits of chinked construction. Even a modest, 1-inch open seam between logs builds in a cushion for these irregularities. You may still have to trim a few limb stubs, but all kinds of uneven log seams can be successfully buried beneath accurate, even chinking. If you have a stack of straight, minimally tapered logs, chinkless construction becomes a more realistic possibility. Full-length V-grooving and scribing for chinkless log walls must be completed before notching. But more on that later. For now, let's work through the saddle notch.

HOW TO SCRIBE
A SADDLE NOTCH

STEP 1. Strip away any remaining bark or bark skin before sighting down the log. Look for the crown and for any major irregularities such as limb stubs.

STEP 2. In addition to marking the verticals on the log ends, for additional reference you can mark a point on each log—say, where a check line, as shown, on the sill log hits the notch log.

STEP 3. The wider you set your scribers, the larger the notch will be. Setting at roughly half the notch-log diameter will sink the notch log roughly halfway into the sill log.

STEP 4. Professional log builders tend to use large, heavy-duty scribers. But you can use almost any kind of compass as long as you keep the points vertical and in contact with both logs while you mark.

1. As shown in accompanying illustrations, start by selecting the top and bottom of the notch log, then mark the top of a centerline vertical with an arrow. This prevents confusion as you roll the log on and off the sill. You have a choice here: to let log crowns dictate top and bottom or to put the smoothest, most even pair of opposing sides top and bottom. The idea is to leave irregularities on outside and inside wall surfaces, not in the way of other logs in the wall. Ideally, you'll get many logs with four smooth and regular quadrants, or logs that have irregularities in the quadrants opposing the line of the crown. If not, you have to shave down the protrusions.

2. Dog or brace the notch log in place, resting it on the sills at each end of the wall.

3. Set your scribe to equal the radius of the notch log.

4. Start the high point of the scribe at the bottommost point of the notch log, with the low point of the scribe directly below, on the outside edge of the sill log. Keep the two points in line and vertical, in contact with the logs at all times. Think of the bottom point as the feeler, the top point as the tracer. If you break the link between them, the trace pattern is just a guess. The better you do this job, the more accurate the contour marks.

5. Move the scribe up and around one side of the notch log, letting the compass drift in from the outside of the sill until you reach the uppermost part of the sill log, at or near its midline. Continue past the midline down the inside of the sill log, letting the compass drift in underneath the notch log until the top point of the scribe meets the midline of the notch log. Half the saddle contour is now complete. Repeat this process on the opposite side of the notch log to finish the roughly oval saddle pattern.

Do not expect to see a smooth-lined, symmetrical outline for the saddle notch. The contour line should be as ragged and as bumpy as the sill log—in fact, exactly as ragged and bumpy. The point is to duplicate a portion of the sill log on the notch log, then remove it so that the two fit together. One last tip may help you get good results. It is almost impossible to have continuous bearing all across the surface of the saddle notch. And for good reason. Remember, you're duplicating the oval contour of the sill log, not all the ups and downs inside the oval. Some of those ups may get in your way. The solution is to make the bearing surface of the saddle notch concave. (See next page.)

STEP 5. You might want to make the first mark lightly, then retrace, following the log contour, but checking the scriber alignment from the side. Try to mark one complete arc on the log without changing position. Working in line with the sill log, you can peak around one side, then the other, and still maintain scriber alignment. The bumps on the sill log are positive space. You have to mark the notch log to make negative space. Repeat this procedure on the opposite side of the notch log.

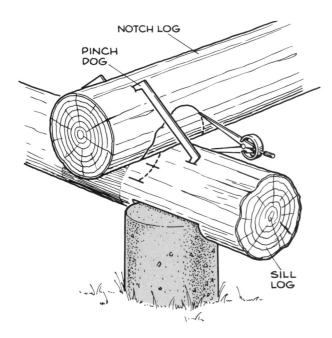

NOTCH LOG

PINCH DOG

SILL LOG

HOW TO CUT A
SADDLE NOTCH

STEP 1. Don't trust fine, pinpoint contour lines. Mark the contour lines clearly, as shown, before starting up the saw. (Photo sequence by the author.)

1. Roll the notch log 180° in from the wall, and dog it in place.
2. Make successive chainsaw kerfs inside the saddle contour, keeping the saw level and square to the log. As the saw bites through the wood, check the progress on both sides of the log, stopping the saw just at the contour line.
3. Using an axe or a mallet and chisel, chop or cut the channeled wood away, leaving a ragged version of the notch.
4. Use a hatchet or hewing axe and work around the contour line to trim a neat edge just to, but not onto, your pencil line. This is not splitting hairs, but it is splitting pencil lines. That's the detailed scale that needs your attention in order to get high-quality notches. This step is similar to "cutting in" when you paint a wall: working carefully with a sash brush where the wall meets the ceiling so that you can use broad and efficient strokes with a roller on the rest of the wall.
5. Working from the carefully cut contours in toward the center of the notch, clear away the remnants of wood channels left from the chainsaw. Since the saw kerfs are at, or slightly above, the contour lines, the remaining wood will hit the sill log before the contours of the notch hit. This leaves an open joint.
6. Leaving the contour edges untouched, chisel a concave shape into the notch. The best tools for this job, as you would expect, have concave shapes. One is an adze—good for the initial concave shaping, but sometimes a little bulky for a saddle shape in small-diameter logs. Another useful tool is a wide-mouthed gouge, which is simply a wide chisel with a gently concave cutting edge. You can safely put a half-inch concave shape, or cup, into the notch.

STEP 3. To clean out the notch on big logs, use larger tools, like this hand sledge and 2-inch paring chisel. For smaller-diameter logs, you can use a carpenter's

STEP 5. The roughly trimmed notch now needs only final cleaning. Some builders tend to this with hammer and chisel or a small, concave-shaped adze. Some builders do the entire job with a chainsaw.

STEP 2. Saw kerfs at the outside limits of the notch must be extremely shallow. As you move toward the center of the notch, approach the contour lines carefully as the chain bites into the wood more deeply.

The more cuts you make, the easier it will be to clean the notch to at least the rough outline of the notch contour.

hammer and chisel. Though messy, the fastest way to clear away wood between multiple saw kerfs is to hit them with a hammer.

STEP 4. The edge of the contour line should be trimmed carefully. A very sharp hatchet can be used to make a clean edge, the edge that will show against the sill.

STEP 6. The inside of the notch should be smooth and slightly concave. It does not have to be as smooth as horse saddle. Clear away all wood that will prevent the notch from seating fully on the sill log, but there's no need to try to make fine furniture.

A good saddle notch should closely follow the contour of the sill log. If you find openings, and the notch is sightly concave, you can close them by trimming just at the edges of the notch where needed. The notch log now sits halfway down into the sill log.

Long-bar, heavy-duty saws have ample power for the job. The trick is to let the saw do the work while you provide the direction. A professional handcrafter can use plunge cuts to shape the notch outline and the notch depth. Noise-deadening earmuffs are standard. Goggles, shown in so many how-to books, are not. Yes, they will help keep wind-blown and chain-tossed wood chips out of your eyes, but I have tried them many times, and found they did more to distort my peripheral vision and general sense of my surroundings. That's bad news working up on the corner of a log home. Ordinary eye glasses serve me better. (Photo by Don Smith.)

Many "handcrafters" are skilled enough with a chainsaw to safely sculpt a smooth saddle notch with the tip of the saw bar. This is potentially dangerous work that invites kickback, where an unexpected kicking force causes your left hand to lose its grip and could cause severe injury. This procedure can be used with reasonable safety *only* by the most experienced craftsman. (Photo by Don Smith.)

Always grip a chainsaw firmly with both hands. A left handguard, as shown, is minimal safety protection. For a little extra money, and superior protection, you can have a left handguard that also serves as a lever-actuated chain brake. This brake stops chain rotation within milliseconds of being tripped forward by your left hand. You may think you would never let go, never expose the hand to the chain. But during a saw kickback, even a slight slip on wet leaves, your right hand may lock on the saw handle (and the trigger) while your left hand instinctively raises up to ward off problems. Since you can't change your instincts, invest in a chain brake. (Photo by Neil Soderstrom.)

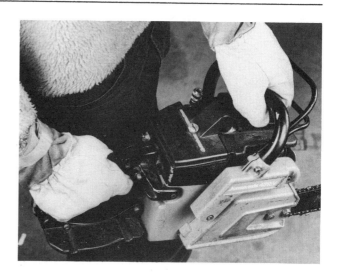

For finish notch work, chisels are safer to use than chainsaws, and also allow higher precision.

If you continue to have trouble, if the contours simply refuse to settle down on the sill log, there is a last-resort solution. Squirt a layer of ground chalk onto the sill log where the notch should rest. Dust the area liberally, using a plastic squeeze bottle. It's like dusting for fingerprints—in this case, fingerprints of high spots on the saddle that are getting in the way. Set the notch log in place over the chalk gingerly, then roll it back off. Chalked areas on the saddle notch show exactly where it is bearing. Shave down these high spots for a close fit. (Then wipe away all traces of chalk and try to look very nonchalant about the perfect fit, like it was no big deal—just a few shots with the axe and it rolled into place.) There is another approach to this problem. You can put a fat inch of concave shape into the saddle and use the excessive clearance to hold some insulation.

CHINKLESS WALLS

You've got to be careful with logs. They are a little like peanuts and potato chips. You get a taste with bark peeling. Then you want a little more, so you go back and hew the sill logs. Then you develop a real taste and want the satisfaction of saddle notches. And finally, with a craving that approaches addiction, you will settle only for the ultimate—chinkless walls.

First, let me play devil's advocate and try to warn you off this addiction. Saddle notches are bad enough.

But scribing contours, hewing up to pencil lines, rolling logs on and off the wall and looking for high spots—why on earth would you want to extend these headaches along the entire length of every log in every wall? Think about the summer turning to fall while you're still working on a cabin with no roof and no windows or doors; it's crazy to spend the extra time, right? And what are the advantages, anyway? Chinkless walls aren't necessarily stronger than chinked walls, or more energy efficient, or more durable. Of course, chinkless walls are always one of a kind, always something special.

Let's take it in stages. Suppose you have a log with an irregular shape, with a bowed section, protrusions and limb stubs on all sides. If you make saddle notches based on the radius at the end of the log, some of the bowed section and all of the protrusions will be in the way. (A 2-inch-deep bump on a log makes the diameter at that point 2 inches larger and the radius 1 inch larger, right?)

So forget about log radius at the ends of the log. Instead, run the scriber along the entire gap between the sill log and the notch log before any notches are cut. This means the top point on the scriber will mark all the protrusions and most of the bowed sections. (Remember, the points must be plumb and in contact with the logs at all times.) Try a few dry runs before laying any marks on the logs. And use your judgment to estimate the largest average distance between the logs before locking up the scriber setting. Now apply

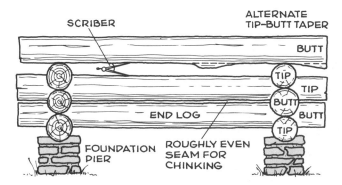

SCRIBER   ALTERNATE TIP-BUTT TAPER

BUTT

TIP
TIP
BUTT
BUTT

END LOG

TIP

FOUNDATION PIER   ROUGHLY EVEN SEAM FOR CHINKING

When logs are peeled and trimmed at the mill, little details like this limb stub are taken care of for you. You can clear these obstructions by placing them inside or outside the wall, or you can use the scriber to discover just how big the protrusion is and trim it off.

This large-diameter, standardized log cabin is from Rocky Mountain Log Homes. Notice the saucer-shaped cove cuts along the bottom of each log. This machined system has a handcrafted look.

this dimension on the ends of the logs to cut saddle notches, and use a chainsaw or hewing axe to clear any irregularities on the log length, working just up to your scribe line on both inside and outside log faces.

This is a rough, on-site method of standardizing your logs. A few manufacturers do this at the mill, peeling and shaving all their logs to the same, smooth-surfaced diameter. This regularity has obvious benefits during construction. It eliminates all the bows and bumps. That's fine for some people. And as you can see, some of these standardized log homes look pretty rugged and pretty darn good. But for some people, this machining is an unacceptable step toward regimented bricks-and-mortar construction—the kind of building they want to leave behind.

Preliminary scribing and cleaning over the full log length will make your logs sit right on top of each other. Sure, you'll be able to see cracks of light—it's not like the doors on a bank vault. But this high-quality fit will let the wood bear against wood—off and on, at least— all along the walls instead of only at the corners. Inevitably, this makes your cabin more of a solid, load-balanced, single structural unit. It also increases your options for wall chinking. You still have room for a little caulking and insulation, but you are not forced, as you

would be with large gaps between logs, to show 3 inches or more of chinking for every 5-6 inches of log. As this proportion starts to approach 40/60 or, worse yet, 50/50, your home may begin to look more like a stucco building with strips of wood added on for effect.

Unless minimizing time on the job is your number-one priority (and I'm guessing that if you're interested in building your own log cabin, this isn't true), I think this extra scribing and trimming is worth the work—for strength and durability, for appearance, but most important, for the pride of taking the trouble to do the best possible job and gaining the satisfaction of refusing to let bumps and bows dictate the quality and craftsmanship of the job.

I did say take it in stages. Well, once you've gone this far (no, I didn't say once you're this far gone), it's not that much further to a chinkless cabin. In the first stage: Lay up logs, forget about trimming, fill in the gaps with chinking later on. In the second stage: Lay up the logs, scribe and trim off the excess to close the gap, and finish with minimal chinking, even strips of molding. In the third stage: Groove the entire bottom of the log, then scribe and trim each edge of the groove so that one log nestles on top of the other.

Most log builders speed up this process by cutting a V-groove along the bottom of the log with a chainsaw. Once the top and bottom are decided and the verticals are marked at each end, here's what happens:

1. Roll the log over 180° so you are working with the bottom up.
2. Mark a V-shape centered on the vertical line at each end, with the legs of the V-shape branching out at roughly 30° angles. (The angle at the point of the V can be 120° to 150°.) The width of the V-groove will vary, depending on the diameter of the log, but commonly does not need to exceed the log radius.
3. Duplicate the V-shape layout at each end of the log, cut tiny notches where each leg of the V hits the edge of the log, and snap two chalk lines to transfer the limits of the V-shape along the log length.
4. Working with extreme care, and only from a secure position with firm footing, use the chainsaw at the angle of your V-mark to cut the groove. Work backward (as long as you know where you're going), with the saw at a relatively flat angle to the log to avoid jamming or burying the tip.
5. Roll the log in place, and scribe the contours of your sill log onto the edges of the V-groove on your notch log.
6. Roll the log back off the wall, and trim the edges of the V-groove to the scribe lines. Use a hewing axe, holding the handle just below the head for the best control.

With this process you create negative space (the groove) that will hold not only all the irregularities of the sill log, but also some insulation, too. In the same way that saddle notches are made concave, trimming the edges of the V-groove simplifies the process of matching and meshing the contours of one log to another. It's very logical. It's very time-consuming. I advise that you experiment with this system before adopting it, because there are a few wrinkles. Some builders use a sharply angled V-groove or a very shallow V-groove, depending on the size and shape of the log. Each log is a little different. Chinkless work is the art of accommodation, not an exact science.

Chainsaw operator's manuals won't show you pictures like these. Burying the tip of the chainsaw is not good practice for novices. However, it is the most efficient way to rough-cut the V-groove prior to scribing each side of the groove to the contour of the preceding log in the wall. This expert is observing a cardinal rule of chainsaw use: keeping the body (including feet and legs) out of the line of cut and potential kickback. (Photo by Don Smith.)

Large handcraft companies like Pacific Log Homes are practical enough to gain advantage where they can from mechanical equipment. This specialized log lifter can pick up and deposit logs quickly and safely. You can custom-rig a forklift or backhoe to gain nearly the same efficiency.

This piece of rigging is called a tip-up jammer and can be site built. There is no single right way to build one. Any tripod-type structure with mechanical assist (in this case, a pulley at ground level) will serve. The ultimate version of this kind of rigging is a gin pole, a tall, guy-wired pole with block and tackle set in the center of the site. The pinch dogs (like ice tongs) release only when the log is positioned and the weight on the steel cable is unloaded.

The purely hand-powered version of chinkless construction requires an adze instead of a chainsaw—more precisely, a gutter adze with a curved head and cutter edge. This tool can be used to hew a shallow trough (a saucer shape) along the bottom of the log in place of the V-groove. The scribing and trimming process is the same. Compared to chainsaw V-grooving, this system is really time-consuming and even more rugged on your back muscles.

RAISING LOGS

No matter how strong you are, there will come a time when you need some mechanical help to lift your logs up on the walls. Maybe you and a helper could pick up 250 to 300 pounds. But lifting that weight in the unwieldy form of a log, then maneuvering along scaffolds and up ladders at the same time, is another story, and not the second story.

Kit companies that use large-scale timbers will most likely arrive with a crane of some sort. If you have a backhoe on the site after excavation, you can chain several logs across the bucket, raise them to the wall height, and leave them there, waiting and ready when you are. Without this heavy-duty mechanical help, your best bet is to take big beams up the wall one step at a time.

Although I almost always worked with a crew, I once built a large timber-frame addition onto a Cape solo. I "walked" the 4 × 10-inch second-story floor joists up a series 2 × 4 angled brackets, put plywood gusset plates on them to avoid any unpleasant surprises, and spiked the brackets into the wall framing. I was able to "walk" the 4 × 12-inch ridgepole up a second set of brackets on the gable-end walls. The real killers were the 4 × 10-inch, 4-foot center rafters. I was able to get them almost in position with a come-along, a hand-ratcheted winch used with a steel cable that has excellent hauling and holding power. But I wound up enlisting help from a fellow builder because I didn't feel secure working at one end of the timber while it was essentially hanging from a steel thread.

The lesson here is to use your common sense about log-raising. You might be able to do everything yourself with scaffolds and braces; you might take the time to build a gin pole, a tall, straight log, guy-wired upright in the center of the cabin, which gives you purchase (with block and tackle) to raise logs and maneuver them into place.

But designing and building scaffolding and rigging is a trade in its own right. You have to resist the temptation (and it is a challenge) to deflect too much of your ingenuity and energy away from the cabin proper and into systems to help you build it.

On chinked walls, you can take advantage of the spaces between logs by inserting anchor bolts or an equivalent piece of threaded hardware to which you can bolt scaffolding. Use scrap lumber and build to suit your site conditions and building loads. Load up your scaffold on the ground first for a test.

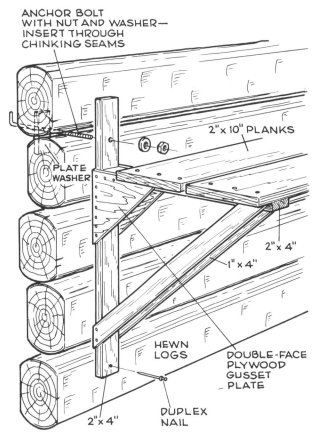

ANCHOR BOLT WITH NUT AND WASHER—INSERT THROUGH CHINKING SEAMS

2" x 10" PLANKS

PLATE WASHER

2" x 4"

1" x 4"

HEWN LOGS

DOUBLE-FACE PLYWOOD GUSSET PLATE

2" x 4"

DUPLEX NAIL

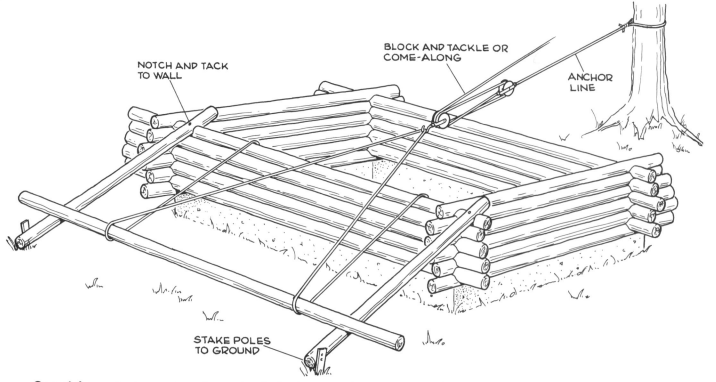

NOTCH AND TACK
TO WALL

BLOCK AND TACKLE OR
COME-ALONG

ANCHOR
LINE

STAKE POLES
TO GROUND

One of the easiest ways to gain mechanical advantage is to use an anchored block and tackle and log skids. Two points: Make sure you have a brake built into the system (even a simple rope tie-off), and make sure the log won't take anybody with it on the way back down if something snaps.

One of the time-honored methods of gaining mechanical advantage, this system is called getting a whole lot of people to come over and help push the great big sucker of a ridgepole up over the eaves.

On most sites, however, there is a reasonable middle ground—skidding the logs into position. Two or three 5- or 6-inch-diameter saplings can be angled between the log wall and the ground. The saplings should be debarked, staked at the ground for security, spiked, chained, or otherwise secured temporarily at the top of the log wall, and greased or soaped to minimize drag. This is the traditional way to gain mechanical advantage.

Make a two-point yoke for the logs, say, with a chain hitch for each end of the log running to an O-ring in the middle. If you pull with one line from the center of the log, you'll lose it. I would stick with steel cable instead of rope. And no matter what kind of mechanical edge you get with pulleys, I would not rely on hand power to haul and to hold. Something may happen,

even something silly like a sneeze, that, just for an instant, makes you loosen your grip. Insert a come-along in the system to gain even more of a mechanical advantage and to gain a safety factor. The ratchet gear will hold what you haul.

Start experimenting with skids while the walls are still at a reasonable elevation. And try not to skid at more than a 45° angle. If there is a high and low side to the cabin site, you might think about getting the log pile set up on the high side, where your skids can be a little flatter and a little shorter.

Many builders use a crowbar to get leverage and wedges to keep the log in place instead of heavier, metal pinch dogs. (Photo by Don Smith.)

TOP LOG NICHE
FOR SAW TIP

WINDOW SILL LOG
BEVELED
FOR DRAINAGE

2" x 6" SAW GUIDE
TACKED TO LOG WALL

Ideally, openings cut after the walls are completed will leave at least half of the header-log diameter intact. Remember to save a few log ends from the cuts for eave walls, firestops between rafters, and other small pieces of the cabin.

If you are sure of the openings you want and don't mind giving up the option to change your mind after the walls are finished, you can prenotch the opening to avoid plunge cuts with a chainsaw later.

OPENINGS

Before I get you too far up the wall, you should know that there are two different ideas about planning for windows and doors in your cabin. One school says to build your walls nonstop. Then you go back and literally cut the window and door shapes out of the solid log walls. Another school says why waste all that good timber? If you have a short log or a log with a bowed section or other substandard section, why not take the time to match it up with a section of the wall requiring a short log—from the corner in 4 feet to a window, from the window another 6 feet to the door.

There are pros and cons to both methods. Full-length logs will give you a certain security at openings. You will know that what you find on one side of the opening will also appear on the other side. The full-log system also lets you change your mind. Many owner-builders look at the huge stack of logs and the spacious numbers on the blueprints only to feel claustrophobic when the walls are up. Where one small window seemed to make sense in the planning stage, reality may suggest a couple of 3 × 6-foot double-hungs. Another bonus of the full-log system is unified cabin appearance, inside and out. If logs at the same elevation in the wall change diameter, tone, and texture from door to window to corner, inevitably the facade will appear a little choppy.

The openings-as-you-go system offers outstanding lumber economy and something else that's pretty basic—a way in and out of the cabin. Without it, you'll

do a lot of climbing over walls, a few feet higher every day.

Whichever system you settle on, the keys are to prevent the logs from shifting (or literally falling out of the wall) between the time you cut the openings and the time you get the openings framed; then to frame with a system that allows for log shrinkage and settling. If your 3 × 7-foot door is attached to a 7-foot-high stack of logs that later becomes a 6-foot-11-inch stack of logs due to shrinkage, something will have to give.

If your wall logs (either full-round or hewn-square) are well seasoned, shrinkage at openings is not a serious concern. The same goes if you take forever building the cabin and let the walls sit there for a year before you get around to closing in the place. In chinked construction, you can use wedge supports (shingle shims work well) in the gaps between logs for vertical support, with scab boards (any long, flat, sturdy piece of wood) spiked to the faces of the log walls for lateral support. If the idea of nail holes in your beautiful log walls bothers you, apply the scab board on the exterior face only; better yet, use a pair of scabs back to back with bolts and washers through the gaps to maintain vertical alignment.

On chinkless walls, where the logs already bear on each other, a single scab should provide enough stability. In any case, don't underestimate the forces in a log wall. If you make a cutout and the logs do shift, you'll have a heck of a time unloading enough weight

WEDGES

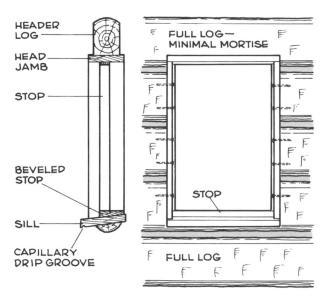

HEADER LOG

HEAD JAMB

STOP

BEVELED STOP

SILL

CAPILLARY DRIP GROOVE

FULL LOG— MINIMAL MORTISE

STOP

FULL LOG

On chinked walls, logs will float at openings unless they are supported. If you use full-length logs, wedge and brace before you cut openings. If you use logs cut to length, you must wedge and brace as you go. These timbers are from Appalachian Log Structures.

Making your own frames takes extra time, but it may be the best way to have trim and finishing details match the ruggedness of the cabin walls. Leave all wedges in chinkless walls in place until the frame has been well secured.

so you can put the pieces back where they belong.

The width of windows and doors doesn't matter much as long as you keep openings several feet away from interlocked corners. (Make sure that at least one of the openings is wide enough to provide access for, say, a big kitchen cookstove.) Height does matter. You don't want the sill jamb or the head jamb to fall in the gaps

of chinked walls (in the chinking material), or just at the edge of a log, or almost all the way through a log. Ideally, the sill and head frames should fall somewhere in the first third of the logs above and below the window. This is particularly true for the load-bearing log across the top of the opening, which acts as a lintel or header.

Individual logs in the wall are secure only as long as they are tied together. Scab boards like this one will keep the pieces together and can serve as braces during assembly of tapering gable ends.

A straight 2 × 6 or 2 × 8 scab board will help to stabilize the gable-end wall during construction. But if you plan ahead, it can do more. Set it at the midline of the wall, as shown, as a brace for a ridgepole.

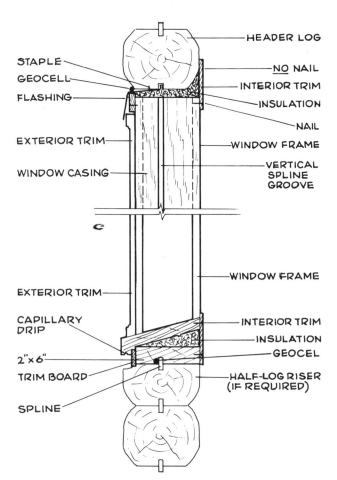

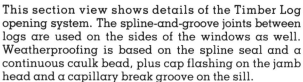

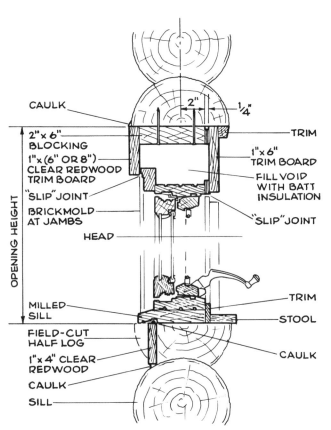

Left diagram labels:

STAPLE
GEOCELL
FLASHING

EXTERIOR TRIM

WINDOW CASING

HEADER LOG
NO NAIL
INTERIOR TRIM
INSULATION
NAIL
WINDOW FRAME
VERTICAL SPLINE GROOVE

WINDOW FRAME

EXTERIOR TRIM

CAPILLARY DRIP

2"x 6"
TRIM BOARD

SPLINE

INTERIOR TRIM
INSULATION
GEOCEL
HALF-LOG RISER (IF REQUIRED)

Right diagram labels:

CAULK

2" x 6" BLOCKING
1" x (6" OR 8") CLEAR REDWOOD TRIM BOARD
"SLIP" JOINT
BRICKMOLD AT JAMBS

OPENING HEIGHT

MILLED SILL
FIELD-CUT HALF LOG
1" x 4" CLEAR REDWOOD
CAULK
SILL

2"
1/4"
TRIM
1" x 6" TRIM BOARD
FILL VOID WITH BATT INSULATION
"SLIP" JOINT
HEAD
TRIM
STOOL
CAULK

This section view shows details of the Timber Log opening system. The spline-and-groove joints between logs are used on the sides of the windows as well. Weatherproofing is based on the spline seal and a continuous caulk bead, plus cap flashing on the jamb head and a capillary break groove on the sill.

Beaver Log Homes provides this section view of window detailing in their construction manual. You can see that a different log configuration calls for different detailing. Without a spline-and-groove seal, this system relies heavily on trim boards and caulking. A piece of cap flashing wouldn't hurt.

This picture could be taken only on a kit-built site. On this Timber Log model, all windows are set and braced over the first few full log courses. Individual precut sections are fitted between the openings using a spline-and-groove system for weatherizing.

If you're building everything from scratch (splitting your own shakes, too), you can tailor window and door elevations to the logs as they fall. If you're selecting windows and doors from a catalog, check out possible opening locations before ordering. It is usually possible to find and stick with window elevations that meet the first-third requirement. Most firms offer heights in small increments.

Here are some of the ways to frame and install windows and doors. First, don't expect conventional jambs made for stick-built homes to work well in heavy timber walls. The slightly chintzy, barely ¾-inch-thick jambs will look out of scale—in fact, will be underbuilt for the situation—and won't come through wide enough to breach the log wall in any case.

It is tempting (if you have the money) to buy everything prehung; i.e., hinged and trimmed and ready to go, usually with finger-jointed jambs. However, they won't look so hot on a log cabin, unless you paint them. Stain will make obvious the fact that these jambs are made of scrap lumber with different grains and tones.

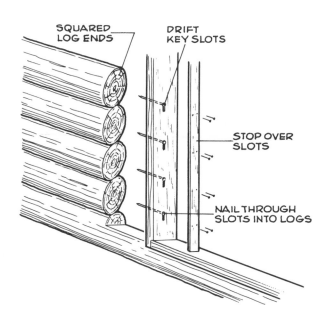

Drift key slots (shaped like keyholes) cut into window and door jambs permit shrinkage and settling to occur without damage. Without some provision for movement, frames can split or logs may become hung up on spikes, which exposes chinked seams and chinkless joints.

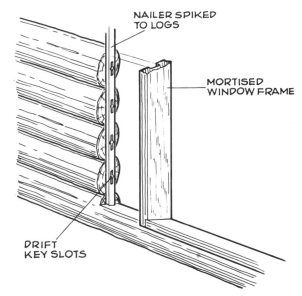

With this drift-key modification, a grooved frame is weather-sealed with a large-scale spline. Only the spline is drift-keyed.

By the time you get done with shimming and adding jamb extensions, you might as well make your own casings, to suit your cabin. Here's how.

For the basics, if you mill your own stock, do it early on in the job so the wood can air-dry. If you use dimensional timbers like 2×8s or 2×10s, buy KD (kiln-dried) stock. Rabbet the corners, use diagonal braces to lock up a square layout, predrill the side jambs with pilot holes to avoid splitting, glue and spike the frame together, and set the frames aside until the glue sets and hardens. This basic, solid frame could be spiked directly to the logs (with some insulation in between) and would hold up fairly well, at least for a while. But I would like to convince you to do more than that.

1. For door bucks, replace the frame sill with hardwood (oak is used commonly), and bevel the exterior face of the sill slightly up to the plane of the door to encourage water runoff. This gives you better weather protection and of course greater durability.

2. For windows, either duplicate this sill bevel on the outside up to the sash line or, better yet, cut a roughly 15° angle on the base of each side jamb to create a sill that is fully beveled against the weather. First-floor frames, even under a substantial overhang, can use this built-in protection against leaks.

3. With either flat or beveled sills, use a plow plane or router to cut a drip groove—technically, a capillary break—on the underside of the sill just in from the outer edge. Even a small, 1/8-inch-wide by 1/4-inch-deep groove will prevent undersill leaks caused by the capillary action of water. And, yes, water can climb back up a slightly beveled sill.

4. At the very least, give the wall logs a chance to settle without overtorquing the frame by cutting small drift slots for the frame-to-log spikes. You can make this small vertical slot with a hand auger or power drill, cleaning the edges of multiple holes with mallet and chisel. Spike through the frame

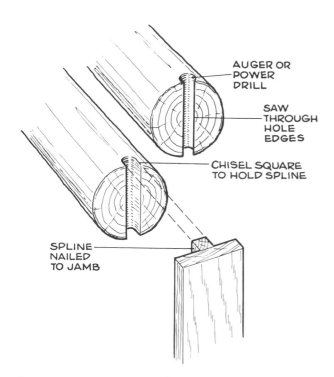

AUGER OR
POWER
DRILL

SAW
THROUGH
HOLE
EDGES

CHISEL SQUARE
TO HOLD SPLINE

SPLINE
NAILED
TO JAMB

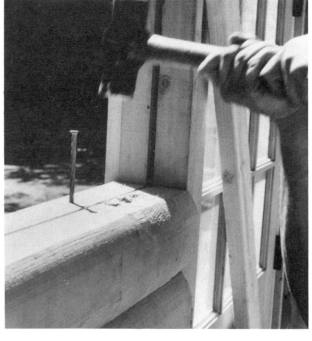

Drift-key slots can be made quickly with a drill. But more time and more care are needed to mill grooves into the end grain of logs at openings. An auger and chisel can create the groove to accept a spline attached to the frame.

Spikes driven through sills can be concealed with frame and trim. Many builders drill a hole through the logs in each new course so the logs can settle without binding on the spikes. (Photo by Timber Log Homes.)

and into each log, setting the spike at the top of each slot. This will secure the frame and keep the logs aligned. As the logs shrink and settle, the spikes can drift downward along with them without splitting apart the frame.

5. To allow for shrinkage and settling and, at the same time, to get a positive seal against the weather, think about one of these two spline-and-groove systems. The weaker version calls for a drift-keyed, slotted spline (a 1×1 or $1 \times 1\frac{1}{2}$ strip, for instance) spiked to the log ends, and a corresponding groove cut into the side jambs. The stronger version calls for grooves in the log ends and a corresponding spline added to the jamb. Dimensional 2×2s (measuring $1\frac{1}{2} \times 1\frac{1}{2}$ inches) make hefty spline stock. The easy part of the job is applying a liberal bead of exterior glue and nailing through the face of the spline into the jamb to make a stubby, very solid T-shape.

6. With scab boards in place for security, you may spike the wall logs together, working in about a foot from the opening. Spikes may be driven directly through softwoods, but they usually need pilot holes in hardwoods to prevent bending. Some builders avoid spikes, which may hang up the logs.

7. Use a chainsaw with a vertical guide board to help you lay a groove into the log ends. The width

should be just slightly wider than the spline minus the thickness of two chainsaw kerfs. This way, when you cut on each side of the guide board, the final groove will be just wide enough for the spline.

8. Tack the guide board to the log ends so that it is plumb with the centerlines and centered in the wall.

9. Working very carefully (this is another potentially dangerous plunge cut) and at a flat angle to the log ends (to minimize the risk of saw kickback), run the chainsaw kerfs down each side of the guide board to a depth just slightly greater than the depth of the spline. Some builders make themselves a clamp-on depth stop to help control the saw tip.

10. Clean the groove with mallet and chisel, and slide the frame down into place. Pack the space between the frame and the logs, and up to the spline, with insulation before trimming.

You may have to cut your own openings. But if hand-built frames, splines, and grooves stretch you to your limits, consider ordering these units from log home kit manufacturers. They may not be thrilled that they cannot sell you an entire home, but a dozen windows and two doors might be a sale that some of these manufacturers would like to make.

ROOF SYSTEMS

Roofing can be the most difficult part of a log cabin. For one thing, all the cutting and trimming and fitting you are used to doing with your feet on the ground now must be done way up there in the air. If working up there bothers you, don't go up there. Unlike the old Henny Youngman joke (I told the doctor it hurts when I do that, and he said, "So don't do that"), this simple advice should be taken seriously.

If the inevitable precarious perches make your heart beat a little faster, make your hands clammy, your throat dry, and your knees wobbly (you know best about your own personal set of anxiety signals), you will probably have a tough time up there. The quality of your work will suffer, too.

I do not want to scare you off this part of the job unnecessarily. But you've got to be honest with yourself. Enough said, except that this advice is here because as hard as you labor, deep down you've got to love the work—you've got to enjoy yourself.

Again, let's run through the terminology so we can conjure up the same pictures. On a cabin with a peaked roof, the walls carrying the rafter tails (usually the long walls) are called the eaves. The opposing walls (usually the short ones) are called the gable ends. The key beams, aside from the rafters, are: the ridgepole, which runs along the roof peak between the rafters; the collar beams, which may be used to turn the inverted V-pattern rafter into a stronger A-pattern; and the plate logs, which are the top logs on the eave walls specially prepared to carry the rafter tails.

Before you decide what kind of roof system to use, you must settle on where the roof should start. Until you know that, you don't know where the wall height should stop. On full, two-story cabins with attic space, the rafters should meet the log walls at, or one log above, the second-floor ceiling joists. A few more logs and you're on the way to building a log skyscraper. On single-story cabins, I urge you to lay enough floor joists for a partial loft, at least, and 3 to 6 logs (depending on log diameter) above the joists.

Extending the eave walls by 3 to 4 feet rescues nearly dead space left in that narrow and unusable triangle where the roof meets the wall. Those extra logs also enlarge the walkable area in the loft. You may not want to sleep up there, but if you are like most homeowners, there will come a time when you'll appreciate the storage space.

If you like the idea of looking at logs, it would be a real switch at this point to go for dimensional timber ceilings, stuffed with insulation, and buried under wallboard, which you are not going to find growing along with your logs in the forest. I've seen it done, with moderate success, using tongue-and-groove cedar

Interiors of these Timber Log homes illustrate two ways to build additional loft space into cabin design. Photo above: You can extend cabin sidewalls several courses above the joists so that rafters meet the walls above floor level. Photo below: You can save some space by laying floor joists in line with girders, using beam-hanging hardware.

or redwood or pine on the ceiling inside and to box in the 2×8 or 2×10 rafter tails outside.

At the other end of the spectrum, you might use 6-inch logs with half-lapped tips at the roof peak, with rafter tail butts notched over the plate logs. Wood shakes over 1×4-inch lath nailed across the pole rafters can complete a solid, good-looking, weather-resistant, authentic, and cold roof. It's nice for a vented storage attic if there is insulation and a vapor barrier on the attic floor.

Between these two systems are countless variations of half-lap rafters, round-log rafters and ridgepoles, log trusses supporting pole purlins and planking, insulation in log rafter bays, rigid insulation sheets over planking—you name it. My experience is that many owner-builders have the same kinds of problems with roofing no matter what variation they use. Here are some of the stumbling blocks, with ways to get over them.

I have never been sold on the conventional rafter layout process. It goes something like this: Measure run, rise, and angle; use slope ratios on a 24-inch framing square to "step off" rafter length, angle at the ridge, multiple angles for a bird's-mouth notch at the wall plate, and rafter tail plumb cut at the overhang.

First, when you measure a long length in a lot of little pieces (the way you lay out with a framing square), you increase the margin for error every time you restart the measurement from scratch. Second, when you cut a flock of identical rafters from a single template, they will work only when the rest of the structure is absolutely uniform. This kind of overall precision is difficult enough to achieve in standardized, stick-built homes, much less in one-of-a-kind log homes, where blindly following a template dimension only duplicates problems and inconsistencies that have crept into the structure.

As you look ahead to compare different roof systems, think about these small points that may make the job work out better for you, particularly if you are an owner-builder who is long on owning experience but a little short on building experience.

Use a ridgepole. Try to avoid full-round ridgepoles requiring carefully sculpted notches that are extremely difficult to adjust up on the ridge. Use ridgepoles with plumb-cut bearing pockets or timbers hewn flat on the two bearing sides. Use plate logs with at least a flat bearing surface hewn on the top face, unless you are willing to saddle-notch each rafter and do a lot of piecework filling in the bays between rafters directly over the log walls.

As you raise gable-end log walls, leave the log ends untrimmed. Periodically string a line or lay a long sapling from one gable to the other, in order to define in space the limits of the roof. What seems to make sense

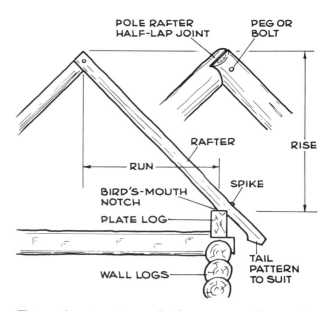

This authentic rafter-and-ridge system relies on the selection of straight trees of similar diameter. A simple half-lap pinned with a dowel (or a bolt if you want) can be used without a ridgepole. Purlins applied perpendicular to the rafter pairs tie the roof frames.

on paper may not make sense when the area is physically laid out on the job. Not everyone can imagine with great accuracy all the blueprint details in three dimensions.

Lay planking over the floor joists to get a feel of the space. Find out firsthand where your head would be bumping into a rafter. This may spur you to add one final course of logs along the eaves, for instance. But when you settle on ridge height, you can select a ridgepole carefully: one gently crowned along the center with enough length to cantilever past the gable ends. This extension carries an extra pair, maybe two pairs, of rafters for your roof overhang. And remember to make this same cantilever with both plate logs. Otherwise, the tails of your overhanging rafters won't have anything on which to sit.

With gable ends in place and securely scabbed (there are no more interlocked corners), you can finagle ridgepole bearing pockets to get the ridge level and running straight down the centerline of the cabin. This divides the roof space in half and sets structural boundaries you can work with.

I don't think you have to cut each rafter as a special case. But I would lug one of them up into place, so as to get the angle cut against the ridge right on the button. If the ridgepole is set plumb, it is safe to duplicate this angle on several, not necessarily all, rafter pairs at the same time.

Again, with the rafter angle cut and scabbed in place against the ridge (you can tack a 2×4 onto the rafter and extending beyond the tip to bear on the ridge), adjust the plate-log notch. Many builders are not able

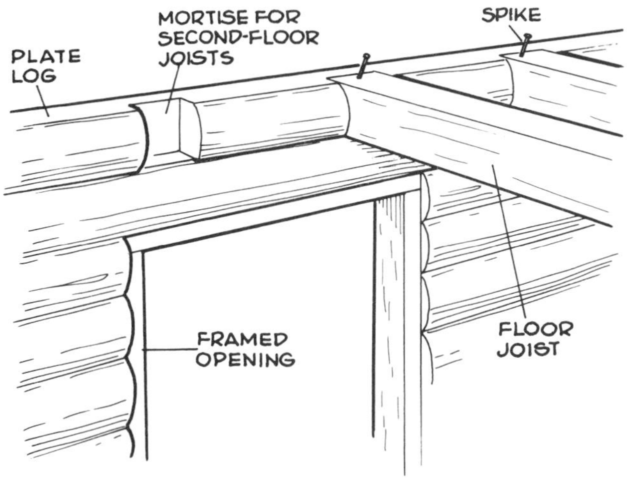
PLATE LOG
MORTISE FOR SECOND-FLOOR JOISTS
SPIKE
FRAMED OPENING
FLOOR JOIST

Like conventional western framing, where each level is decked fully before walls are tipped up for a second or third story, second-floor joists here are set into mortised pockets. Note that at least one full log course covers openings, acting as a header.

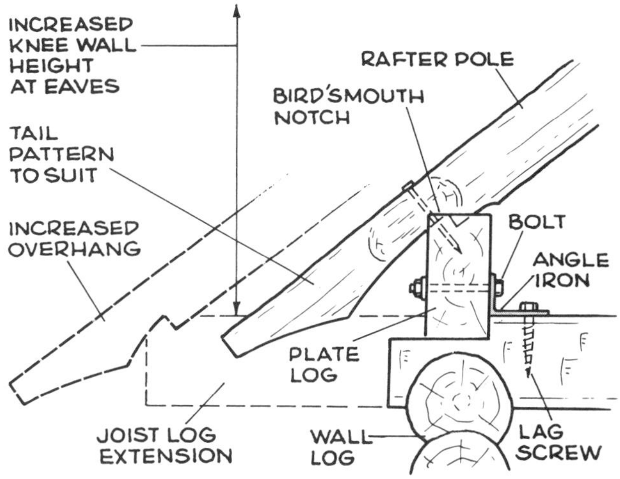
INCREASED KNEE WALL HEIGHT AT EAVES
RAFTER POLE
BIRD'S MOUTH NOTCH
TAIL PATTERN TO SUIT
INCREASED OVERHANG
BOLT
ANGLE IRON
PLATE LOG
JOIST LOG EXTENSION
WALL LOG
LAG SCREW

Lateral (sideways) thrust would tip over this plate log if it were held in place only by notches at the corners of the cabin. On small-diameter logs, spikes can pin the timbers. Another option is to add angle brackets at mid-span to keep the log straight and rigid.

With conventional and alternative building, there is frequently a rush to get the roof on and get protection from the weather, even though you may plan to use the second floor only for storage. During construction of this Berkshire Hills Vermont Log home, sturdy second-floor joists support the roof scaffold system while you work on rafters and roofing. (Photo by Neil Soderstrom.)

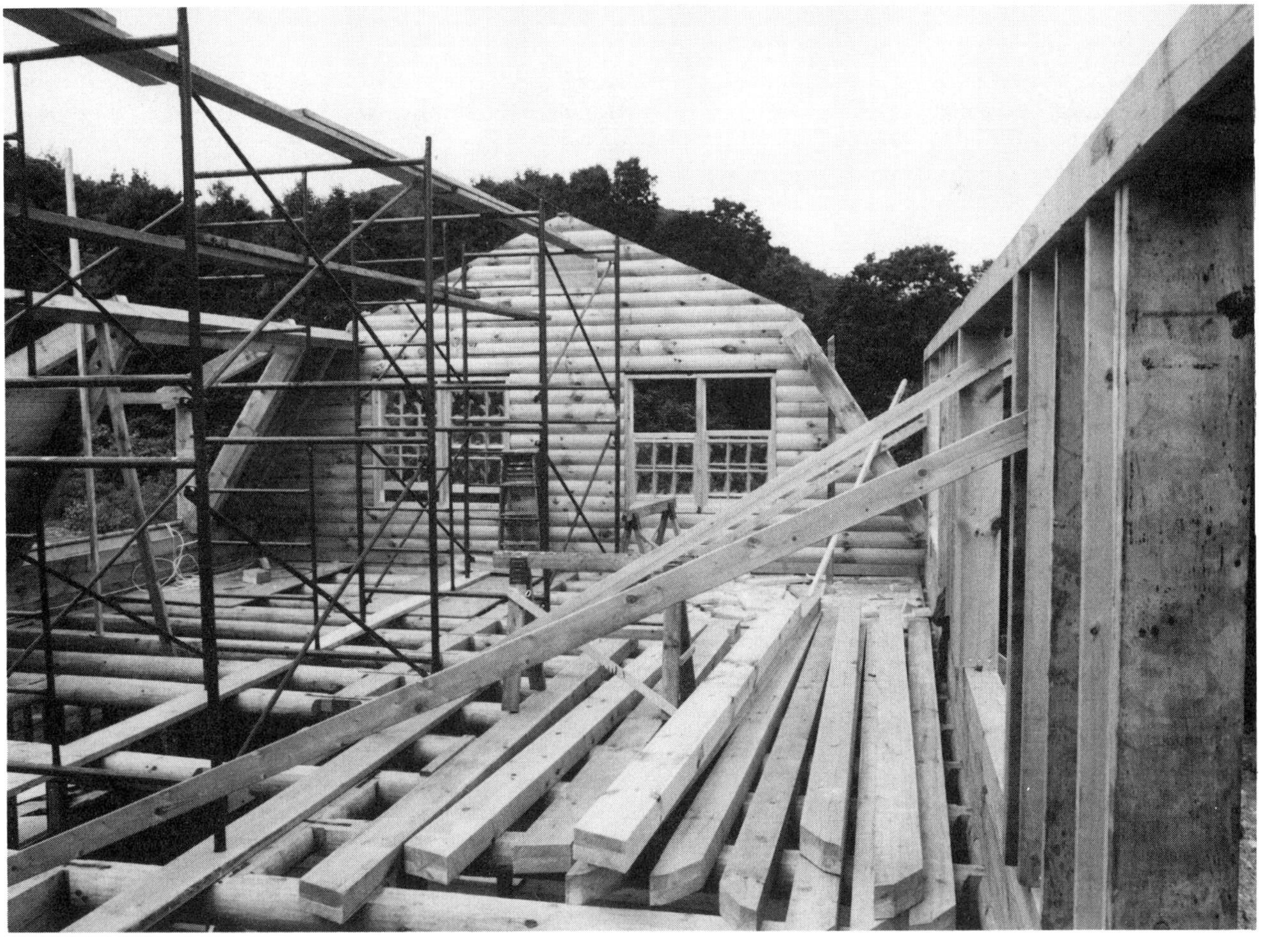

RAFTER PLAN

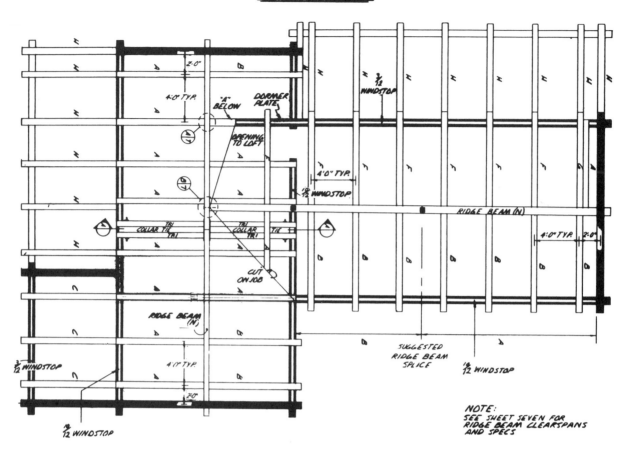

New England Log Homes supplies this rafter plan in its construction manual. In this scheme, 4-foot centers could be bridged with 2×6 T&G decking or a system of purlins. Note that even on precut log homes, some areas (where the two rafter systems join creating valleys) require a personal touch, noted CUT ON JOB.

LOAD-TO-SPAN RATIOS FOR ROUND LOG TIMBERS

Diameter (in.)	Weight (lb./ft.)	Load (lb./lin. ft. at span [lin. ft.])							
		10 ft.	12 ft.	14 ft.	16 ft.	18 ft.	20 ft.	22 ft.	24 ft.
6	5.5	113	65	41	27	19	—	—	—
7	7.5	209	121	76	51	35	—	—	—
8	9.8	357	206	130	87	61	44	33	25
9	12.4	572	331	208	139	98	71	53	46
10	15.3	872	505	318	213	149	109	81	63
11	18.5	1,277	739	465	311	219	159	119	92
12	22.1	1,809	1,047	659	441	310	226	169	130
13	25.9	2,492	1,442	908	608	427	311	234	180
14	30.3	3,352	1,940	1,221	818	574	419	314	242
15	34.5	4,417	2,556	1,610	1,078	757	522	414	319
16	39.2	5,719	3,309	2,084	1,396	980	714	537	413
17	44.3	7,288	4,217	2,655	1,779	1,249	911	684	527
18	49.6	9,160	5,301	3,338	2,236	1,570	1,145	860	662
19	55.3	11,372	6,581	4,144	2,776	1,950	1,421	1,068	822
20	61.2	13,962	8,080	5,088	3,408	2,394	1,745	1,311	1,010

to hit both these joints perfectly the first time around. Realize, though, that the bird's-mouth notch on the first rafter may not work farther along the line, where your log wall bows out just a bit. On stick-built homes, it is relatively easy to gather in a bowed stud wall and make it fit the rafter. But a 10-inch-diameter log may not be so willing to move.

Finally, no matter how much success you have working from a rafter template, do not precut the rafter tails. When the cabin is complete, there is only one strong eyeline reference point for this line of rafter tails, and that's the log wall behind the overhang. Common sense dictates that you use this reference to establish the tails. If they wobble in and out according to the rafter length or ridgepole, you'll see it, and it will bother you every time you come home. So measure out from each end of the wall, run a chalk line from the first to the last rafter tail, and cut all the tails between them to make an absolutely straight line. This is one of the last jobs to do on a roof, just before you begin shingling.

Before looking at the individual roof systems, the table on page 60 will give you a sense of what you're in for—specifically, how massive the timbers must be, depending on roof design. Given the log diameter, the table shows suggested span limits at different allowable loads. But no span limits should be followed without reference to local conditions, such as snow and wind loads, and to structural conditions like the weight and relative strength of the rafter, purlin, and joist wood species.

Although this simplified table provides useful and safe parameters, you should know that there are other ways to size joists and rafters. Several references list characteristics of different wood species, among them fiber stress in bending and modulus of elasticity. When these are known, conventional span-limit tables from the National Design Specification may be used. These tables list different joist and rafter sizes and show span limits based on various conditions: 30-, 40-, even 50-pound live loads on floor joists limited by deflection to $\frac{1}{360}$ of the span; ceiling-joist span limits, which are typically based on only a 10-pounds-per-square-foot live load when the area is not occupied and not anticipated for storage; rafter spans for low-slope designs at 20-, 30-, and 40-pounds-per-square-foot loading, which are typically figured at a reduced deflection limit of $\frac{1}{240}$ of the span; and another set of numbers for medium- or high-slope rafters at various loads per square foot, where the deflection limit is further reduced to $\frac{1}{180}$ of the span.

If you are going to use tables, use a lot of them, and make sure the figures you get meet your specific situation. Otherwise, I would rather see you use your common sense, which would mean, most likely, that

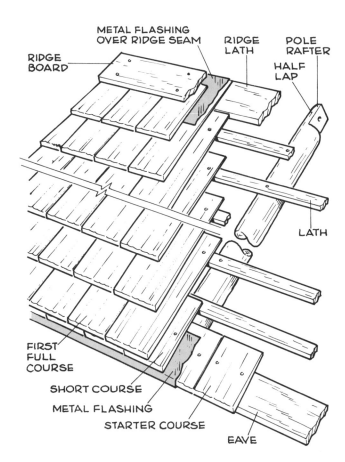

This is the prototypical purlin-and-shake roof used on so many early-American homes. It looks great and will last a moderately long time, keeping out the wet but not the cold. You can leave the system as is for cold-storage attics, use rigid insulating board over the purlins, or insulate between rafters and close in the ceiling with finished planking to the inside.

you would overbuild rather than underbuild the framing system.

If you are turned off by tables like this one, relax. All it tells you is that bigger logs are stronger and, therefore, can carry more weight over greater distances. On this simplified load/span table, allowable loads include the beam weight, and deflection (bending over the unsupported span due to the load) is limited to $\frac{1}{360}$. This means the load/span ratio is limited so that the timber will not deflect a distance more than $\frac{1}{360}$ of the unsupported span. This very stringent requirement builds in a hefty safety margin on log structures. Finally, the table uses a modulus of elasticity of E = 1,200,000, which reflects most types of fir, hemlock, and spruce and some types of redwood, cedar, and pine.

If you decide to use the NDS tables (and they do go on and on), first verify the modulus of elasticity for the species of wood you are using. Technically, this is the ratio of stress (force per unit area) to strain (deformation per unit length). In plain terms, it is a natural

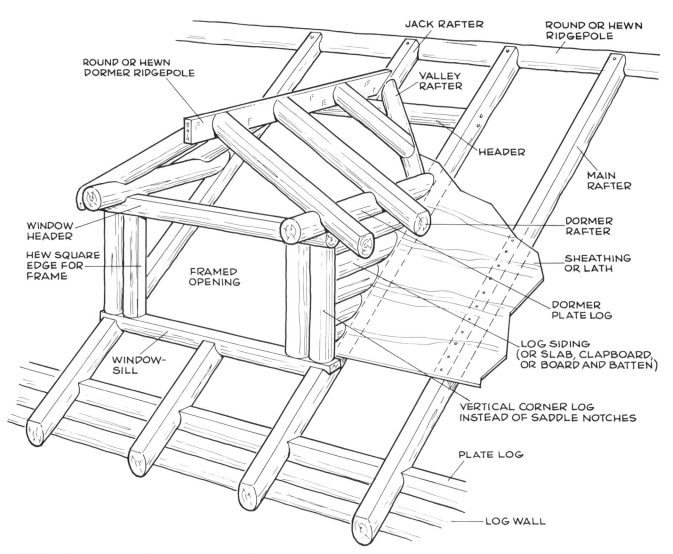

JACK RAFTER

ROUND OR HEWN RIDGEPOLE

ROUND OR HEWN DORMER RIDGEPOLE

VALLEY RAFTER

HEADER

MAIN RAFTER

DORMER RAFTER

WINDOW HEADER

SHEATHING OR LATH

HEW SQUARE EDGE FOR FRAME

FRAMED OPENING

DORMER PLATE LOG

LOG SIDING (OR SLAB, CLAPBOARD, OR BOARD AND BATTEN)

WINDOW-SILL

VERTICAL CORNER LOG INSTEAD OF SADDLE NOTCHES

PLATE LOG

LOG WALL

Full-log dormers can place excessive loads on rafters. The extra weight may require heavier or doubled rafters and hardware to secure the joint between rafter and ridge. The saddle cut shown is a difficult joint to make on the job and is certainly not the strongest, even though it duplicates the custom connections of saddle-cut corners used elsewhere in most handcrafted cabins.

characteristic of wood that changes from species to species; simply, its resistance to downward bending (deflection) when loaded from above.

With that ammunition, here is a look at many of the possible roofing systems.

Half-lap and lath. Working on a roof frame like this one, you will appreciate a solid platform level with the top of the eave wall. Working on that deck, you can cut half-laps at the rafter tips, based on the diameter of the rafter logs, and secure the lap with a bolt or, for authenticity, a hardwood peg. It is sensible to add a temporary cross brace before tipping the rafter pair up in place. Brace at the ridge line, and spike through the rafter into the plate log; 1×4, even 1×3-inch furring strips (lath) are then laid perpendicular to the rafters, roughly 10 or 12 inches on center, depending on shingle length, with a wide board (1×6)

across the rafter tails and on both sides of the ridge. If you cut into an older colonial house, this is the roof construction you are likely to find.

Log rafter and ridge. Think about cutting a nice tight saddle notch up on the ridgepole—kind of tough without elaborate scaffolding. And this joint, while unusual and an elegant match to the corner saddle notches, is not that strong. On horizontal interlocked corners, the full saddle cut bears and closes on the log beneath it. But up at the ridge on a steep angle, only the top of the notch bears on the ridge. The bottom part of the saddle is only cosmetic, so you must rely on a bolt or spikes. You can improve this joint considerably by shaving a shallow saddle on the rafter or a shallow bearing pocket matching the slope angle on the log ridge, then running the rafter pair up above the ridge to meet in butt angles or half-laps.

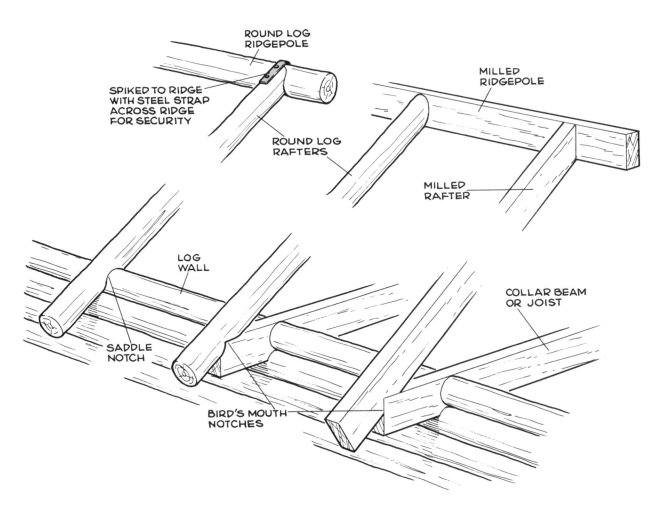

ROUND LOG
RIDGEPOLE

SPIKED TO RIDGE
WITH STEEL STRAP
ACROSS RIDGE
FOR SECURITY

ROUND LOG
RAFTERS

MILLED
RIDGEPOLE

MILLED
RAFTER

LOG
WALL

SADDLE
NOTCH

BIRD'S MOUTH
NOTCHES

COLLAR BEAM
OR JOIST

When you set a round log against a square ridgepole at an angle, the section of the log becomes an oval. Joining the sidewall, the round rafter must be saddlecut or notched to sit on the overhang of a floor joist. To avoid these curvilinear cuts and joints, many builders use round wall logs but switch to square-hewn rafters and ridgepoles.

Whether round or square-hewn rafters are used, you will have an easier time cutting and fitting, and get a stronger joint, by raising at least a portion of the rafter above the ridgepole. At the roof peak, the ridgepole can be beefed up without sacrificing headroom, so as to carry a good portion of the load.

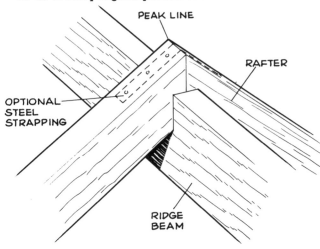

PEAK LINE

RAFTER

OPTIONAL
STEEL
STRAPPING

RIDGE
BEAM

Log rafter and hewn ridge. Even though the cabin interior may not display any hewn surfaces, a steeply angled rafter pair flush cut against a flat-hewn ridge makes a strong and simple joint. On high-slope roofs, you can lock up the rafters and the ridge by drilling through all three members, parallel to the deck, and installing a ⅜-inch-diameter machine bolt with large, semicountersunk bearing washers. Alternatively, you can spike in metal strapping over the ridge and about a foot down each rafter. I've done that on several post-and-beam jobs, so that after I have made absolutely sure that the 4×10s are in the right place and will stay there forever, I can forget about them. You can always camouflage the hewn wood by leaving the bottom of the beam rounded.

Log truss and purlin. This is a good system for cabins with long eave walls. The idea is to divide the long roof into manageable sections, maybe 8 to 12 feet apart. The log truss is a reinforced combination of loft-floor joist and rafters in the same vertical plane, including a vertical king post set between the center of the joist and the rafter peak, and usually at least two branch supports running from the bottom of the king post to the center of the rafter span on each side of the ridge. Collectively, these timbers are exceptionally strong. Purlin logs with minimal taper can be run perpendicular to the trusses and past the gable end for overhangs. They serve as the main supports for planking and shakes or another efficient match for this system, corrugated-steel barn roofing.

I am partial to terne (metal roofing with standing metal seams) and corrugated metal for log homes. To me, they fit in almost as well as shakes and considerably better than asphalt—even modern, strip shingles that are supposed to look like shakes. The metal is not natural, like the logs, but is just as utilitarian.

The temporarily supported log trusses in this Pacific Log Homes design are counterbalanced at the side walls by extensive, built-up cantilevers that greatly reduce the span. Round truss members must be sculpted to a concave shape to seat properly.

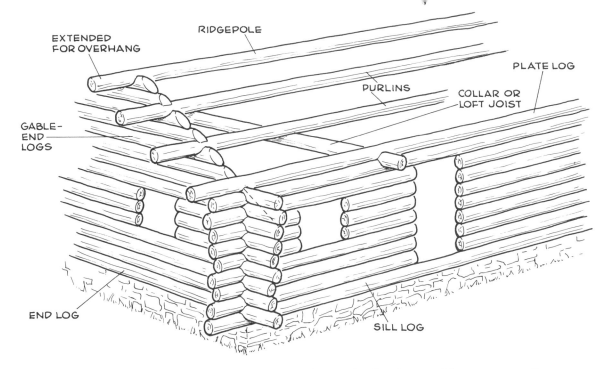

This traditional roof-frame system uses structural purlins and collar beams. This is an excellent way to get extensive overhangs on the gable ends. Depending on cabin length, intermediate truss frames that mirror the angle at the gables may be needed.

The end pair of trusses on this Rocky Mountain Log design show the extremely secure structural detailing at the roof peak: a beefy, square-hewn, V-topped ridgepole under the rafters.

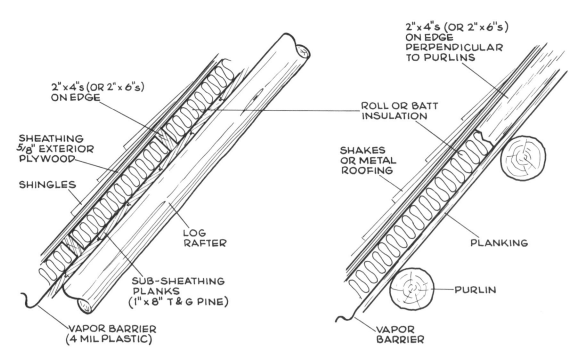

2"x4"s (OR 2"x 6"s) ON EDGE

SHEATHING 5/8" EXTERIOR PLYWOOD

SHINGLES

LOG RAFTER

SUB-SHEATHING PLANKS (1"x 8" T & G PINE)

VAPOR BARRIER (4 MIL PLASTIC)

2"x 4"s (OR 2"x 6"s) ON EDGE PERPENDICULAR TO PURLINS

ROLL OR BATT INSULATION

SHAKES OR METAL ROOFING

PLANKING

PURLIN

VAPOR BARRIER

Using purlins that run parallel to the sidewalls (perpendicular to the gable end walls), you can cover the frame with planks running from sidewall to ridge, then install rigid insulating board and roofing; or you can install 2 × 4 sleepers on edge, then sheathing and roofing. This system is the opposite of conventional rafters running from sidewall to ridge. However, neither method compares favorably (in terms of time and material) with 2 × 6 T&G decking directly over the frame.

While web-pattern trusses create space that is not habitable, you may not be able to leave the space unobstructed and bridge the entire cabin span with a single ridgepole. If a splice is required, make sure the vertical line of support underneath is continuous, all the way down through the frame. Even though the logs on this gable end are spiked together, the tapered wall is braced from both sides.

Rest assured that it is okay to use power tools on log homes, even handcrafted log homes. This overhang detail is on a home custom-built by Pacific Log Homes, which routinely incorporates sculptural shapes into the frame, either by cutting a curve to extend canti-levers in one log course or by gently tapering several courses, sanding the final log to a thin taper. (Photo by Don Smith.)

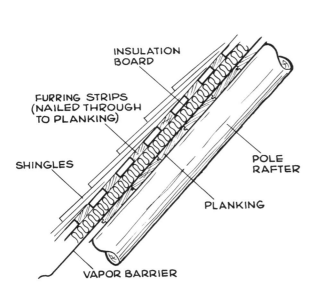

INSULATION BOARD

FURRING STRIPS (NAILED THROUGH TO PLANKING)

SHINGLES

POLE RAFTER

PLANKING

VAPOR BARRIER

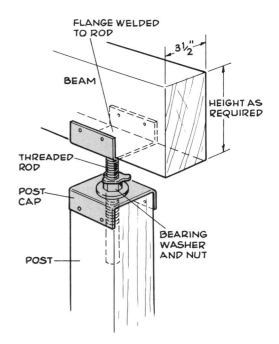

FLANGE WELDED TO ROD

BEAM

3½"

HEIGHT AS REQUIRED

THREADED ROD

POST CAP

POST

BEARING WASHER AND NUT

Rigid insulating board over finished ceiling planking and under furring strips offers a reasonable combination of roofing systems. Furring strips spread the nailing load through shingles or shakes and allow for ventilation of moisture.

It is reasonable to anticipate settling and twisting at this critical seam. Of all the elaborate joints possible at this juncture (half-laps, laps with mortise pockets, angled laps with holding teeth, and more) this simple piece of hardware, while not the most elegant, is easy to work with and secure.

Although energy questions are detailed in the following section, it is helpful at this stage to see how insulation can be included in these roofing systems.

Planks and sleepers. Solid logs have reasonable insulation value. But planking run between rafters won't do much to keep the heat in. So this system calls for a sandwich of sheathing. You'll see the bottom layer but not the top layer, so you could even use (dare I say it?) plywood. Between the sandwich skins, 2 × 4s or, for more insulation, 2 × 6s are laid perpendicular to the rafters 24 inches on center, providing bays for the insulation.

Planks and panel. As much as "the pink stuff" is pushed on television by the panther, fiberglass batts are not terribly attractive. Also, fiberglass loses its effectiveness when compressed. So if you put it under the roof planks, you have to cover it; if you put it over the roof planks, you have to build a secondary structural system to keep it from getting squashed by the shingles. Rigid-foam insulating panels solve this problem. They offer a good solution for people who want to see all that beautiful work they put into the ceiling.

Different foam configurations are rated up to an R-value of about 6.2 per inch. They are lightweight and easy to handle, and while you could easily snap them in two, they have surprising compressive strength over planking and tar paper and under shingles.

You should spend some time on, and derive great satisfaction from, planning and executing overhangs. On a sloped site you might think about extending rafters 6 feet past the eave wall to protect half of a 12-foot porch. You might run plate logs 8 feet past the gable end to shelter a massive wood pile. This is one time when you get to cash in on the massive, strong construction of the full-log walls.

And there is one part of the roofing job that should be pure fun: designing and cutting rafter tails. In the "Introduction" to this section, I wrote that you might carve one into the shape of a raccoon's face, but not if you are too quick to trim. Log ends at corners, for example, can hold planking and scaffolding during construction. You can chop them off as you go, but then you have firewood. Better yet, you can chop the ends of the rafter tails plumb, hang a gutter, shave them down to blunt points, round them, or notch them, turning one of the most prominent details on your cabin into your calling card—just like an artist signing his painting.

CHINKING

Straw, moss (dry sphagnum moss), clay, and almost any other malleable material could be used to chink the gaps between logs. Moss, in fact, is still used by some builders. But with fuel at a premium, it pays to insulate as much as possible.

If you use chinkless construction, you must take the opportunity to fold some insulation into the V-grooves as you go. If you use chinked construction, wait as long as possible before chinking, to let the logs shrink and settle. Try not to chink in mid-summer heat and humidity. And brace yourself as the other shoe falls: All that time you saved, leaving the gaps and irregularities on the logs, gets eaten up with chinking.

Each log is a little different, so that each saddle notch is made following the same routine but always with a few unique twists. Chinking, although it must link up all the different bumps and twists, is a repetitive, boring job. And after you complete the outside, you have to come around and do every inch of it over again on the inside. So wait until the fall if possible, but start chinking in stages—like half a day on chinking and half building the fireplace; half a day chinking and half on the loft floor planking.

There are three basic steps involved in chinking log walls: fill the center of the gap between logs with insulation; tack on a binding system like metal-mesh plaster lath; apply the chinking mixture. Rigid foamboard, and fiberglass batting or fill, all work well. The loose fiberglass is more accommodating to irregular shapes, but I like a combination of the materials, applied as follows.

1. Lay loose fiberglass in the heart of the gap, leaving some excess material on both sides. In very large gaps, where the batting wants to slip off the log as soon as you let go, spread on a thick varnish to grab the glass strands.
2. Cut ½-inch rigid foamboard in strips, using a mat knife at roughly 45° angles. Since the material is reversible, you do not have to double-cut edges; just turn the strip over.
3. Fold up (don't squash) the excess fiberglass under the foamboard strip, which, with its beveled edges, will fit snugly into the gap. If a knot prevents the strip from seating against one or both of the logs, break away a small piece of foam.
4. Keep this assembly in place by tacking (every 5-6 inches) continuous metal lath over the foamboard. Once the central fiberglass is in place, the foam and lath must be added from both sides of the cabin.

Unless the gaps between logs are uniform, don't use corner guard or another narrow, precut lath. Thin-

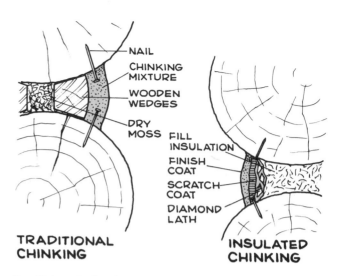

TRADITIONAL CHINKING

NAIL
CHINKING MIXTURE
WOODEN WEDGES
DRY MOSS

INSULATED CHINKING

FILL INSULATION
FINISH COAT
SCRATCH COAT
DIAMOND LATH

Traditional chinking used moss as an insulator, wooden wedges or thin splits as a backer, and a variety of mixes for the surface coating. The modern equivalent uses the same series of components with modern material substitutions: foamboard, fiberglass, and metal lath.

gauge diamond lath can be tapped into deformations with a hammer, then easily trimmed with a tin snips. You can use short, galvanized roofing nails, which won't eventually send rust stains through the chinking and which have large heads for holding the lath securely. In a pinch, you can fabricate your own fine pattern lath by folding over a few layers of chicken wire.

Now you get to choose from the chinking-mix recipe list. There are dozens of possibilities. But with a well-insulated and reasonably well-sealed gap behind the lath, the prime concern should be resistance to the weather and to cracking. Here are a few of the better-known recipes.

1. Equal parts of sand mortar and clay mixed with chopped straw.
2. One part each of clay, fireplace ashes, and lime, with three parts sand and chopped straw.
3. Equal parts of clay and sand with a ¾ part of cement.
4. One part mortar mix and three parts sand.

There are other mixtures. And as it is with cooking, many owner-builders feel a touch of this or a dab of that really makes the difference. In any case, make small test batches and experiment—early on in the job, if you can—to get an idea about how the mix weathers. You may not like the stark contrast of warm wood tones and grayish masonry chinking. If it bothers you, add powdered color pigment to the dry mix. Sift it very thoroughly to avoid two-tone chinking. You should need no more than one ounce of powdered coloring per five shovels of mix, probably less.

WOOD CHINK METHODS

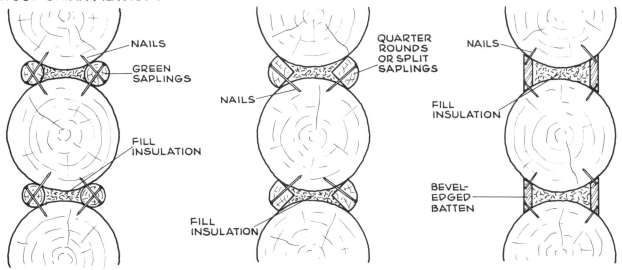

Three traditional chink facings use wood to cover caulked and insulated seams between logs. Green saplings, split quarter rounds, and flat splints all make a somewhat ragged and flexible closure against logs.

The chinking process is similar to applying stucco, except you have to apply it in a confined area, where the log edges keep getting in the way of your smooth, long, even trowel strokes. Work with a small pointing trowel and literally anything else—a soupspoon, for instance—that helps force the mix onto the lath without smearing it on the wood. It pays to work from the top down, cleaning drips off the logs as you go.

Different objects or professional mason's jointing and shaping tools can be used to put a curve, V-shape, or combination of bevels on the chinking. This can help to minimize the lifeless appearance of wide bands of chinking. In all cases, leave a reveal on the underside of any flat-hewn timbers. (A reveal is simply a small setback that avoids a clumsy butt joint where two pieces or materials meet.) If you chink flush up to the edge of hewn timbers and the chinking opens up a crack, it will catch water, funneling it into the wall instead of shedding it.

Chinking does keep you pinned on the walls for some time. One person, moving right along, might easily take seven full days to chink both sides of a 16 × 20-foot cabin. This may increase your curiosity about alternative chinking systems. Even though the heavy work is behind you, this is an important choice that can dramatically alter the appearance of your log home.

1. Saplings. After packing the gap with fiberglass or oakum (a weather-resistant hemp or jute roping impregnated with tar), small-diameter, full-round saplings are spiked to bear against the logs on each side of the gap. This very fast system has draw-

backs. The top of the sapling can funnel water into the wall, and irregularities in the sapling against irregularities in the logs can leave gaps.

2. Splits and moldings. This is about the same process used with saplings, but it's a little slower, with more efficient use of materials. Milled quarter-rounds offer an interesting, even bizarre contrast to one-of-a-kind logs. The trick is to find molding of just the right size—big enough to cover the gap, but not so big and stiff that it cannot bend up and down a bit to accommodate crowns and tapers. Quarter-split saplings are a rustic version of quarter-round moldings, offering a better weather seal than round saplings. You can get a somewhat sloppy but equally effective closure with splits, which are thick shavings (or thin strips of molding, depending on how you look at it). Cedar splits are commonly used, if available, for increased protection against insects. None of these systems is particularly waterproof on its own.

3. Beveled batten. This is another version of wood chinking, the custom-made equivalent of split saplings. The idea is to use ¼-inch stock cut into strips with edges at 45°. These wafer edges can be planed or filed on the spot to work around obstructions on the logs. Oakum, which swells slightly when wet, makes a good, if expensive, backup for all wood-strip chinking.

I would rather have you try moss chinking than one of the modern options, polyurethane foam. But some energy-conscious builders use it, particularly when the gaps between logs are tight, although not completely closed. Even though the material needs special tools for application, must be painted or stained, and releases highly toxic fumes in a fire, it does have a high R-value per inch and does make a positive seal between irregular logs. I think it looks pretty incongruous with logs.

I have saved for last what amounts to a true revolution in chinked construction: a weathertight, waterproof, elasticized chinking material that does not harden, does not crack, and does not pull away from the logs as they shrink and settle. The material goes by two trade names, Perma-Chink and Weather-All. Perma-Chink Systems Inc. told me this is not a new material, even though the application to log homes is brand new.

The material is polymer-based, modified slightly from a material first used in Germany 20 years ago as a replacement or alternative to stucco. Some sand is

Replacing traditional chinking mixtures, Perma-Chink can be applied to new log homes and over existing log seams with existing chinking material that is dried and cracked. You might feel a bit like a baker icing a cake, but the squeeze-bag application method turns out to be a quick way to force the material into place neatly. Tests on this stucco-derived product show impressive results: It does not support mildew growth; it has passed 60 cycles of extended submersion tests; it has undergone 2,000 hours of accelerated weathering tests (simulating about 10 years of wear and tear) with no effect on color, texture, or structural integrity. Full test results are available from Perma-Chink Systems, PO Box 2603, Redmond, WA 98052.

mixed in, and the polymer really does look like mortar chinking. It is applied from a squeeze bag (like adding decorative icing to a cake), then troweled smooth. The release agent is denatured alcohol. On new work, a ¼-inch-thick layer is applied over backer rod or styrene panels. Metal lath is not required. And some of the best news about this product is for log-home owners, who can rechink for the last time, using the polymer directly over old, cracked chinking.

I have not been able to uncover any drawbacks or side effects of this too-good-to-be-true product. It has been used by the U.S. Park Service over existing mortar and has undergone tests on Park Service sites in West Yellowstone. The U.S. Forest Service of the USDA has also tested the product successfully. Neither agency, of course, endorses the product in any way. But the fact that they are trying it lends some legitimacy to the claims.

CAULKING

If you pack saddle notches with insulation, add splines to fit into potential air infiltration and weather gaps between jambs and log walls, mortise rafters for snowboards fitting between rafter tails at the eave walls, and generally design in weather-resisting details, you should not need much caulking. In the few places where it can make a real difference (over header jambs on windows and doors, for instance, even if you install a drip cap), there are two types that work well. Try either a brown-tinted butyl or a clear silicone (unfortunately, about the most expensive type on the market). Both remain supple in most kinds of weather, resist hardening and cracking brought on by seasonal variations, and stretch and shift just enough to move with the logs without losing adhesion. Butyl may stay a little gummy for years. Silicone starts to form a rubbery outer skin almost immediately after application. It does not offend the eye because it is clear; you just don't notice it.

SEALING

On conventional homes, with all kinds of joints between brick and stucco and shakes and clapboards and trim (sometimes all on the same house), caulking is critical. On log homes, the simple structural system and the consistency of material makes sealing critical. It is so easy to concentrate on construction, then neglect preservation. And what a mistake that would be! Even though your cabin seems indestructible, it can be attacked by fire (most likely from a mistake inside the cabin) and by water. But your cabin can deteriorate before its time from sap stain, a fungus that makes wood more permeable and receptive to other wood disease. Wood is also susceptible to brown rot and white rot, both fungi that destroy wood strength, and to mold and mildew. Any of the diseases, if unchecked, could destroy the sturdy timbers.

Most forms of decay flourish in moist conditions with temperatures between 75° and 90°. If that sounds a lot like the southeastern United States a good part of the year, you're right. With many areas receiving an annual rainfall of more than 65 inches, the Southeast is the region for caution: from Florida to southeast Texas, north to the Virginias, and east to the Atlantic. That's where substantial overhangs help ward off effects of rain on the walls.

The good news is that there are many ways to make your log home survive these environmental attacks. But a caution: Many of the protective treatments use chemicals. I am neither a chemist nor a doctor. But while I was building mammoth pole barns with 2×6 T&G penta-treated sheathing, the backs of my hands turned red and burning. And then after I wiped sweat from my forehead, the penta so irritated my eyes that in desperation I flushed them with cold coffee from the thermos.

All preservative chemicals available to the public have advantages and disadvantages. I know of none that is completely benign. You should investigate the efficiency and toxicity of any mix you choose. And I hope that you refrain from Mr. Wizard experiments— inventive concoctions that tend to smoke a little before you paint them on the log. Use treatments containing chemicals with current EPA (Environmental Protection Agency) registration numbers, treatments that you test early on during the job on scrap logs.

If you could keep well-seasoned logs from soaking up moisture during three weeks of 90-percent relative humidity, keep them out of the rain, and keep insects away, they might last hundreds of years. Some species (cedar, cypress, and redwood, for example) give you a head start. Almost all applications offering an extended lifespan for your logs will alter their appearance, at least a little.

Creosote, the preservative commonly used on telephone poles, is effective, but because it stays oily and keeps that unique and potent smell so long, it has few practical applications in log cabins. I have used it on girders and joists near ground level supporting exterior decks; out of direct contact and outside the house, so the odor eventually dissipates.

Proprietary clear sealers and waterproofers, like Thompson's Water Seal, Woodlife, and Sikkens coating offer reasonable protection with only slight yellowing and darkening, but require frequent resealing. A 2×4 fir deck I built in central New Hampshire long years ago (before Wolmanized became nearly a household

See wood preservative note on page 360.

word) is now silver gray but still solid because the owner followed my lead and recoated the deck yearly.

Aside from these well-known applications, there are several types, both proprietary and custom-batched at the site, that can be used successfully. All applications will penetrate the wood more deeply if the wood is dry. The raw wood will change tone a little as it seasons, but the extra penetration is worth the wait. As a rule, you should concentrate on rough, exposed wood fiber (at notches, for example) and on end grain, where the fiber is longitudinal and more permeable. With many of the surface applications you should expect to maintain protection by resealing frequently in severe climates; in other areas, on average, every seven years.

Clear urethanes. This group gives reasonable protection with frequent recoating, with the least change in wood tone.

Working with sealants course after course, you could go through hundreds of expensive, individually packed caulking tubes. Commercial equipment used by Timber Log Homes, including this oversize, refillable caulking gun, and commercial quantities like this 5-gallon can of sealant, are investments that save you time and money over the job.

In the standard log builder's tool kit caulking guns, saws, hammers, and drills are king-size. (Photo by Timber Log Homes.)

See wood preservative note on page 360.

Penta. This is the trade slang for the fungicide chemical pentachlorophenol. Most penta solutions leave a reddish tint and darken the wood somewhat. I would be wary of surface applications at the site. Some builders will not use penta in habitable areas. I would willingly use it on any covered timbers (girders and floor joists, for instance). With some hesitation, I would apply it outside—but not inside. Again, this judgment is based on my experience. I would avoid possible direct contact but would not worry about fumes. Penta may be used as an additive with many proprietary mixes; however, always check with the manufacturer before you arbitrarily throw some in the pot. (Note: Because penta is highly toxic, use extreme care in handling it. Consult the EPA for their most recent evaluation before you apply it.)

Cuprinol 10 and 20. C10 has a copper naphthanate base, which gives a greenish tint to the treated wood. C20 is a clear mix with zinc naphthanate. Neither contains penta. I would use either in areas of probable direct contact. Test C10 on the end grain of a scrap log to see for yourself just how green it gets; it may be too much for you.

Oil waterproofers. The U. S. Forest Service builds and maintains all kinds of log structures. They commonly use the following preparation: 2 quarts boiled linseed oil, 1 quart paint thinner, and ½ cup penta. It's a potent mix with excellent fungicidal and water resistance. It will darken the wood markedly, usually with a yellow-red tinging. But it is highly toxic (see note above). A slightly jiggled mix without penta

could be made as follows: 1 part boiled linseed oil, 3 parts thinner (or turpentine) and ¹⁄₁₀ pound paraffin per gallon of mix (that's 1 pound per 10 gallons). This has to be a hot batch, heated gently up to about 100°, and combined with the liquefied paraffin. You play the chef, follow the recipe, and melt the paraffin in a double boiler. And I'll bet you never imagined you would need a double boiler along with your broad axe and chainsaw. For more money you can add a fine, lighter tone to this mix by substituting tung oil for the linseed oil. If you want to go the other way and add more tone (and many owner-builders like the look of dark logs), you can add 2 quarts of oil-base colorant and ¼ pound of zinc stearate per 10-gallon batch. As always, test by making a mini-batch, following the proportions religiously, in order to minimize surprises when you scale the ratios up to large-batch mixes.

No matter what preparation you use, as your logs check, the checks expose raw, unsealed, unprotected wood. The only way to improve on surface applications is to pressure-treat the timbers. This is not a do-it-yourself job, although there is an in-between stage: dipping and soaking. This sounds good, and there are all kinds of pamphlets on the subject, intended mainly for site work, where factory pressure-treating is not available. I think by the time you get done fabricating 20-foot-long dip tanks out of old oil drums, you'll regret the decision.

In pressure-treating, preservatives are driven deep into the wood grain while the log is inside a sealed, pressurized cylinder. Pressure-treated penta timbers do not have to be sealed at the site, even as they settle and check, although they may be stained for the sake of appearances. Another interesting system is CCA salt pressure-treating. The water-borne preservative, chromated copper arsenate, is infused under pressure, and as the water evaporates, it binds to the wood fibers. Pressure-treated CCA logs appear somewhat darkened but still in the family of natural wood tones, as opposed to an artificial, chemically induced apple green or cheese-puff orange. You can find CCA logs supplied with a 25-year guarantee against rot and insect infestation.

If your building is subject to rugged environmental conditions, pressure-treated logs or hewn timbers are a good long-term investment. Instead of hewing oversize logs down to heartwood for sills, pressure-treated CCA or penta logs are the cautious, more expensive, and longer-lasting alternative. You can consider using them throughout the cabin, but particularly for girders, vertical girder supports, sills, and joists over a crawl space. But heed my cautions with regard to toxicity in foregoing text.

ENERGY EFFICIENCY

INSULATING VALUES

As an apprentice learning to build houses, I often got stuck with some of the most repetitive and boring jobs on the site—highly skilled work like digging holes for deck piers, setting floor bridging, cutting firestops, nailing off sheathing. One day I was up on scaffolding, fitting ½-inch plywood around large roof beams and around openings for fixed glass frames. It may have been crummy-looking sheathing, but I was approaching my work with the precision of a cabinetmaker. The general contractor asked me, a little sarcastically, what I was doing. I thought he would really appreciate the fact that his sheathing job had everything but dovetail corners. He didn't.

"It looks nice, but how is the house supposed to breathe if you build it like a coffin?" That scene flashes into my mind when I see plans to cut special ducts required to bring in enough fresh air to support combustion in a furnace; when I see 2×4s exchanged for 2×6s so the wall can hold more insulation; when I see puny 2×4 roof trusses swathed in a neoprene vapor barrier and buried in a sea of fiberglass. I'm not suggesting that you leave all the windows open during a snowstorm. But I question the sense of saving a few dollars on fuel oil only to spend it on a dehumidifier, plastic sheeting, fans, ducts, and more.

If you believe wholeheartedly in the tight-house theory—in altering structural systems, even sacrificing some structural durability, and generally increasing building costs in order to save energy—a log home, particularly a handcrafted log home, is not a good choice for you.

Some log-home manufacturers go to great lengths trying to turn log homes (or homes that look a lot like log homes) into superinsulated houses. Some companies cut inside and outside faces off the log, mill T & G edges on them, use the milled inner core for joists and other concealed structural timbers, then reassemble the log faces around an insulated foam core. That's a lot of work (and cost) to force one idea on, or in this case inside, another. And it may not be necessary.

To start with, wood, fiberglass, cellulose, and other building materials all have some insulation value. The accompanying table by the U. S. Forest Service lists many log species according to R-value, the conventional measure of thermal effectiveness, or the resistance to heat transmission.

You can compare these R-values to fiberglass (about 3.4 per inch), cellulose (about 3.6 per inch), and other insulators. The comparison would not favor log-wall construction. For instance, a 3½-inch bay between

R-VALUE OF WOOD SPECIES

Log species	R-value/in.
Northern white cedar	1.41
White pine	1.32
Balsam fir	1.27
Basswood	1.24
Cottonwood	1.23
Aspen	1.22
Jack pine	1.20
Ponderosa pine	1.16
Hemlock	1.16
Spruce	1.16
Yellow poplar	1.13
Western red cedar	1.09
Cypress	1.04
Red pine	1.04
Eastern red cedar	1.03
Redwood	1.00
Douglas fir	.99
Black ash	.98
Maple	.94
Yellow pine	.91
White birch	.90
White ash	.83
Sugar maple	.80
Beech	.79
Red oak	.79
White oak	.75
Black locust	.74
Shag hickory	.71

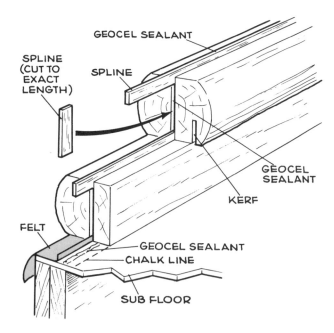

SPLINE (CUT TO EXACT LENGTH)

GEOCEL SEALANT

SPLINE

GEOCEL SEALANT

KERF

FELT

GEOCEL SEALANT

CHALK LINE

SUB FLOOR

One of the many proprietary weatherizing systems, this one from New England Log Homes uses splines set into grooves filled with Geocel sealant at all horizontal and vertical seams between logs.

studs in a stick-built home, filled with fiberglass, gives an R-value of 12. To get the same R-value with white oak (.75 per inch), you would have to build with 16-inch-diameter logs, or 12-inch-diameter redwood, or 9-inch-diameter white pine. In northern climates requiring, say, 5½ inches of wall insulation giving an R-19 rating, the equivalent white pine log would have to increase from 9 to 14 inches in diameter.

But it turns out that logs have more going for them than R-values. All that strength and bulk give energy benefits beyond the natural insulating value of the wood species expressed as an R-value. It's called thermal mass. In log homes, this mass can make a dramatic difference, particularly in spring and fall heating and cooling costs.

Two recent reports have confirmed this finding. A five-year study conducted by Muir Publishing Company in Canada (they publish the *Log Home Guide*, so they have a stake in the issue), titled *The Energy Economics and Thermal Performance of Log Houses*, looked at eight different types of wall construction in two climatic zones—Ottawa, Canada, and Miami, Florida. The study took issue with the "indiscriminate" use of R-values alone to determine energy efficiency, and called for performance-oriented evaluations instead.

This report notes that R-values were developed by ASHRAE (American Society of Heating, Refrigeration, and Air Conditioning Engineers) to estimate the size of heating and cooling equipment, not to measure the overall thermal performance of a building. Here are some of the findings:

1. From tree felling to finished construction, log-home manufacturing has a cost margin 3 to 14 percent better than brick, wood-frame, and concrete-block construction.
2. Solid-log walls have a thermal performance 54 to 74 percent more efficient than conventional construction. The report specifies that "uninsulated log walls outperform any and all of the other walls, whether they are insulated or not."
3. R-values do not account for moisture content, air infiltration, and thermal mass, the ability of a material to store and reradiate heat.

A second report, *A Field Study of the Effect of Wall Mass on the Heating and Cooling Loads of Residential Buildings*, was conducted over seven months on six 20×20-foot test homes by the National Bureau of Standards (NBS) for the Department of Housing and Urban Development. The six test houses included a log home with 7-inch-square lodge-pole pine logs and compressible foam backer rod with caulking applied to interior and exterior joints. Also tested were an R-11 wall wood frame, an uninsulated wood frame, insulated and uninsulated masonry block walls, and an insulated block plus facing wall. The R-value of the log home wall was 10.3, two to three points lower than that of conventional insulated buildings. All the homes were identical and were measured under controlled conditions. Wall construction was the only variable. Here are a few of the findings:

1. During the 14-week winter heating period, the log home and the insulated wood-frame building used roughly equal amounts of energy.
2. During the 11-week summer cooling period, the log home used 24 percent less cooling energy than the insulated wood frame.
3. During the 3-week spring heating period, the log home used 46 percent less heating energy than the insulated wood frame.
4. Overall, in a 28-week test period, the R-10 log home "performed as well as the insulated wood frame home with its R-12 walls (approximately 17 percent higher value) but without the benefit of the thermal mass of the log wall."

Both reports are available to owner-builders (see "Information Sources" on upcoming pages). If you want to play the energy numbers game, these reports provide all the formulas, facts, and figures you need. If you are sick to death of crunching energy numbers, what you have just read should convince you that in addition to building a beautiful, rugged, long-lasting home, when you build with solid logs you are building an energy-efficient home as well.

However, if you live in northern Minnesota and you have a limited or exceptionally costly fuel source, you might be able to do as well or better with 2 × 6s, 24 inches on center, with 5½ inches of fiberglass plus 2 inches of rigid styrene sheathing, which, with other materials in the structure, could provide something like R-35 in the walls. In severely cold climates, look closely at white pine and other species that resist heat transmission most effectively. And look at the up-front costs of large-diameter logs. They may be difficult to find and harvest in your own woodlot, but you can buy them. And a few manufacturers of log-home kits offer several large-diameter options. For example, Rocky Mountain Log Homes in Hamilton, Montana, uses white or lodge-pole pine in 7- to 12-inch diameters and offers, on special order, 18-inch-diameter logs.

This is an absolutely pivotal issue for log-home manufacturers, owner-builders, and all who are interested in log construction. There are still areas where energy standards based solely on R-values exclude log structures. You can imagine how this could cut into the potential market of a log-home kit manufacturer, how it could change the plans of a family who wanted a log home but could not get their plans approved.

Well, in addition to the two energy-performance reports, there is more good news for owner-builders on this subject. The Canadian government has recently initiated the Government Industry Task Force on Log Building. Along with inevitable red tape attached to any government undertaking, this source also adds legitimacy to log building, may become involved in construction standards, possibly even handcrafter certification, and will make the log industry a little less of an adventure, a little more accessible, and probably more reliable as well.

Also, questions of thermal performance have been helped by systems like the Hercules HBR backer rod (a closed-cell polyolefin foam tubing) and the brand-new Perma-Chink, polymer-based chinking material. Even so, it is only recently that the state of California accepted the benefits above R-value of thermal mass. California's *New Residential Building Standards Energy Conservation Manual* establishes an elaborate point system with mass factors that are good news for log builders and buyers. Write the California Energy Commission, 1516 Ninth St., Sacramento, CA 95814.

VAPOR BARRIERS

This section is particularly for owner-builders in cold climates. Some of the most practical information on cold-climate housing comes from the Department of the Army and from the Cooperative Extension Service attached to the University of Alaska. The Army commonly takes its own cold-weather structures along with its personnel. But the Alaska Extension Service has done a lot of work with log homes on construction, insulation, and vapor barriers.

Extension Service reports pinpoint a second, crucial function of vapor barriers. The first and most obvious purpose of 4- or 6-mil polyethylene sheeting is to stop, or at least minimize, migration of water vapor into framing cavities, where it can cut the effectiveness of insulation and eventually rot framing timbers. In near-arctic climates, vapor barriers are intended to stop the free flow of water stemming from frozen vapor in roof cavities and attics that melts in the spring thaw.

Extension Service recommendations include the last-resort step of furring out the log walls for insulation and vapor barriers, but with a refreshing sense of reality, they go on to say that this is probably the last thing you want to do. Instead of covering up your logs, they suggest the following:

1. Close any open joints between logs with nonhardening caulk such as silicone rubber.
2. Close large cracks with wood-fiber plaster held to the logs with fiberglass sill sealer laid between logs during construction.
3. Vaporproof all interior log surfaces with a clear urethane or epoxy coating.
4. Paint the chinking with a clear waterproof sealer such as liquid silicone.

MECHANICAL ACCESS

One of the things you do give up with solid-log construction is the concealed space between framing members—the bays filled with insulation and accessible for water-supply lines, drainpipes, electrical wiring, and recessed receptacles and switches. This is not as big a problem as it first appears to be.

Whether you use dimensional lumber or log timbers for the floor frame, you have all kinds of room for pipes and wires. Supply and drain lines can pass beneath flooring, between joists, in cabinets, cupboards, and closets. In bathrooms, exposed, standing pipes for water supply and drainage can be quite elegant and utilitarian (a style that is in harmony with most log homes) in bright copper that is steel-wooled and lacquered and in brass.

In two-story cabins, concealed access should not be a problem if you design kitchens and baths back to back on the first floor and stacked directly on top of each other from first to second floor. No matter what framing system you use, it is economical to run a single vent stack, a single waste line, and a single set of supply lines, then tap into them as required from floor to floor.

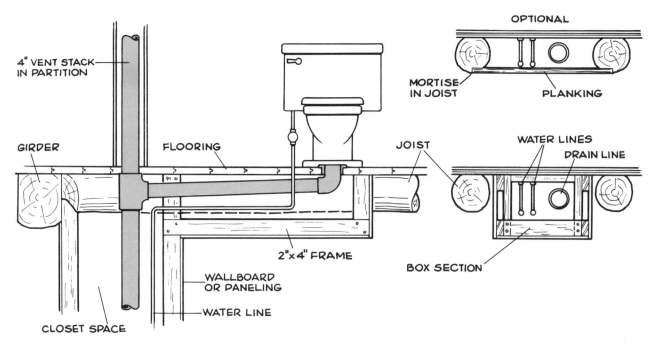

4" VENT STACK
IN PARTITION

GIRDER

FLOORING

JOIST

OPTIONAL

MORTISE
IN JOIST

PLANKING

WATER LINES

DRAIN LINE

2"x4" FRAME

WALLBOARD
OR PANELING

WATER LINE

CLOSET SPACE

BOX SECTION

Supply, waste, and even vent piping is easily concealed in partition walls—unless they, like the exterior walls, are made of full-diameter logs. Aside from exposing bright copper piping, the most sensible way to route utilities where you need them is to build pipe and wiring chases (channels).

Recessed duplex receptacles at conventional wall height can be fed only by drilling through several log courses, either in one shot with a long auger or by drilling aligned holes in each log course as it is set in place. Recessed floor outlets are a good alternative.

Electrical lines also can run just about anywhere you want beneath the first floor. I would try to route them up to the second floor along with the plumbing lines. Without stepping on local electrical codes, I would not quickly follow the advice in many log-building books, which take you through elaborate steps to route duplex receptacles into the log wall. Some suggest boring holes in wall logs as you lay them in place so the holes line up. That's a bit tricky. The idea is to create a channel through which you can pass cable. At the top of the channel, you have to mortise for a gem box. I don't understand the need for this struggle when you can use code-approved floor outlets.

Switches are a different matter. Switches just inside doors and at the top and bottom of staircases, for example, must be convenient to be safe. Running cable in surface-mounted raceways is the easiest way to solve

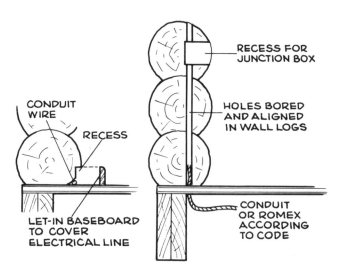

CONDUIT
WIRE

RECESS

LET-IN BASEBOARD
TO COVER
ELECTRICAL LINE

RECESS FOR
JUNCTION BOX

HOLES BORED
AND ALIGNED
IN WALL LOGS

CONDUIT
OR ROMEX
ACCORDING
TO CODE

this problem. If metal raceways bother your eye, make your own with wood molding.

To run cable in the log wall, you can try to auger large-diameter holes in line with each other as you build (don't try it with less than a 1-inch drill) or run cable with spline boards (or in their own grooves) inside door jambs, then along the wall with insulation in the horizontal V-grooves.

INFORMATION SOURCES

No single book can give you all the definitive answers. And even if such a book existed, you would make a mistake by reading only that book. Log building is not like a mathematical equation; there is more than one answer. There are a great many options and even more differing points of view on their merits.

In the spirit of this book which will not lecture you on the one and only way to do it, I encourage you to select from this source list.

LOG-BUILDING BOOKS

Build Your Own Low-Cost Log Home, Roger Hard (Garden Way). This large-format, profusely illustrated 200-page book covers handcrafted and kit log construction. Coverage is very broad at the expense of some detailing, and the text is a little impersonal. But you will find information and several unique solutions to stovepipe installations, log stair construction, and such.

Building with Logs, B. Allan Mackie (Log House Publishing). A classic, basic, 76-page book on log building with quality photographs by one of the premier log-building teachers. This book is used as the course guide for sessions at Mackie's log-building school.

Your Log House, Vic Janzen (Muir Publishing). This practical, learn-as-you-build book by a former chief instructor at the Mackie school offers guidance suited best to first-time owner-builders. The drawings are a bit primitive, but the thorough text starts with cabin design and works through mechanical and finishing jobs.

Hand-Hewn, William C. Leitch (Chronicle Books). One of the friendliest books on the subject, with a unique collection of photographs, the text has a lot of commonsense help for owner-builders. Unfortunately, the presentation is all too brief, frequently sending you after other sources to get necessary details. Still, this is an easy-to-read, very enjoyable introduction to the subject.

A Logbuilder's Handbook, Drew Langsner (Rodale Press). An illustrated case history of the author's cabin. It covers only hand-hewn, square-faced con-

struction. This specific, well-detailed look at a specific subject is worth reading, even though the author seems to take it all very, very seriously.

The Craft of Log Building, Hermann Phleps (Lee Valley). This monumental, profusely illustrated book offers a complete, distinctively European view of building with logs; a fascinating historical text rather than a how-to book.

In Harmony with Nature, Christian Bruyère and Robert Inwood (Sterling Publishing). The title tells you the point of view, although this practical book never goes overboard (warning you off chainsaws, for instance). Wonderful, realistic, annotated illustrations (five small drawings sum up V-grooving) make this book a treat.

Building a Log House in Alaska, Cooperative Extension Service, University of Alaska. Don't miss this compact, inexpensive, incredibly practical book that includes and explains a lot of technical data. Some of the arctic climate details may not apply, but the construction details are applicable and first rate.

Complete Guide to Building Log Homes, Monte Burch (Popular Science Books; Stackpole Books distributor). The author has covered all the bases from foundations to roofing, maybe even too much as big chunks of the 370 pages (and over 840 illustrations) are eaten up with modest treatments of conventional building subjects, such as laying asphalt roof shingles and installing water pipes. While the book is a bit short on the nitty-gritty, learned in the trenches, practical tips and log-working skills, it is long on laying out the constantly growing log-building options: kit homes, different layouts and styles, and such. One of several log books worth reading before finalizing your plans.

TECHNICAL BACKGROUND

Why Buildings Stand Up, Mario Salvadori (McGraw-Hill). This is one of my favorite building books, and I recommend it to anyone interested in understanding why buildings work as they do. It is not easy reading, and it is about all kinds of construction. But the slow, winding path leads you to interesting places usually reserved for architects and structural engineers.

Design of Wood Structures, Donald E. Breyer (McGraw-Hill). A totally technical, detailed examination (practically a textbook) of wood construction; to be used with pencil and calculator in hand.

Homebuilding, Mike McClintock (Scribner's). Yes, it is immodest to mention one of my other books, but I do so because I frequently refer to the first 100-page section on footing and foundation construction. I know how much time and trouble went into organizing and presenting the endless technical details of mix ratios and curing strengths and such, in a format that can be used by owner-builders.

National Design Specification for Wood Construction, National Forest Products Association. The *NDS* is a nationally recognized guide for wood structural design. Filled with grade distinctions and stresses and formulas and tables of all kinds, the book leaves it to you to do all the extrapolating.

Pressure-Treated Southern Pine, Southern Forest Products Association. Representative of many of the fine publications available from wood industry groups, this 16-page booklet covers specifications for pressure-treated timbers. Also ask for: *Technical Note on Preservative Retentions* and *Technical Note on Construction Poles*.

A Field Study of the Effect of Wall Mass on the Heating and Cooling Loads of Residential Buildings, Burch, Rennert, Krintz, Barnes (National Bureau of Standards). A detailed test of solid-log walls and both masonry and wood frame construction is available from the NBS, Washington, DC 20234.

The Energy Economics and Thermal Performance of Log Homes (Muir Publishing). A study of log wall thermal performance compared to other forms of conventional construction. A condensed consumer edition is available from Muir Publishing, Gardenvale, Quebec, Canada H9X IBO.

BASIC BACKGROUND

Wilderness Adventure, Ralph and Riki Plaisted, (Dillon Press). One family's start-from-scratch experience building a three-room log cabin in the wilds of northern Saskatchewan. Not the book for log-building details, but an enjoyable, chatty, diarylike account (complete with home recipes for moose stew) of the solitary, silent life at the edge of a lake, where spare parts for the portable mill came in by seaplane.

Practical Homesteading, John Vivian (Rodale Press). All the nitty-gritty details about the practical side of living on your own homestead are explained and thoroughly illustrated; everything from keeping livestock to putting up preserves.

One Man's Wilderness, S. Keith (Alaska Northwest Publishing). This book continues to be difficult to locate, but it is worth the effort. It is the diarylike account of Richard Proenekke, who hand-built a cabin in the Alaskan wilderness. Not a book about log-home carpentry, this story is about the problems, the emotional and mental struggle, and the incredible rewards of building a cabin in the wild.

TOOLS

A Museum of Early American Tools, Eric Sloane (Ballantine). Sloane's carefully crafted illustrations and folksy text offer a nice introduction to log-building and homesteading hand tools.

Country Craft Tools, Percy W. Blandford (Funk & Wagnalls). A very thorough treatment of all kinds of handcraft tools, including handmade tree saws, bow-string-powered shoulder augers, felling axes, and more.

Ancient Carpenter's Tools, Henry C. Mercer (Horizon Press). An in-depth, historical text, ideal for authenticity mavens with time enough to research every detail of cabin construction.

Bailey's Discount Mail Order Logging Supplies Catalog, Bailey's, PO Box 550, Laytonville, CA 95454. A complete look at, and source for, every kind of logging tool and piece of equipment.

Barnacle Parp's Chain Saw Guide, Walter Hall (Rodale Press). An extremely thorough, basic guide to buying, using, and maintaining the crucial piece of log-building equipment.

ASSOCIATIONS

LOG HOMES COUNCIL, National Association of Homes Manufacturers, 6521 Arlington Blvd., Falls Church, VA 22042.

NORTH AMERICAN LOG BUILDERS ASSOCIATION LTD. (NALBA), PO Box 369, Lake Placid, NY 12946. NALBA includes log-home builders, manufacturers, distributors, and dealers.

CANADIAN LOG BUILDERS ASSOCIATION (CLBA), PO Box 403, Prince George, BC, Canada V2L 4S2. CLBA writes and updates log-building standards at their two meetings every year; the organization is oriented more broadly to include individual builders. There is a modest annual membership fee.

PERIODICALS

The Mackie School Newsletter, Log House Publishing, PO Box 1205, Prince George, BC, Canada V2L 4V3; modest fee for 6 annual issues. Updates school programs and includes help-wanted and log-builder notices.

Log Home Guide, Muir Publishing, Gardenvale, Quebec, Canada H9X 1BO; 4 issues per year plus a Winter Directory. A complete look at the log-home industry, handcrafters, and some tools and equipment; stays current on key issues like insulation standards, code compliance, and more. The Winter Directory is an excellent starting place for kit buyers. Write for their Book-Log review flyer.

LOG-BUILDING SCHOOLS

B. ALLAN MACKIE SCHOOL OF LOG BUILDING, PO Box 1205, Prince George, BC, Canada V2L 4V3. Courses range from one-day seminars to six-week general building sessions, including tent space or cottage.

RESTORATION, Drawer G, Free Union, VA 22940. Charles McRaven teaches a log-building program running five days and covering foundations to chinking and hewn-log building.

PAT WOLFE SCHOOL OF LOG BUILDING, RR 1, McDonalds Corner, Ontario, Canada KOG 1MO. This school offers one- to eight-week courses year round.

MINNESOTA TRAILBOUND SCHOOL OF LOG BUILDING, 3544½ Grand Ave. S., Minneapolis, MN 55408. Ten-day courses are held at the school's base camp deep in the Superior National Forest. The group courses also stress community wilderness skills.

LEGENDARY LOG HOME SCHOOL, PO Box 1150, Sisters, OR 97759. This school conducts three-week courses and a special two-week beginner's course. Instruction begins with site selection and includes several log wall options.

There are other schools and sources, but it should be clear from the number and variety listed here that interest in log homes is growing. More people are becoming interested in timber-frame building, pole building, and the other types of construction. And there are new and growing periodicals and associations covering restoration work on Victorian homes, row houses, and more.

It's as if home consumers buying, building, and maintaining conventional housing have been doing a slow boil that has finally started to bubble over. Increasingly unsatisfied with fancy wrappings over puny structural systems, more home consumers are looking for alternatives. Log building has become one of the most visible and most appealing options. This means that many firms in the field are relatively new. You have to examine them and their products carefully, since they may not have extensive track records.

Despite increased energy efficiency, quality cutbacks in new home construction may have contributed to this renewed interest. But there are other factors. A log home symbolizes the rugged, early-American homestead. And maybe even more important, a log cabin symbolizes the tightly knit, self-sufficient, early-American family, unfettered with daily commutes to work, aluminum siding, lines at the gas pump, and rising real estate taxes.

Life on early-American homesteads was taxing in its own way, to be sure. But many of the most pressing dangers and concerns are no longer concerns today. What remains is the positive side of the picture.

Maybe log owner-builders are only resisting all the impersonal technology of the brave new world. Maybe they have found something in the past, that seems to work as well today, that offers an unusual peace and quiet and a pride in craftsmanship and accomplishment that seems timeless.

ILLUSTRATION ACKNOWLEDGMENT

The following contributed illustrations to this part of the book:

ADIRONDACK LOG HOMES, PO Box 369, Lake Placid, NY 12946

APPALACHIAN LOG STRUCTURES, PO Box 86, Goshen, VA 24439

AUTHENTIC HOMES, 14-636 Clyde, B.C., Canada V7T 1E1

BEAVER LOG HOMES, PO Box 1145, Claremont, OK 74017

HUSQVARNA SAWS, 520 Lafleur St., Lachuto, Quebec, Canada J8H 3X6

NEW ENGLAND LOG HOMES, PO Box 5056, Hamden, CT 06518

PACIFIC LOG HOMES, PO Box 80868, Burnaby, B.C., Canada V5H 3Y1

PERMA-CHINK SYSTEMS INC., PO Box 2603, Redmond, WA 98052

ROCKY MOUNTAIN LOG HOMES, 3353 Hwy. 93 South, Hamilton, MT 59840

STIHL SAWS, 536 Viking Dr., Virginia Beach, VA 23452

TIMBER LOG HOMES, Austin Dr., Marlborough, CT 06447

TIMBER-FRAME

HOUSES

Brochures prepared by the big lumber companies usually have a full-page illustration showing how their trees are sawn into endless shapes and sizes. The texts stress efficiency, explaining how a tree is divided into plywood skins, trim, 2 × 4s, joists, beams, girders—all the wooden parts of a modern house.

This jigsaw millwork is impressive. Even the sawdust is scooped off the floor to make pressboard panels and fake logs for fireplaces. Conservationists are placated. Modern technology gives us "full forest utilization."

Proponents swear by the practical, economically efficient system. Others swear the system is needlessly complicated, involving endless transportation, distribution, and middle men. But the lumber jigsaw puzzle is a crucial part of the hot concept in stick-built construction today—the "engineered" frame. The essence of the idea is to get more structural value from less wood by slicing the tree into all kinds of special-purpose timbers, then reassembling the pieces very scientifically.

"Engineered" is a reassuring word, as though instead of a simple house frame you are getting a suspension bridge. But depending on your point of view, the frame you get may not be called engineered, but fully utilized, minimal, underbuilt, or even rinky-dink.

A log builder would want that long, straight tree before it goes to the mill. That is the option at the other end of the spectrum;

Photos by Richard Starr.

using trees "as is," or almost "as is" after clearing limbs and bark, to make solid wood walls. Working with logs, you minimize material preparation but maximize labor to and on the site. And transportation can be an ordeal. Taking an 8- or 12-inch-diameter tree from the forest to the building site is just about impossible without help, whether animal or mechanical.

A lot of labor is expended on the preparation of dimensional timbers, even to the point of dressing (that's what the industry calls it even though "undressing" would be more accurate) 2 × 4s so they measure a half-inch scant in both directions. But 2 × 4s, and even 2 × 12s for that matter, are very workable. All that preparation means a hardy owner-builder can get almost all dimensional timbers in place in the frame without help. And unlike logs, which are all one of a kind and must be sculpted individually to fit into cabin walls, dimensional timbers are prepared to be uniform and to conform easily to a building system, to modular measurements, to an engineered frame.

Knowing about both ends of the framing spectrum will help you understand timber framing, because it is right in the middle. The timbers are not sticklike (as in stick-built) and are not whole logs. But they are substantial. The timbers are not as easy to handle as dimensional studs and joists, and they are not as impossible to move as full logs. But several helpers plus block and tackle are standard with timber framing. The timbers are not dimensionally uniform, like conventional studs and joists, and are not all one of kind, like logs. But while the timbers are hewn square and the bents, or repetitive members of the timber frame, are similar, timber framing does require a lot of custom work. And this kind of work, the craft of timber framing, is very different from log and from stick work.

All building systems require planning and design skills. And all require the skills of accommodation, possibly the most important and least-talked-about skills needed to bridge the wide gaps between clean, theoretical, building systems in textbooks, and water-logged 2 × 4s, tapered logs, and crowned timbers that reflect the naturally imperfect state of the real world. But that's where the similarity among systems ends.

With log work, structural engineering skills are reduced (stacking one log on another is not the most complicated notion), but hand skills (the traditional skills of carpentry) are concentrated on the logs. And the level of skill shows, both inside and out. With stick-built frames, structural engineering skills are in-

If you are used to setting one measly little 2 × 4 stud at a time, prepare for some heavy-duty work. Individual bents, even on small timber frames, need a lot of bodies, block and tackle, a crane—all the help you can get. Notice how this bent is clamped and tied together with come-alongs during the tip-up at Cornerstones, a building school in Brunswick, Maine.

Below is the modern version of timber framing: minimal 2 × 4 timbers woven together with gang-nail plates into roof and floor trusses. Stand them up one at a time just like bents on a real timber frame, and you get a fairly typical, unexciting, but low-cost home. This one is a test house built by the U.S. Forest Service.

creased. The completed frame on even a conventional home products a maze of woodwork: headers, belts, jack studs, built-up corners, nailers, firestops, cats, bridging, and more. Selecting the right piece of the puzzle for the right location, even pricing and ordering material for the project, can be quite complicated. But since the pieces are dimensional, fixing them in place is not that complicated. In fact, one of the prime ideas behind this system is to speed construction.

What about timber frames? They require high-quality carpentry work, classic skills like mortising, and making tenons that fit into their pockets without play. Cut the wrong side of a pencil line, and the tenon is loose as a goose. Cut the wrong side in the other direction, and the tenon won't fit in, not even when you "persuade" it with a sledge. Cuts on stick-built frames rarely require the same precision as timber-frame work. And remember, this does not say that one system is better than another or that you have to be a better carpenter to work with one material as opposed to another. It says only that the emphasis is different.

Compared to the "engineered" 2 × 4 truss timber frame, the real thing (left and below) is reassuring. Stark, severe lines of the raw frame (and massive gunstock corner posts) can be exposed inside a timber-frame home. The large timbers allow for open interior spaces and an atmosphere of quiet strength and security. (Photos by Richard Starr.)

Timbers may be massive and a bit rough. But if you take the time to make high-quality joints, you wind up with a sound structure and a beautiful interior from one operation.

A timber frame can be incredibly strong and durable. It can also be beautiful, and you don't have to bury it. In this house built by Ed Levin in Greenwich, Connecticut, all conventional trim is painted, with the walls highlighting the elegant structural skeleton. (Interior photo by Richard Starr.)

If the oldest house in town isn't made of stone, it is probably a timber frame. The classic saltbox with the typical appendages of ell, porch, barn, and privy is the John Adams house in Quincy, Massachusetts.

As you can see, completed timber frames are generally handsome inside and out. Yes, they must be covered on the outside. But no, they do not have to be covered, at least not completely covered, on the inside. Oddly, some builders do just that, adding wallboard, a three-coat taping job, and paint over the red oak or some other hardwood frame. I know I couldn't do it, even though it would be satisfying to know the frame would likely last 200 years or more. Drive through an old New England town, spot a house that carries enough layers of paint to smooth the seams between clapboards, and a number over the front door like 1685, and underneath you would likely find a hardwood timber frame locked together with mortise, tenon, and peg.

The longevity is admirable. How could there be a negative word to say about such building; about materials selected for strength and durability; about construction joints tight and neat enough for living-room furniture; about carpenters who had the time and the

This cape, ell, and log barn is the birthplace of Daniel Webster in Salisbury, New Hampshire.

At the Farallones Institute Rural Center in Occidental, California, the community raising is a learning experience. Instructors and students push and lift beams with pike poles, tighten joints with come-alongs, and generally heave, shove, wrestle, and guide 8 × 8s.

this would be reversed, looking at the short term, and invoked against vinyl siding, garbage landfills, nuclear energy, and the endless list of expedient products and services—expedient against a backdrop of 200 years or so.

For all the work and longevity, I would want to see some of those 4 × 8s over the rim of my coffee cup in the morning. It's nice to know the frame is solid, that maintenance will be minimal, that the structural skeleton will be clean and elegant, that the frame is suitable for compartmentalized floor plans or wide-open spaces, and, down near the bottom of the long list of advantages, that it will last a very long time.

Unlike pole construction, probably the least publicized building system in this book, timber frames are available in many forms: from custom builders, stock plan building firms, restoration firms that rescue and recondition beams from dilapidated barns and even import complete barn and house frames from the English countryside, and kit manufacturers. And, of course, you can start from scratch.

All these possibilities are covered in this part of this book. But I urge caution with the last option: planning, designing, milling, and assembling your own. Log building was often accomplished by one family, specifically one man (possibly with the help of a horse for hauling) all alone in the wilderness. But the traditional timber-frame barn-raising was more often a community event. Neighbors gathered partly out of the goodness of their hearts, but also because timber frames required the strength of numbers.

Allowing for exceptions, the social side of these building systems shows that log construction, taxing as it is, was and continues to be successful as a solitary operation. It's not often that you see the facades of the buildings on Main Street built with logs. Timber frames, more suited to community building, generally require more tools, more finesse, more practice, and more bodies.

Now that I've done with the devil's advocate questions, we can concentrate on one of the most interesting and elegant ways to build a house. (I try to give you a few reasons not to use the system up front, figuring that if they really hit home, it should register that this may not be your best alternative.)

Timber framing is challenging but not unfamiliar. In fact, it may be the most recognizable system. Think of it this way: Pick five everyday objects normally made of wood—say, a bed frame, blanket chest, rocking chair, bookcase, even a sailboat. The materials, joints, and tools used to put these objects together have practically nothing to do with saddle notches and solid log walls, or with poles embedded in the soil on concrete pads, or with earth-sheltered, sod-covered roofs. But furniture joinery and timber-frame joinery are very much alike.

skills to put the pieces together? Well, there may be one question worth asking, particularly when you discover just how much time and skill are needed to build a frame that may well last for 10 generations—a house for your great, great, great, great, great, great, great grandchildren. Will those descendants want it, understand it, or even know what it is? Will they really be able to live in a frame house with timbers hewn from the forest when the number over the front door is 1985 and the year is 2185?

I wonder, sad as it seems now, if the idea of a single-family home complete with dug well, backyard, fireplace, garage, doghouse, and a row of petunias out front by the picket fence won't be a thing of the past by then.

You would get the wrong idea if you think this is a knock against timber framing. I am all for strength and durability. And well-built frames are as pleasing as any sculpture. The question goes more to our pride, our self-importance and sureness, the pleasure we take, sometimes to the point of blindness and pure pigheadedness, in our own rightness. Normally, a question like

EARLY TIMBER FRAMES

Timber frames—simply, trees hewn to specific shapes—were used in construction 2,000 years ago. None of those examples still exist. Most homes, temples, and other ancient structures that have survived exist as stone shells minus the roofing, which was wood-framed. But some European timber frames dating from the 14th century have survived. Some of the oldest existing timber frames are in Japan, where the tenacious attention to detail, craftsmanship, and quality construction that we now associate with their automobiles was applied to housebuilding.

The joinery illustrated in accompanying pages may stretch your woodworking skills to their limits. But it is apprentice level compared to the catalog of some 400 joints devised by Japanese builders over the centuries. Many of them are extraordinarily complex but structurally redundant, including intricate twists that do not add to the value of the joint. But this joinery typifies the full exploration of heavy timber-frame work; the highly developed state of the art that has produced the hammer beam truss, the English braced box frame, and barn ceilings of exceptional beauty with 60-foot clear spans.

Like fine furniture now, timber frames began as modifications of very basic shelter designs. The heart of many primitive, hut-type shelters was called a cruck, a naturally angled tree split in two, with one half reversed against the other to form an arch. This was a simple way of using naturally preformed timber to make walls and roof frame in one shot.

This is an ingenious and economical system, provided you don't run out of splittable trees with natural bends in the right places. When you do run short of naturally curved wood, the next step is obvious: use two straight timbers, joined at an angle, to divide the arced, one-piece wall-roof into two legs—one for the vertical wall, one for the sloped rafter. And this step, more likely dictated by materials than purposeful design, led to another element in the frame: Collar ties, placed at the weak link, the joint between the two legs, ran from one wall to the other.

Previously, loads traveling smoothly down the arced wall-roof of a cruck stressed the structure evenly. But loads traveling down the rafter on the two-leg system

If you like puzzles, you'll appreciate this tying joint (post to plate to girt) on one of Ed Levin's timber frames. Photographed from varied angles, these pieces made one joint—one of the most complex you would be likely to encounter. (Photos by Richard Starr.)

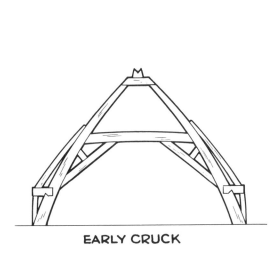

EARLY CRUCK

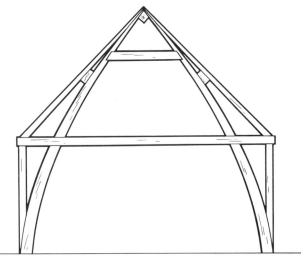

CRUCK IN BOX

applied a lateral force against the vertical side wall. To resist the sideways pressure, collar beams tied the walls together. At the same time, they provided the frame for a second floor.

Now we are used to vertical walls, linear designs, and floor plans with a lot of 90° corners. They stem from our linear building materials. But traces of curved cruck construction survive as braces, sometimes straight but frequently curved, built into the timber-frame bents.

Early timber frames in this country have European, particularly English, roots. In England as well as in Japan, the quality of construction resulted not from ingenious owner-builders but from professional craftsmen who apprenticed and practiced within a highly organized guild system. Japanese guilds, organized along family lines, produced distinctive, intricate, and diverse results. The European guilds, organized more for business purposes only, produced less intricate, more practical frames—right in line with the needs of colonists struggling in new settlements along the seacoast.

It is not surprising that early-American timber frames reflected European frames, complete with wattle and daub or other masonry filler between members of the frame. Conventional wisdom says that settlers always bring housing systems, furnishings, and more along with them. It is remarkable that the colonists did not insist on imposing all the elements of these frames on the new environment. They adapted quickly. There was more cold and more wind in the northern Colonies than in England, more than the traditional fillers between frames could handle.

The environmental adaptation to the European timber-frame system was an exterior skin, a sheathing of clapboards, generally presenting 4 inches or so to the weather. Underneath was the same massive frame.

The early English cruck was incorporated into a box frame that surpassed the curvilinear cruck. Remnants of the 14th-century design linger in the roof frame of a meeting house in Massachusetts, in the trusslike system called a hammer beam, and even in the short, gently arced knee braces in an otherwise rectilinear timber frame.

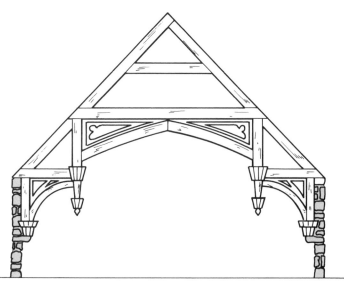

HAMMER-BEAM TRUSS

But instead of the frame and filler facade generally described as Tudor, colonial frame houses took on the classic, protected look of small windows, minimal trim, and clapboard walls—clean-cut shelter against the weather.

Before water-powered saws made the job a lot easier, timbers were pit sawn—tiring work, especially for the guy at the low end of the saw down in the pit. Clapboards (originally called clayboards because they covered the clay-type filler in the walls), wideboard flooring and sheathing, wainscoting, and other pieces

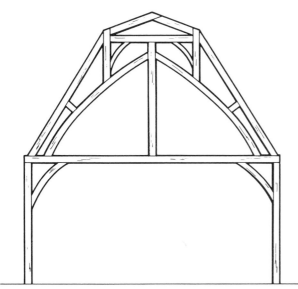

OLD SHIP MEETING HOUSE

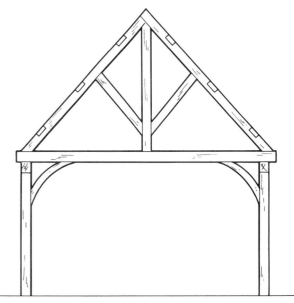

ENGLISH BOX FRAME

of woodwork were cut from the ample supply of large-diameter, tall, straight trees, and cut in enough quantity for exports to England. Shipping records indicate lumber exports as early as 1621.

Also, there is evidence that frames were exported within the Colonies—that is, complete timber frames were fabricated with mortise-and-tenon cuts, even prefitted clapboards, at an established settlement where sawyers, carpenters, and others could work efficiently (and safely), then shipped to more remote sites. (And you thought *knockdown* was a recent invention, allowing shoppers to take bed frames home in the backs of their cars.) In 1633, records of the Plymouth Colony indicate that a house frame including "boards to cover and finish" was shipped up the Connecticut River to a new site in Massachusetts, where it was "clapt up."

American timber frames took shape during hard times. They were and continue to be fundamentally English, even though the new environment tempered the traditional designs. Aside from the addition of clapboards, timber frames in the Colonies were, at first, very similar to their English counterparts. Both established the first floor at or close to grade; both relied structurally and socially on a massive fire column; both used minimal story heights, creating strikingly low ceilings by today's standards. This helped to increase the usable percentage of heat produced at the hearth (heat that was reradiated through the night due to the mass gain of the masonry—supposedly, a modern innovation in building), but was largely an unnecessary holdover. Low story height was an English feature designed to protect the filling between timbers from the weather by increasing the shading effect of the overhang. The style was transplanted to the Colonies even though timber-frame fillings like cob (one of the words

describing the combination of clay and chopped straw) were protected by a wooden exterior skin.

There is a lot about us in timber frames, a lot about our past. That's probably why they are so pleasing to us. Looking way back to the change from huts and bent-cruck construction to hewn timber frames, we can recognize a fundamental change from primitive to civilized life—civilized in the best sense of the word. Just think of the difference between simply using what you find and shaping what you find to be useful. There is a monumental difference—and several stages of cultural evolution—between the two. When building materials no longer determine where and how you live, even how long you can stay there, structures may be built to suit personal and even community needs. The ability to use the properties in material instead of the material as you find it, to imagine the potential for construction, to mold the potential to conform to a preconceived design, requires skillful tool use, a degree of permanence and stability, and communities where existing is part of life but no longer all there is to life.

Looking back only as far as early-American timber frames, there are less-momentous but even more-recognizable characteristics that mirror our history. Colonial frames were not ornamental or flamboyant. There simply wasn't time to spend on nonessential decorative detail. Even though the weather, the need for security, and the endless pressures of starting a new life might have caused expedient structural shortcuts to be dealt with as maintenance and repair later on, durability was preserved.

No wonder timber framing is perhaps the most popular and most accepted (by mortgage bankers and building inspectors as well as homeowners) alternative to conventional modern construction. Perhaps it touches our subconscious, primal drives for security

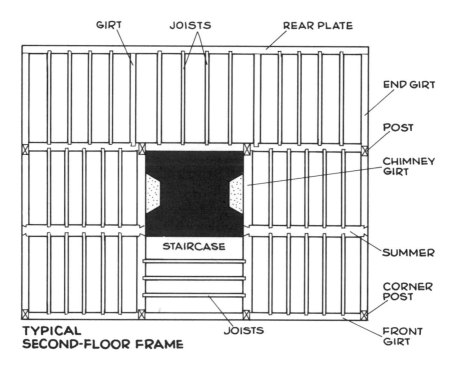

TYPICAL SECOND-FLOOR FRAME

Labels: GIRT, JOISTS, REAR PLATE, END GIRT, POST, CHIMNEY GIRT, SUMMER, CORNER POST, FRONT GIRT, STAIRCASE, JOISTS

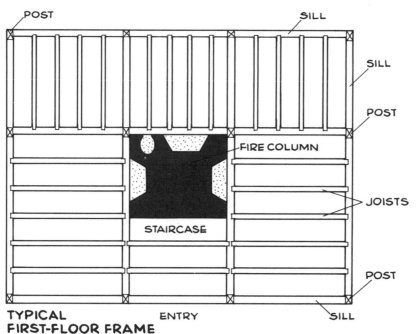

TYPICAL FIRST-FLOOR FRAME

Labels: POST, SILL, SILL, POST, FIRE COLUMN, JOISTS, STAIRCASE, POST, SILL, ENTRY

Although it is difficult to select the average, typical colonial frame, the plans on these two pages come close. The central hearth served rooms to each side of the entry hall and, with the addition of a built-in oven, the kitchen. Around the massive hearth a small number of large timbers carried loads from floor joists.

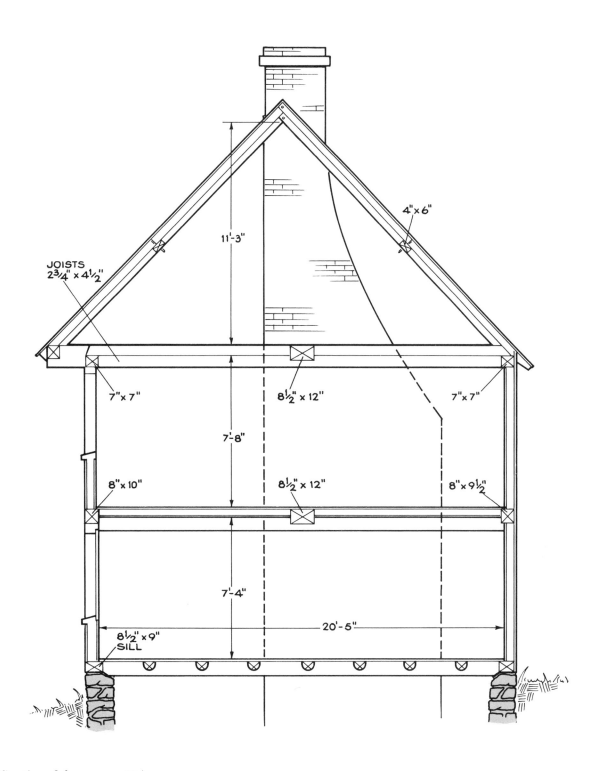

JOISTS
2¾" x 4½"

11'-3"

4"x 6"

7"x 7"

8½" x 12"

7"x 7"

7'-8"

8"x 10"

8½" x 12"

8"x 9½"

7'-4"

20'-5"

8½" x 9"
SILL

(Continued from page 89.)

and permanence and for exerting our will and our plans despite (not in spite of) the limitations of our surroundings. More clearly, it starts us thinking about traditional values, almost a hackneyed phrase by now. But with timber framing the traditions ring true: community spirit, simplicity, efficiency, durability.

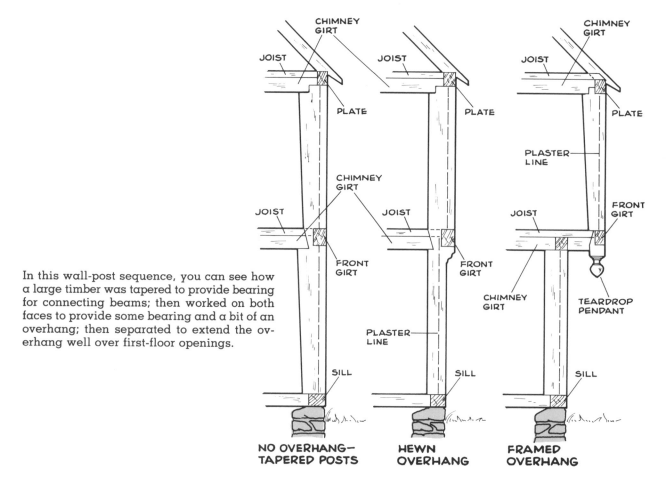

CHIMNEY GIRT

JOIST

PLATE

CHIMNEY GIRT

JOIST

FRONT GIRT

PLASTER LINE

SILL

CHIMNEY GIRT

JOIST

PLATE

JOIST

FRONT GIRT

PLASTER LINE

SILL

CHIMNEY GIRT

JOIST

PLATE

PLASTER LINE

JOIST

FRONT GIRT

CHIMNEY GIRT

TEARDROP PENDANT

SILL

In this wall-post sequence, you can see how a large timber was tapered to provide bearing for connecting beams; then worked on both faces to provide some bearing and a bit of an overhang; then separated to extend the overhang well over first-floor openings.

NO OVERHANG— TAPERED POSTS

HEWN OVERHANG

FRAMED OVERHANG

Timber frames are compact. And although some joints may be complicated, the idea of the frame is not. Each piece of wood is substantial, in a specific place to do a specific job.(Photo by Timberpeg.)

DESIGN PRINCIPLES

Following faithfully the two components of timber-frame construction—the material and the design—presents a problem for owners and builders today. Designs are well documented from the most basic single room with hearth to much larger central-hearth plans with full lean-to additions. An owner-builder can follow historically accurate floor plans and elevations that span the complete design evolution of American timber-frame homes. Materials are another matter.

Oak is the traditional frame material. But unlike building ideas and designs that can proliferate on paper, hardwood does not regenerate so easily. To put it simply, straight, select oak, a finite resource, was used for frames, flooring, and (early on) even for clapboards until about 1800, when oak was surpassed by white pine and other softwoods. Many custom builders still provide frames completely in red oak, which has roughly the same design values (compressive strength, modulus of elasticity, fiber stress in bending, etc.) as white oak, and even Douglas fir. Performance factors for Eastern and Sitka spruce and pine are reduced by about one-third in every area, which must be taken into account when you calculate loads and spans.

Choosing timbers is the easiest decision, because selection and availability are limited. Choosing a design is the hard part, because there are so many beauties. To give you an idea of the possibilities, and to demonstrate a distinct advantage of timber frames—that they can be built successfully in stages—look at the development of the Thomas Lee house (next page).

The skeleton timber frame defines living space more clearly and succinctly than any other residential system. (This home was built by Timberpeg, a firm in Claremont, New Hampshire.)

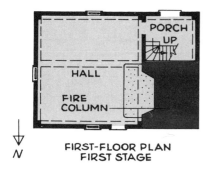

FIRST-FLOOR PLAN
FIRST STAGE

The most basic colonial plan, seen in the Thomas Lee house in East Lyme, Connecticut, enclosed one room on the first floor next to the "porch" and fire column.

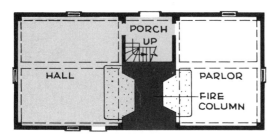

FIRST-FLOOR PLAN
SECOND STAGE

The second stage of timber-frame design, which occurred as colonists had more time for more than subsistence living, added a parlor.

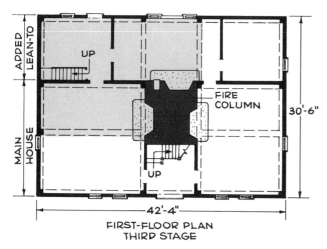

FIRST-FLOOR PLAN
THIRD STAGE

The Thomas Lee house was further expanded with a shed addition off the back of the house, creating the classic colonial center-chimney, timber-frame house.

The house was first framed in 1664 and typified the earliest, most basic colonial timber frames. In the first stage, one small room (box-framed with a central girder) faced south. The room was exposed on three sides, with most of one wall occupied by a massive chimney. Tucked in next to the chimney was a kind of entryway, called a porch, with stairs to the second floor.

A typical second stage, added to the Lee house in 1690, was a mirror image of the initial room reflected on the opposite side of the chimney. The back-to-back hearths were now enclosed on three sides as the stair porch became a pass-through from one main room used for cooking and eating to a slightly more formal parlor.

A third addition to the Lee house increased its first-floor space by about one third with a lean-to addition across the long side of the rectangular layout opposite the stairway porch. The finished dimensions were 42 feet 4 inches across the front of the house, and 30 feet 6 inches along the gable end wall including the lean-to. This third-stage plan was a product of growing families and of relative security and prosperity, which meant less time spent on sheer survival and more time spent on and in the home.

The fourth stage of early timber-frame design incorporated the lean-to addition into the initial building plan, generally in homes built after 1700. Later, a final stage altered the sweeping roof line from peak to wall to lean-to wall practically to the ground, and included the slightly dead space above the lean-to in the full-height, usable living space on the second floor.

There are so many variations: Capes, brick end walls with frame and clapboard between, two-chimney plans with a central hall running the depth of the home, and more. But whether the extended frames are original or add-ons, whether you build large or small, the basic framing method, the tools, the joinery, and the skills are the same.

The grain in oak runs in tightly compressed fibers. That's what makes it a hardwood. Naturally, it is more difficult to work than a softwood. Chisels must be sharper, saw teeth finer, mallets heavier. But if you use highly sharpened high-quality tools and have the patience to work the fibers one small piece at a time, the results are outstanding. I suppose riveted steel is stronger, but heavy timber frames will support a structure a very long time in a fire, and when properly pinned and braced, they are almost impossible to knock down. In his book *Building the Timber Frame House*, Tedd Benson describes his first experience with the frames—a case of meeting the immovable object—as he tried to tear a frame down. He wound up having to dismantle it one piece at a time.

The overall strength of timber frames is due as much to the joinery and the framing system as to the heavy wood. And this is not only because wood-in-wood joints

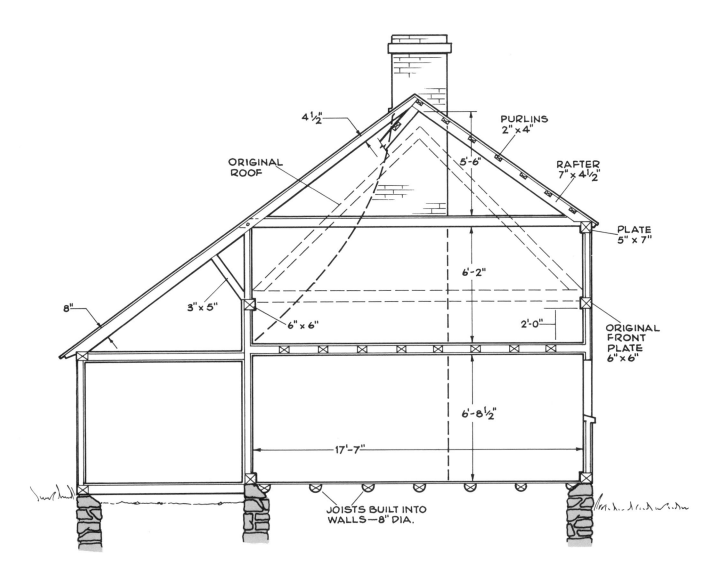

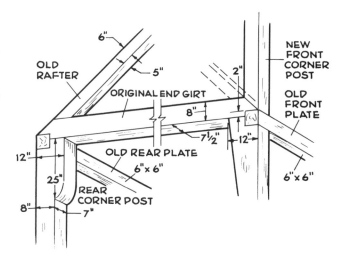

On well-preserved colonial timber frames, exposed framing shows how the original end girt was recut to make way for corner-post extensions that carried a new raised roof.

have more holding power than eightpenny and ten-penny common nails in a butt joint. On a stick-built home, most wood-to-wood connections are simply one on one—for instance, where a stud meets a shoe or plate. This is not always true in balloon framing, the system of full-height wall studs devised by one G. W. Snow (an efficiency-minded building contractor) in the mid-1800s that effectively ended timber framing as the conventional building system in this country. In balloon frames the full-height, sill-to-rafter studs provide structural continuity and at several points form combined lap and butt joints with more than one other structural member.

In timber frames, where major timbers are widely spaced, most joints involve three structural members, and many join four timbers. Instinctively, without worrying about thrust lines, fiber deformation, and other engineering technicalities, a three-part joint should conjure up the ideas of triangulation and truss construction, and suggest an obvious improvement in strength over a one-on-one right angle.

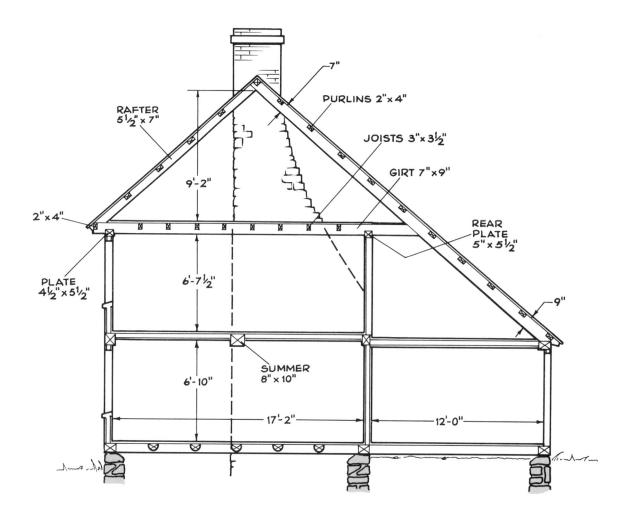

RAFTER
5½" x 7"

7"

PURLINS 2"x 4"

JOISTS 3"x 3½"

GIRT 7"x9"

9'-2"

2"x 4"

REAR
PLATE
5"x 5½"

PLATE
4½"x 5½"

6'-7½"

9"

SUMMER
8" x 10"

6'- 10"

17'-2"

12'-0"

Typical extensions to existing timber frames included an extended second-floor girt to provide headroom in the shed extension (above), and a broken-back saltbox roofline (next page) covering a two-story and single-story shed.

Joints with four members are common along the front and back walls of a timber frame; between vertical front chimney posts on the first and second stories, the front girt running parallel to the sill between stories, and the chimney girt running perpendicular to the joint from front to back wall. In fact, many of the remaining one-on-one joints in a timber frame are triangulated with a short brace mortised in as the hypotenuse of the conventional right-triangle joint.

The strength of the framing system (not only the strength of the timbers) is obvious to most people. But that added strength also requires more time-consuming labor compared to that for dimensional lumber. Builders of dimensional lumber homes always order extra studs to allow for waste, which is often figured at an alarming rate of 15 percent, and also to serve as bracing during construction. On these stick-built homes, braces are everywhere, even after plates are doubled up with lap joints to lock corners together. Let-in diagonal wind bracing at corners, a feature unfortunately absent on most new homes, helps to make frames stable if not

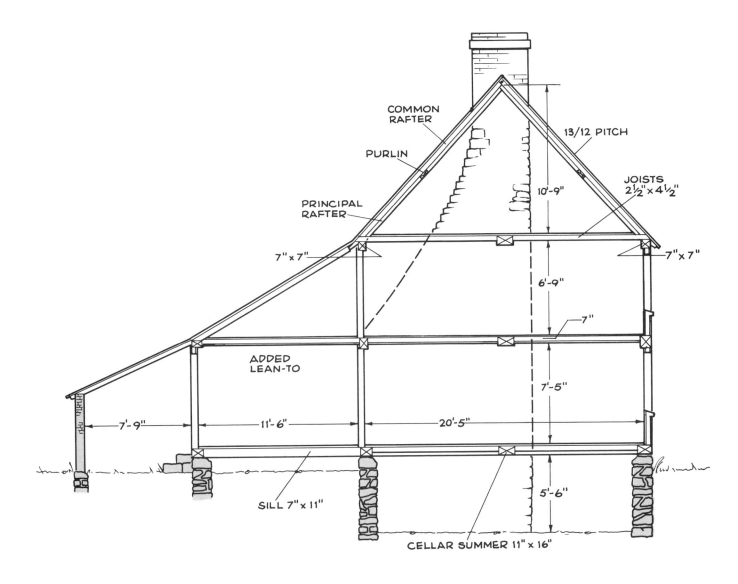

COMMON RAFTER

PURLIN

PRINCIPAL RAFTER

13/12 PITCH

JOISTS 2½" × 4½"

10'-9"

7" × 7"

7" × 7"

6'-9"

7"

ADDED LEAN-TO

7'-5"

7'-9"

11'-6"

20'-5"

5'-6"

SILL 7" × 11"

CELLAR SUMMER 11" × 16"

On the plate-to-post joint (below), three timbers are depositing loads on the post. Three braces take some of the weight and keep the system square and rigid. Photo left: Braces set into the corner post on this Timberpeg home are locked with hardwood pegs. Imagine trying to pull this joint apart!

At the Cornerstones School, Tedd Benson (far right) shows how purlins are set into rafters between bents. Purlins plus temporary braces lock up each bent as it is placed.

truly rigid. But a stick-built house can be racked (shifted out of square) at any stage before sheathing is applied. I've seen it happen all at once from a sudden, unexpected load, and happen gradually from lack of bracing and cumulative errors over the course of the job. And I've seen the process reversed, so that a frame built slightly out of plumb is racked back to 90° before plywood is nailed in place. Any engineering-standards book tells in numerical detail the dramatic increases in strength and rigidity when a modular frame is sheathed and face-nailed to architectural specs.

While a stick-built site is littered with angled braces inside and out, most timber-frame sites are remarkably clean. When bents that include wall posts, floor girt, rafters, and a collar tie are raised into place, then joined to adjacent bents with purlins, there is simply nowhere they can go. You might think that the size of the timbers, the great overall mass of wood in one of the walls, is what keeps the wall from racking. But a typical timber frame has roughly 20 percent (several analysts say 25 percent) less wood mass overall than a stick-built home. It's not the wood. It's not the fancy joinery, although that helps. The strength is in the building system, in the principles of lumber-efficient, durable, timber-frame design.

Full frames making up this Timberpeg structure are so completely interconnected that they brace themselves.

DESIGN DECISIONS

There are many frame and floor plan designs, including both traditional and highly energy-efficient, very modern adaptations complete with superinsulation and large panels of fixed glass. A tour through only a few of the books listed in the information sources at the end of this section would likely provide more choices than you'll want to see. But aside from flow patterns and social operations within a given floor plan, and aside from the completely personal choices about appearance and architectural style, consider some of the consequences that flow from these basic design decisions.

RESTORATION

This is not work you can do yourself. It takes training, the right connections, and a lot of scrounging around on back roads. Only a few companies handle these structures, dismantling and reassembling truly antique timber frames. Each one is special case. The frames didn't all pop out of the same shop, after all, even though the firms doing restoration usually provide a single, illustrated inventory (for a few dollars) of their diverse holdings. Generally, you are responsible for the site preparation: access, utilities, and foundation

This magnificent English barn restoration and conversion is one of many such enterprises pulled together by David Howard, Inc. Working with frames that are centuries old presents specialized engineering, transportation, and construction problems. As you can see, when such a project is accomplished successfully, the results can be monumental, beautiful, and unique.

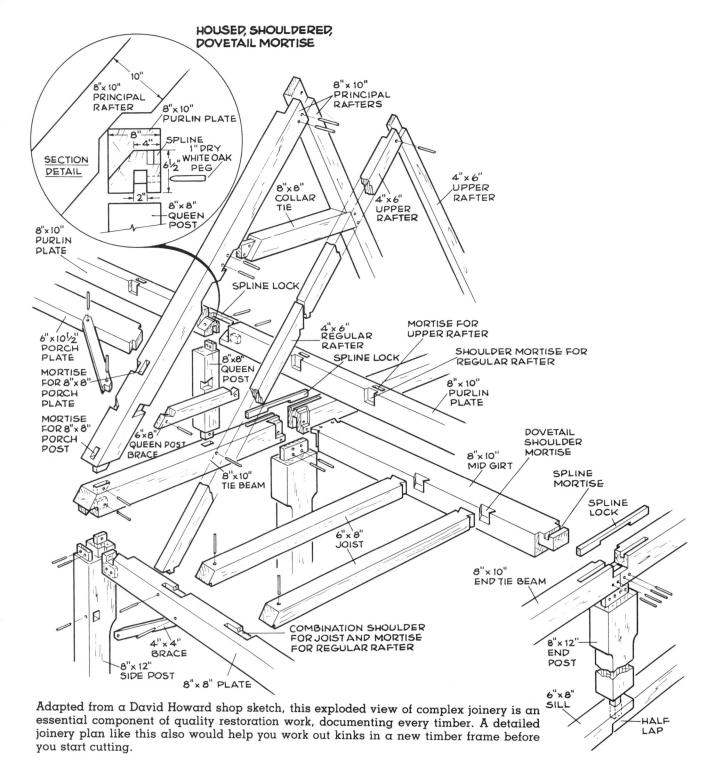

HOUSED, SHOULDERED, DOVETAIL MORTISE

10"
8"x 10" PRINCIPAL RAFTER
8"x 10" PURLIN PLATE

SECTION DETAIL
8"
4"
SPLINE
1" DRY WHITE OAK PEG
6½"
2"
8"x 8" QUEEN POST

8"x 10" PURLIN PLATE

8"x 10" PRINCIPAL RAFTERS

8"x 8" COLLAR TIE

4"x 6" UPPER RAFTER

4"x 6" UPPER RAFTER

SPLINE LOCK

4"x 6" REGULAR RAFTER

MORTISE FOR UPPER RAFTER

SHOULDER MORTISE FOR REGULAR RAFTER

6"x 10½" PORCH PLATE

SPLINE LOCK

8"x 8" QUEEN POST

8"x 10" PURLIN PLATE

MORTISE FOR 8"x 8" PORCH PLATE

8"x 10" MID GIRT

DOVETAIL SHOULDER MORTISE

MORTISE FOR 8"x 8" PORCH POST

6"x 8" QUEEN POST BRACE

SPLINE MORTISE

SPLINE LOCK

8"x 10" TIE BEAM

6"x 8" JOIST

8"x 10" END TIE BEAM

4"x 4" BRACE

COMBINATION SHOULDER FOR JOIST AND MORTISE FOR REGULAR RAFTER

8"x 12" END POST

8"x 12" SIDE POST

8"x 8" PLATE

6"x 8" SILL

HALF LAP

Adapted from a David Howard shop sketch, this exploded view of complex joinery is an essential component of quality restoration work, documenting every timber. A detailed joinery plan like this also would help you work out kinks in a new timber frame before you start cutting.

ready for frame erection. Theoretically, you or a local mason could build an adequate foundation. But the masonry must be matched to a sill and frame that may have a few peculiarities after standing somewhere else for a hundred years or more. You may also have to worry about the effect of an environmental change on the old timbers; from a dry, windy hill in New Hampshire to a somewhat soggy and humid piece of bottomland in Virginia, for example. And then there is transportation.

However, if you can overcome these potential prob-

lems through frequent and accurate communications with the restoration company, you can wind up with a jewel. Be sure the firm you deal with tests the structure and tells you more than you really want to know about it. A thorough treatment includes measuring and diagraming the frame in place, gently dismantling it, replacing structurally deficient sections, wirebrushing or cleaning by some other means, and reassembling the frame on your site. Instead of having a frame built as closely as possible to the way they used to be built, you'll have the real thing. That's something special.

Top photo: Tom Page, a Cornerstones instructor, takes students through at least some of the fine, small-scale points essential to timber-frame joinery. At Farallones owner-builder classes, students also get a taste of bent raising (as shown), one of the not-so-fine points of building and assembling a timber frame.

CUSTOM BUILDING

The decision to start from scratch by yourself or working with a custom builder raises the two choices with the most uncertainty. Both are a little scary. On your own, you need every detail and picture and drawing and numerical table in this section, several of the books listed later, probably a course at one of the timber-framing schools, a lot of good advice, a reasonably cooperative building inspector, a lot of bodies available to push and pull with you when needed, a modest amount of money, and a lot of time. That's all. A run of decent weather wouldn't hurt either, but there's not much you can do about that.

Working with a small custom-building firm or individual can be a rewarding if obviously more expensive alternative. When you get into the thick of books and articles, and write to the schools, and ask around, names of people like Tedd Benson in Brunswick, Maine; Ed Levin in Canaan, New Hampshire; and others are likely to surface from independent sources. There are many names, particularly in the Northeast, of people who love the work, who know their trade, who have credibility. As in all trades, this does not apply to all the names.

Realize that there is a certain back-to-nature, in-touch-with-your-roots, *real*-experience trendiness associated with alternatives like timber framing. And it has produced some dilettantes—folks who are more wrapped up in the idea of being a timber framer than in the time on the job details of the craft itself. This is something to beware of with alternatives that, to many, imply a radically alternative lifestyle as well. Sometimes the idea of an alternative overshadows the substance.

Personality instead of hard knowledge can foul you up. But the wrong personality (incompatible with yours, at least), no matter what the knowledge, can be nearly as bad. If you don't like the builder, don't really trust him, can't make sense out of his plans and descriptions, you'll have trouble in spite of every possible credential.

This Cornerstones timber-frame class learns traditional techniques and poses proudly.

STOCK PLAN BUILDING

A growing number of companies produce a limited range of brand-new, traditional (I know, the words don't mix very well) timber frames. The description

Ed Levin has stretched traditional techniques in a few extraordinary homes with major pieces of curvilinear joinery. And you can imagine what is going through his mind here: If I insert tab A into slot B. . . . Yes, it is a bit like a giant-scale puzzle.

"stock plan building" is a bit unfair, because many of the firms will alter dimensions of Plan Number 9 from their catalog in order to sell you a house. They would be dumb not to. Established companies give you a lot to look at and a lot in writing that you can use to comparison-shop. Don't overlook the age of the business. This criterion works against fledgling firms that may have new ideas to offer, but a track record is important, particularly with firms that have worked past custom building and stock plans to kit and prefabricated frames. Unless you feel you can tackle timber-frame joinery yourself, stock plans make a lot of sense.

Inside this Native Wood Products home, the bays between ceiling joists are closed in to cover the plumbing and the wiring.

Timberpeg offers stock plans for many traditional designs and a series of solar-efficient timber frames. Like custom frames, stock plans can deliver unbroken interior space that lets you showcase the interior structure.

This Timberpeg Solar model has a gambrel design modified to increase southern glass exposure. Inside is an elaborate system of solar air collection and distribution ducts tied to a washed-rock, hot-air, mass storage below the first floor. Timberpeg estimates that this plan provides a 65-percent solar contribution to annual heating requirements, leaving the owner to provide only 10.4 million BTUs, or about 110 gallons of fuel oil.

ENERGY-ADAPTIVE BUILDING

For lack of a better phrase, this option will involve you with firms that have married the traditional and the modern, the old and the new, the durable-frame and the superinsulated, solarized, passively oriented, mass-gained, triple-glazed, hermetically sealed—wait for it—energy-efficient envelope. Some of these marriages are successful. Some are headed for a divorce because the two parties stuffed together in the same place have little in common.

Listen closely for the difference between firms that explain the marriage as a single entity, with integrated, complementary components, and firms that are really selling two products: a timber frame and a unique, energy-saving system. The products may have great merit independently but little to offer in combination.

Before you turn over the job to a professional builder or order up a stock plan from a catalog, I wish you would get hold of a piece of hardwood, even a fir 4×4 will do, and try one of the mortise-and-tenon joints shown on accompanying pages. If that one works out pretty well, maybe you could try one of the three-member joints some Saturday morning. Then maybe you could hunt around for a scrap piece of oak 8×8. Then when a neighbor asks what you're doing with these weird T-shapes, you might find yourself saying, "Ah, I guess I'm thinking about building a timber frame house."

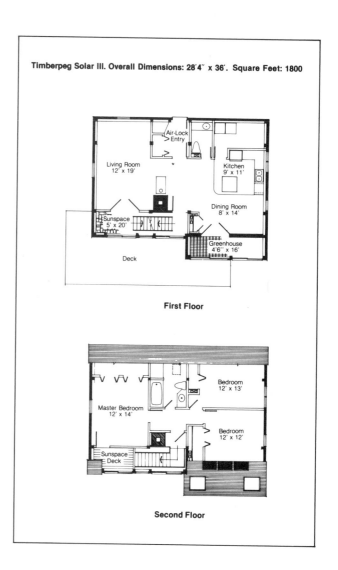

Timberpeg Solar III. Overall Dimensions: 28'4" x 36'. Square Feet: 1800

First Floor

Second Floor

PREPARATION FOR CONSTRUCTION

SITE SELECTION

Timber frames can be built on virtually all foundation systems. Concrete, block, even AWWF (all-weather wood foundations) construction are structurally and

Aside from a tight roof and weatherized building skin, the best way to keep water out of, and away from, the wood frame is to build on an elevated site with a significant slope at grade away from foundation and sill. It wouldn't bother me to build a traditional timber frame on a modern block foundation (as shown) or on reinforced- concrete. Structurally a modern foundation makes great sense, although visually it looks incongruous with timber frame and wants siding run almost to grade.

cosmetically suitable as long as you keep exterior siding relatively close to grade level. Yes, I've left out the perfect match—the natural fieldstone foundation.

I hesitate to introduce another trade, another set of tools, and another set of skills, but most of the timber frames that have survived in harsh climates for more than 100 years did not sit on wooden ground sills (the most elementary method) but on full-height stone foundations or shorter stem walls.

The availability of local stone can be a factor in site selection if you are determined to go whole hog and re-create a timber-frame home literally from the ground up. An interesting alternative is to form and pour a concrete wall with an integral stone facing.

There is not a lot to consider about site selection and building orientation (other than the points applying to almost all building types discussed on page 15) that is peculiar to timber frames. However, common sense dictates that because heavy timbers are the heart of the home in every sense, siting for energy advantage and environmental protection should be considered, with the highest priority given to potential effects on the wood.

Water and moisture. Select an elevated site if possible, or try to create a minimal slope away from the foundation wall during excavation. Completely eliminate contact between the timbers and filled-to-capacity natural runoffs and drainage-collection points. And whatever energy envelope you use, please try to avoid a skin so tight that condensate drips down on the sills. This is one of the new causes of an old problem. Records of town meetings in New Haven, Connecticut, as early as 1656 describe the diagnosis of the meeting house, which needed repairs: ". . . viewed by workmen and finde it verey defective, many of the timbers being very rotten, besides the groundsells."

Fire. Everyone knows that it is difficult to start a fire without kindling. If one 8-inch log rests on two 4-inch logs that rest on four 1-inch saplings that rest on a box pattern of ½-inch splits, a few sheets of newspaper will get things going. Even then, it will be a

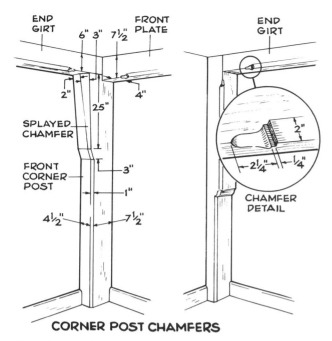

CORNER POST CHAMFERS

Chamfering the fine edge of timbers is a traditional practice. The colonists may not have been thinking about childproofing (a hard edge does more damage) or fire retardation (a hard edge burns more quickly), but only of an elegant finishing touch, bringing the chamfer out to a hard edge in a teardrop pendant just before the joint.

After waterproofing a masonry foundation, take care with backfilling. Avoid burying wood debris that could become termite food. Also, wood, masonry block, and large rocks will induce uneven settling. In addition, create enough of a clearing so the house and frame can breathe, superinsulated or not.

while before the large log starts burning, even longer before its fibers give way. Hardwood lasts a long time, longer than softwood. And heartwood lasts longer than sapwood. Also, it turns out that one of the traditional timber-framing details, chamfering the hard corners of exposed beams, delays ignition by removing the feature of every timber that is most like a piece of kindling—the thin, fine edge, which is most likely to burn first.

Pests. In addition to installing tar-sealed termite shields, do not bury construction debris prior to back-filling. First, bank run and topsoil will settle into the debris, creating a swale by the foundation that traps water instead of shedding it. Second, even short timber cutoffs may provide bracing material, firestops, or, at the very least, excellent firewood.

Exposure. Remember that most properties have an orientation, a kind of weather emphasis, even before you site a house there. It is impossible to alter. Your building site orientation can enhance the emphasis of the property or work against it. Be cautious of energy systems and sites that can produce damp, dark, poorly ventilated conditions that deteriorate timbers. Let the darn thing breathe a little. Stories about homes with 2 feet of insulation in the walls, heated by light bulbs and having some bodies walking around at 98.6°, scare

me to death. I know air exchangers are supposed to do the job (when the power fails, you better throw open a few windows), but I just can't see having only one complete air change per day. (You have to hope no one in the house breaks a sweat.)

Solar gain. Large-scale timbers have ample room for insulation and mechanical systems. Because the beams are massive, they also have room for heat—stored heat. Don't be too eager to bury timbers or to eliminate vertical-frame components in open spaces by increasing header size. Exposing most of the wall frame behind a greenhouse addition, for example, will provide significant mass-gain benefits.

PRELIMINARY PLANS

The initial planning stage should be roughly the same whether you are building a custom home or selecting a manufacturer's model. Of course, if the planning nets you a particularly atypical home, and if you can't find even a facsimile in the company catalogs, you'll be going the custom route.

Unless you are absolutely smitten with a very specific architectural style (a basic Cape, a saltbox, or a gambrel roof, for instance), it is usually more productive to think first about the clear spans, timber limitations, and possible floor plans they permit. Consider the operational home. Ideally, you would think about form and function simultaneously. But specific frame styles will limit your design options. Keep them open for a while. Experiment on paper at least with full, center hall plans and other possibilities. Then examine basic architectural styles.

Cape. The most basic, easy-to-frame, space-efficient, and pleasing design, the Cape is usually laid out in what I call a soft rectangle—something like 32×26 feet on early colonial homes. A typical early plan at that scale might have a 6-in-12 roof pitch (45°), first-floor walls 8 feet high or even a bit less, some 1,650 square feet of floor space, and about 6,000 board feet of timber in the frame. (See page 111.)

Cape and ell. Take the basic Cape and add a mini Cape (usually a single story with attic space) to the gable end wall to create the ell. Like a small version of the first Cape, ell walls are typically inset from the main house, with roof lines lower but at the same pitch. The basic 32×26-foot Cape with a 12×16-foot ell increases the floor space to almost 1,900 square feet, with about 6,500 board feet in framing. (See page 111.)

Early colonial timber frames were very modest structures. Depending on the size of the timbers and complexity of the joined frame, you can build a modest home, as above, a one-room with loft by Native Wood Products, or a huge structure like that below, which became a vacation complex on Martha's Vineyard.

The Manning House, designed and built by Ed Levin, demonstrates the use of a traditional method to achieve an innovative, contemporary result. Straightforward linear joinery is highlighted with some very slick arch additions between greenhouse posts. (Photos by Richard Starr.)

An extreme version of this plan, a kind of endlessly telescoping Cape like a main house with six or seven ells attached in succession, has become the "in" design, housing the rich and famous. One *Architectural Record* house of the year was a telescoping Cape with only a foot or so indentation between frames, the same basic plan used on Jackie Onassis' estate on Martha's Vineyard.

Saltbox. For some reason I always associate this Cape-plus-shed design with Nantucket and a sleek roof line to the ocean weather. There are three distinct types. In the pure saltbox, the shed addition along the back of the house on the first floor has a roof line that is a direct extension of the Cape roof. In the broken-back saltbox, the shed extension roof is flatter than the main roof, thus the broken-rafter line and the "broken-back" name. My absolute favorite is the pure saltbox and overhang. In this plan, the second-floor offset protects first-floor openings. It's practical. And the overhang sharpens the sense of complementary

Four three-post bents capped by rafters with collar ties create a classic Cape design. This straightforward building has only one drawback: a fair amount of unusable space at the attic eaves. (Photo by Richard Starr.)

This classic Cape with ell creates functional floor plans with a little more separation room to room than you get in a basic Cape. The ell also requires valley rafters and more roof flashing.

components in a unified frame: first floor, second floor, shed, and attic comprising a unified whole.

Slight variations in the roof pitch and story height can dramatically alter the shed dimensions and headroom. You will discover by making some of these small alterations on paper that this enigmatic plan in elevation can, with the wrong proportions, become a bulky, top-heavy home instead of a graceful classic.

Gambrel.
This roof design may be associated more with barns than with homes. The distinctive two-leg rafter system (one steep and one flat) is created by building a second-floor wall frame set back several feet from the perimeter walls of the first floor. The lower leg rafters might have a 60 to 65° pitch, while the upper rafters might be pitched at only 20°—possibly a little flat for areas with heavy snow loads. This design minimizes wasted space in the corners of triangles where rafters meet second-floor joists. By extending the wall height past the second floor joists, this problem can be minimized in Cape designs.

This Ed Levin gambrel frame provides the greatest amount of usable space on the second story and avoids slight changes in proportion of rafter slope and length that could have made the profile squat and unappealing. (Photo by Richard Starr.)

With a straight (as opposed to broken-back) rafter extension, the saltbox profile anchors the frame to the site. It is an ideal frame to combine with solar systems.

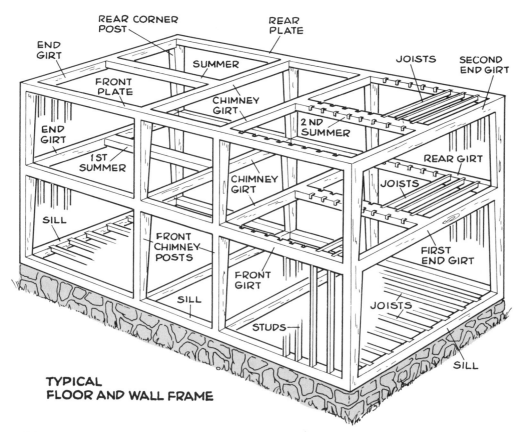

REAR CORNER POST

REAR PLATE

END GIRT

SUMMER

JOISTS

SECOND END GIRT

FRONT PLATE

CHIMNEY GIRT

2ND SUMMER

END GIRT

1ST SUMMER

REAR GIRT

CHIMNEY GIRT

JOISTS

SILL

FRONT CHIMNEY POSTS

FIRST END GIRT

FRONT GIRT

SILL

JOISTS

STUDS

SILL

**TYPICAL
FLOOR AND WALL FRAME**

For approval by local building authorities, you can supply your own nomenclature, and you should supply your own framing plan, backed up by written or, if possible, illustrated notes on joinery throughout. (Plans on these two pages adapted from those by David Howard.)

When your frame plan is established, it pays to work through floor plan, elevation, and section drawings (even isometrics, if you can produce them) to make a joinery plan as well. It may help to cut a few scaled-down models of the timbers you will be using in the frame—to butt them together as they will meet on the job in order to select an efficient joint.

As you work through the plan, you should develop what amounts to a parts list. Every timber need not be precisely dimensioned at this stage. But you should have rough length, the timber width and depth, and the joint system pinned down for all major pieces. You can't discover every possible inconsistency on paper, but any you do find and correct will save time, material, and hassles on the job.

Finally, it is important to catalog the parts of the frame, including the rough order in which each one will be needed, in order to familiarize yourself with the job and to facilitate storage. On extremely remote sites, you may either cut, haul, and hew each timber (a monumental job that is physically taxing, time-consuming, and difficult without accomplished hand skills)

or provide only limited access to the site—for a portable mill instead of a lumber truck. Unless there are indisputable reasons for hand hewing (50 acres on a clear lake for $50 an acre is a good one), it is probably a mistake to undertake this extra work unless you are a stickler for tradition and like arduous work. If you can drive a pickup carrying a portable mill to the site, you can probably haul milled timbers in as well.

Where you stack the timbers may depend on what kind of mechanical help you will use, if any. Gin poles rigged to nearby trees may locate the pile for you. In any case, get the stack of timbers up off the ground, supported every 4 feet or so. Lay strips of lath or a similar spacer between timbers. Provide protection from the weather with suspended tarps or plywood (or a storage barn), but don't create a hothouse environment by wrapping the wood in plastic. Mold, bugs, and fungus that decay the wood thrive in this kind of environment. Spaced timbers protected from the rain will continue to air dry. Also, spacers permit easy access for a belt lift, and for your fingers if you have to move the pieces by hand.

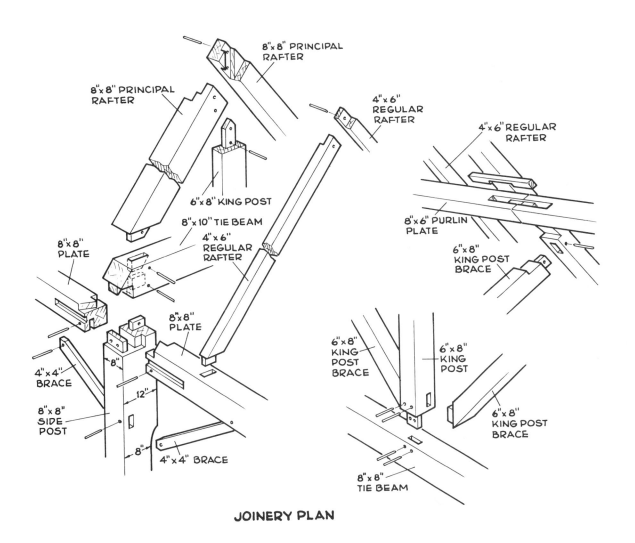

8"x 8" PRINCIPAL
RAFTER

8"x 8" PRINCIPAL
RAFTER

4"x 6"
REGULAR
RAFTER

4"x 6" REGULAR
RAFTER

6"x 8" KING POST

8"x10" TIE BEAM

4"x 6"
REGULAR
RAFTER

8"x 6" PURLIN
PLATE

6"x 8"
KING POST
BRACE

8"x 8"
PLATE

8"x 8"
PLATE

6"x 8"
KING
POST
BRACE

6"x 8"
KING
POST

4"x 4"
BRACE

8"

6"x 8"
KING POST
BRACE

8"x 8"
SIDE
POST

12"

8"

8"x 8"
TIE BEAM

4"x 4" BRACE

JOINERY PLAN

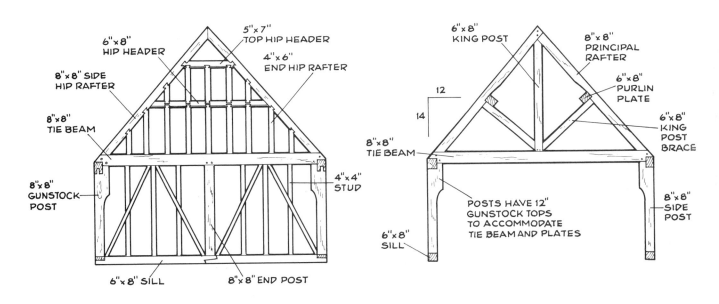

6"x 8"
HIP HEADER

5"x 7"
TOP HIP HEADER

4"x 6"
END HIP RAFTER

8"x 8" SIDE
HIP RAFTER

8"x8"
TIE BEAM

4"x 4"
STUD

8"x8"
GUNSTOCK
POST

6"x 8" SILL

8"x 8" END POST

END WALL ELEVATION

6"x 8"
KING POST

8"x 8"
PRINCIPAL
RAFTER

6"x 8"
PURLIN
PLATE

12

14

6"x 8"
KING
POST
BRACE

8"x 8"
TIE BEAM

8"x 8"
SIDE
POST

POSTS HAVE 12"
GUNSTOCK TOPS
TO ACCOMMODATE
TIE BEAM AND PLATES

6"x 8"
SILL

CROSS SECTION

The more you build, the more you develop personal preferences, say, for a retractable steel tape over a folding box rule. But at timber-frame scale you should use a rafter square to mark accurate crosscuts, saving a try square for detail work. Use sturdy marking gauges for large-scale, repetitive layouts and a punch or scratch awl to scribe joint details. (Photo by Richard Starr, courtesy of *Fine Homebuilding*.)

Many builders use cardboard templates (center). The hinged, metal template is difficult to fabricate but can travel unaltered from job to job. Ed Levin uses both, but never a marking pencil. He uses a knife or awl to mark cross grain and a marking gauge holding a sharp nail to mark with the grain. That way chisels and saw teeth start in the right groove, literally. Inset photo: Complete timber frames are test-assembled, adjusted, temporarily pegged, and marked before disassembly and shipping. (Template photo by Richard Starr, courtesy of *Fine Homebuilding*; inset photo by Timberpeg.)

CONSTRUCTION

TIMBERS

Unless you're going to see the home team, and you know the batting average, homerun count, and shoe size of everyone on the field, it's true—you can't tell the players without a program. From this point on it's important that we talk the same language, so the timber-frame program, complete with words and pictures, follows.

Don't let the terminology throw you. And there may be disagreements about the names of certain timbers from source to source. No matter. Using these traditional names saves you from reading gobbledegook like, "It fits into the big piece about halfway between the timbers over the front wall and the back wall with mortise pockets in the sides facing the front and back of the house." Whew! The timber is called a summer. That's so much easier.

Sill. It is possible to set a timber frame on a conventional deck made of joists and plywood over a block or concrete foundation. Some builders opt for this construction on parts of the house that will be covered inside and out—that is, the traditional system isn't so important if you can't see it. And there's nothing so wrong with that.

The traditional center-hearth plan devoted a large share of interior space to the "furnace." At current costs per square foot of living space, modern timber-frame plans turn this precious space back to the inhabitants, losing the mammoth thermal mass of the masonry column.

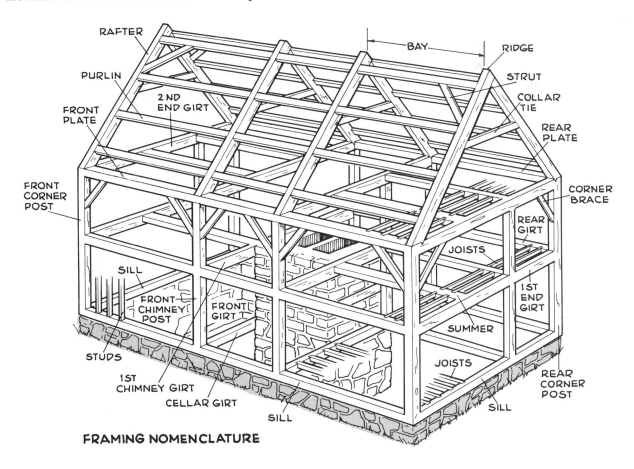

FRAMING NOMENCLATURE

For this Timberpeg frame, the walls are built piece by piece and by assembled wall bent. Then the rafter sections of the bents are set separately.

On full-blown timber frames, sills are generally 6×6 inches or even 8×8 inches, and joined with a tongue-and-fork joint (slotted), a hybrid of that joint including dovetail angles for more security, or a simple half-lap of the sills meeting at right angles.

Girt. This is a major horizontal timber that connects vertical timbers. This description fits the sills, too. But timbers on the foundation are always called sills. In stick-built construction, double 2×4 plates above stud walls are the equivalent of a girt. In fact, many builders call the girts plates when they connect vertical posts on the uppermost story.

Girts are further described according to their location in the frame—for example, on the gable-end walls between corner posts, the timber is called an end girt. One system of nomenclature specifies "first" or "second" for girts at the ends, front, or rear of the frame according to story level. Another system calls the beams "sills" at the foundation, "girts" at the second-story floor, and "plates" at the second-story ceiling.

Joists and studs. Nothing new here. Joists go side to side for floors and ceilings. Studs go up and down for walls.

Posts. The vertical timbers in the frame also are qualified by their location—e.g., corner posts. Chimney posts (even if there is no chimney) describe the major verticals in the front and back walls inside the corners. Traditionally, they framed the stairway porch and enclosed the fire column.

Summer. These timbers can run 6×12, 8×12, or even larger (placed with the longest side parallel, not perpendicular, to the floor), and connect girts—say, between the first end girt and first chimney girt midway between the front and rear walls. They significantly shorten the span for floor joists. Summers are like floor girders in the same plane as the floor joists. Although it may sound a little like a surfing movie, on two-story frames the summer in the attic floor is called the second summer.

Rafter and ridge pole. Nothing new here either. The rafters go up and down, and the ridge pole connects the tops of the angles formed where rafters meet.

Purlins. These horizontal timbers connect rafters. On a modern, engineered barn frame, long 2×4s (nar-

On the U.S. Forest Service test house above, truss frame bents are shop assembled, stacked, and tipped up with pike poles the way it was done centuries ago. However, old-time bents were never this flexible. These are practically pretzels until tied together.

Drawings below: King posts are used on high-slope rafters, while queen posts add support to low-slope rafters. Depending on the overall design profile, these traditional timbers may be offset, as in this Timberpeg home, photo above.

row side down) are set on top of the rafters or trusses. On timber frames, purlins are mortised into the rafters.

Bent. This term refers to all major timbers in the same plane through the frame, as though you sliced through the house, making a cross section, to find a connected series of timbers: two corner posts, a first end girt, second end girt (or plate, if you like), two rafters, a collar tie, and, on most bent designs, short braces set across the right-angle connections between corner posts and girts.

Because timber frames are built by assembling bents, then tying bents together, you can also keep track of all the pieces of the house by assigning them to bent one, bent two, etc., or to the spaces between bents—for instance, summer 1–2 (the summer between the first and second bent).

King and queen posts. A king post is the single vertical timber in a bent running between the ridge and the uppermost girt (or plate). Queen posts are used in pairs, spaced apart on either side of the ridge line, instead of a single king post.

That's about it for the major timbers—certainly enough to get you started.

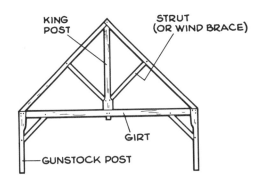

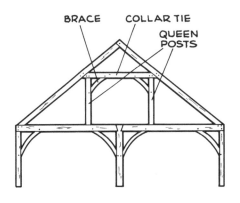

Bents built overall as tightly as this Ed Levin joist-to-girt dovetail can be racked into square if they are only a bit out of square. A come-along can do it. But a tight, hardwood joint like this one is practically immovable. (Photo by Richard Starr.)

SILLS

There are several ways to lay sills on or even into a foundation. None of them work properly if the foundation is out of level. This is a purposeful overstatement. An out-of-kilter sill won't make the house fall down, but it will cause all kinds of trouble. Put more positively, you should strive for perfection here; for splitting the pencil line on a story pole with the cross hairs on a transit; for using bubble-free water levels instead of dime-store line levels.

A level and plumb frame is the product of a specific succession of events—like stacking blocks. Placement in the first few courses is crucial, because any errors are reflected through each block in the stack. By the fifteenth block, a slight misplacement, by itself as close to perfection as humanly possible, may cause a poorly aligned stack to topple.

Sure, you can fudge the margin of sills on foundations, laying truly square corners even if they appear at odds with the masonry underneath. (You can fudge only so far, though.) And, yes, you can add shims here and there and stuff insulation into the gaps. But what a way to start the job—compensating for mistakes with the first timbers you handle.

There is another reason for seeking perfection at this stage (knowing you can only do your best), for measuring diagonals, triangulating corners, checking and rechecking, using every trick and formula even though you are dying to get some wood in place and make visible progress. It is difficult enough to keep a tenon aligned with the walls of a mortise pocket, to calculate angles on a brace and cut them so the shoulders of the brace tighten up flush on four sides, leaving the main

beams fixed like a rock at 90°. If that chimney post, for example, is leaning in just a bit, a true 90° joint (possibly the most accurate piece of joinery in the entire frame) will produce a floor girt that heads downhill. To solve that problem, you have to start laying out would-be square cuts with a bevel square. Each joint becomes an adventure. The process quickly deteriorates into a crapshoot where you are preoccupied with reconciling 87° corners and 93° floor beams that won't quite put the bubble between the level lines.

Reconciling such discrepancies is tedious work to be avoided on new construction. But it may be necessary if you are rebuilding or tying into an older timber-frame building. When I was a kid, my family moved into and renovated an old house. I remember the excitement when we found flaking newsprint stuffed into some of the walls as insulation; the headlines were about Abraham Lincoln (from 1863 I think). We refitted the house inside and out, but left the original frame, floors, woodwork, and such intact.

My second-floor bedroom had a distinct slope on the narrow strip of floor under the dormer. And it became a wonderful place where I could see the Connecticut River in the distance out the dormer window, and play all kinds of games with marbles and balls and tops, which would roll or spin up to the window, then slowly work their way downhill, bumping off the baseboards and return as faithfully as the steel balls on a tilted pinball machine. Thus, structural idiosyncrasies, particularly in houses with old frames, are not all bad.

The most obvious location for timber sills is directly on top of the foundation wall. Surprisingly, many books on timber frames bypass this issue. I would want a conventional set of sill anchors between the wood and the masonry, even though the frame is locked together

Purists would not toenail wall posts to the sill (as shown in these two photos). That defeats the purpose of using the timber-frame system. The close-up view (left) also shows a toenail on a commercial, large-scale frame, used here to pin a shingle shim and draw the sill just slightly onto a layout line. Notice the wedge driven to snug up the post-to-sill mortise and tenon.

as a unit, and even though the great weight presses the sills against the foundation. You may have to use threaded rod, washers, and bolts instead of J-anchors to hold a 6- or 8-inch sill instead of the normal 1½-inch-thick 2×6. Obviously, special and unusual site conditions call for special engineering solutions.

There is another option, which relies on the accuracy of the foundation and cannot be fudged to correct square. The idea is to build a shelf into the top of the foundation, equal to the width of the sill on its bottom and the depth of the sill on its side. Floor girts and floor joists that fit into this L-shaped pocket are then flush with the top of the masonry sill. This design brings the first floor closer to grade level.

There are other twists on this system. On some early timber frames, the sill was independent of the floor frame and served more as a 2×4 shoe does in stick construction—as an anchoring and bearing plate for posts and, if you use them, studs. The historic Bushnell house in Saybrook, Connecticut (built in 1678), had 8-inch-diameter logs (hewn flat only on top to bear against plank flooring) built into the foundation walls.

A similar detail is used on many modern homes for a central girder to carry first-floor joists. To minimize floor height above grade, the beam is set below the sill into the concrete foundation wall. Specs for this detail call for the masonry pocket to be lined with tar or flashing or both. These steps are taken to prevent wood rot from leaks or condensation puddling on the floor of the beam pocket. This possibility presents a strong argument against nestling the sill or individual beams into a masonry ledge.

On timber frames, there are three basic sill joints to choose from. And the one you select may set the tone for the joinery on the job.

At Farallones, teachers and students recycled 8×8s from a pier to frame a barn. Posts have been custom-trimmed to fit snugly between double anchors embedded in raised piers along the foundation. Pulling from above and pushing from below get the bent raised.

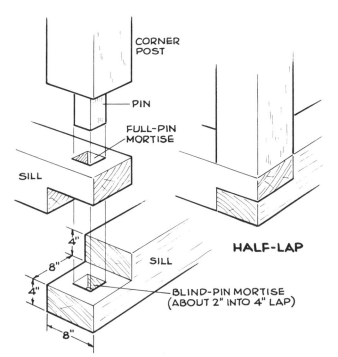

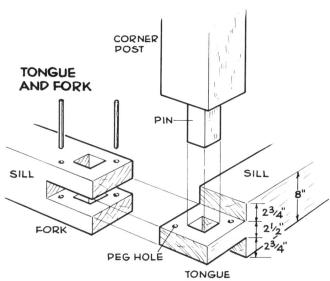

The all-time easiest piece of joinery, the half-lap, does very well on timber-frame sills when it is compressed by the post load and pinned by the post tenon.

The tongue-and-fork sill is like a double half-lap. On 8-inch timbers there is enough meat for three separate overlapping sections. At 5- or 6-inch depths, three layers minus saw kerfs gets marginal.

Half-lap. This is an easy and effective way to join wood. It puts pieces together like cards in a deck. But unless you hold them together, they will separate under stress. The lap can be pinned with a bolt or with a tenon cut into the bottom of a corner post resting directly on the joint. When half the cross section of timbers meeting at right angles in the same plane is removed, the pieces fit together nicely.

Tongue and fork. This joint is like two half-laps in one. The depth of each sill is divided into thirds. On one sill, the top and bottom thirds are removed, leaving a central tongue. On the joining timber, a reverse pattern is cut, and the central fork section is removed. But for all the notching and cutting, each timber can still pull away from the joint. A corner post tenon is used to pin the fork and tongue along with pegs. But this can cause checking and splitting. On an 8 × 8-inch sill, for example, the half-section on a half-lap is 4 inches thick, enough safely to house a 3 × 3-inch tenon extending down from a corner post. But the third section on a fork-and-tongue joint is only about 2¾ inches thick. A 3 × 3-inch hole there would cause problems.

On some designs, the three sections of the fork-and-tongue joint are unequal, producing 3-inch forks and a 2-inch tongue. Even so, these pieces may open up along the end grain under stress, even if you offset the tenon toward the inside corner of the sill. This move will only ensure that the offset side of the sill splits first under pressure.

Dovetail tongue and fork. If you want to go to the trouble to cut the end of each timber into three quadrants, you really should gain some structural advantage from the joint itself, not only from tenons and pegs pinning parts of the joint together. This becomes a more serious issue on horizontal to vertical connections (a girt to a post, for instance), where design loads cannot be handled safely with a small-scale tenon and a few wooden pegs.

The dovetail tongue and fork requires the same number of cuts as the more basic joint, but the cuts are purposefully out of square. While the flaring of the tongue section prevents withdrawal in one direction, it still requires pinning.

Sill table. This treatment can be used with any joint system and can be found on early plank-frame houses. The idea is to rabbet what amounts to a combination ledge and water table into the top outside edge of the sill—something like 2-3 inches high and 1½ inches deep. Traditionally, this rabbet held 1½-inch-thick oak planking (12-16 inches wide). The planks butted against the rabbeted ledge and were secured with two oak pins in each plank. The scale of the sheathing eliminated the need for studs. This ingenious detail has possibilities inside the house, too, where joists are let into a rabbeted sill and even a rabbeted girt, leaving a portion of the sill above the floor line to back up interior planking.

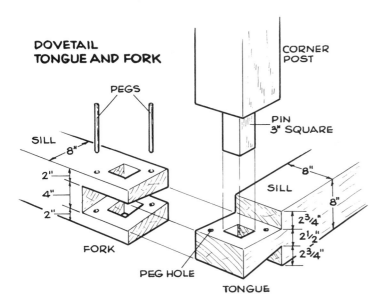

DOVETAIL
TONGUE AND FORK

PEGS

CORNER
POST

PIN
3" SQUARE

SILL
8"
2"
4"
2"

FORK

PEG HOLE

SILL

8"

8"
2¾"
2½"
2¾"

TONGUE

The dovetail tongue and fork is definitely a step up from the straight half-lap. Dovetailing prevents withdrawal in one line, while the post tenon plus the pegs (they are a bit redundant) lock the corner.

Beneath the heavy loads traveling down a corner post, a shallow mortise and tenon, or dovetail mortise and tenon (both details from 17th-century colonial frames), are enough to securely lock the right-angle joint.

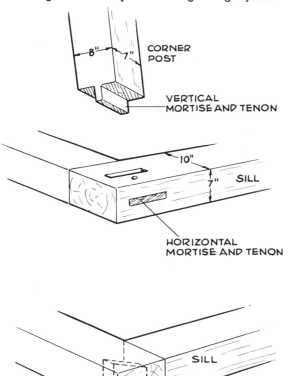

CORNER
POST

8"   7"

VERTICAL
MORTISE AND TENON

10"

7"   SILL

HORIZONTAL
MORTISE AND TENON

SILL

DOVETAIL
MORTISE AND TENON

SILL CORNER JOINTS

One of the pleasures of having a substantial mass of wood to work with is the room to create rabbeted details on the sill, such as this unique, flush siding inset (and water table).

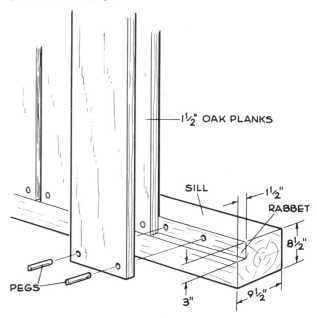

1½" OAK PLANKS

SILL

1½"
RABBET

8½"

PEGS

3"   9½"

From the beginning, think carefully about getting structural reward for your efforts on joinery. Beware of noodling. It's an insidious affliction that keeps you shaving and trimming and sanding and measuring and adjusting and recutting needlessly complex joints, or treating straightforward joints like the Hope diamond. When you get stuck noodling, it's difficult to get timbers off the bench and into the frame.

BENT DESIGN

You can continue the job along traditional lines; that is, designing and building individual bents, assembling them on the first-floor deck, raising them, then connecting the bents with girts and purlins and summers. You can also build and assemble the frame one timber at a time. The advantages to the latter are obvious. Handling individual timbers requires fewer bodies, and you may be able to get by without any mechanical help. Dealing with a timber instead of a multitimber bent is less challenging to many owner-builders. And you get the opportunity to adjust square and plumb step by step at each joint. But this last plus can be a minus.

In any series of small, independent decisions, each one has a margin for error. Problems can arise as the insignificant errors accumulate. There is a good analogy between piece-by-piece framing and laying out studs in a wall. When 16-inch centers are stepped off from a single starting point using a 50- or 100-foot tape, each 16-inch center line is related through the fixed initial point of the tape. But when each center line is meas-ured from the previous line, there is no single reference point. If you tend to reset your ruler so that all of the previous pencil marks are visible, you will gain space and head off center with each new measurement.

So even though building full individual bents (traditionally four of them for a basic Cape) creates some handling problems, it is more likely to produce a plumb and level frame. Four similar assemblies give you fewer chances to go wrong. Four or more mirror-image bents may all be a bit shorter than planned. The rafters may all be a bit steeper. But you are less likely to get a ridge pole running downhill or clapboards that wave in and out along the walls.

Finally, bent designs offer an efficient way to analyze the architectural and structural elements of a timber-frame design. A bent plan can show the usable space inside the frame, the number and size of most timber components (aside from connecting timbers that can be penciled into the plan as cross sections), the joinery to be used, and the pure architectural lines of the house. Several bent designs outlined here are only some of the possibilities.

To raise a rafter truss smoothly and safely, it is good practice to have the builder (or other experienced worker) act as overseer, coordinating the efforts of pushers and pullers. Here the rope tied to the rafter ridge would better have been secured to a solid support. (Photo by Richard Starr.)

Many firms build timber frames piece by piece, while most individual builders take the traditional handcraft approach and build bents. Sometimes, piece-by-piece building (shown below) includes lumberyard 2 × 4s and other common stock in the frame creating a kind of mutant.

Cape and collar. This is the most basic bent: eight timbers, including knee braces between wall posts and connecting girt, and 11 joints, including tenons at wall posts into sills. All right-angle corners in the bent are triangulated for exceptional strength. On small-scale, single-story frames, timbers and rafters may be assembled on the deck before raising. Usually, larger three-post and two-story bents are assembled and raised without the roof timbers, which are set later, from above, with a crane or some form of gin-pole rig. (See next page.)

Cape and king post. This bent design alters only the roof-support system of the basic one-story Cape. An extra set of braces (small timbers that are difficult to set so that they supply their full structural potential) adds considerably to assembly time. The elegant design cuts usable loft space but is capable of supporting heavy snow loads on long rafters with only a moderate pitch. (See next page.)

Three-post colonial. This classic design doubles the first-floor space of the basic Cape with a third, central post. Loads on the long rafters (triangulated with a collar tie) are carried in part down queen posts to the connecting girts. Between the eaves a portion of the second story has full headroom. (See next page.)

High-wall Cape. The designs described above can be altered without much structural work to greatly increase their usefulness. The alteration is simply an extension of the wall posts a few feet past the connecting girts. First-floor space stays the same. Usable space on the second story is increased because the eaves are raised, and the normally useless space where rafters and girts come together is eliminated.

Also, a high-wall Cape introduces a beautiful structural element—the gunstock post, flared out just like the broad stock of a gun beneath the connecting girt. And a high-wall Cape can be extended to include a small shed-type addition at a later date, using the broken-back shed rafter or straight extension to create a traditional saltbox.

Two-story Cape. Taken in steps, it is not that far from a single-story Cape to a high-wall Cape to a full two-story Cape. For extended posts, a second girt, and two more braces, you can double the floor space. The two-story bent is a handful to raise, and it can be difficult to set rafters and assembled roof trusses without a crane.

Two-story Cape with overhang. By extending the connecting girts between stories, you get

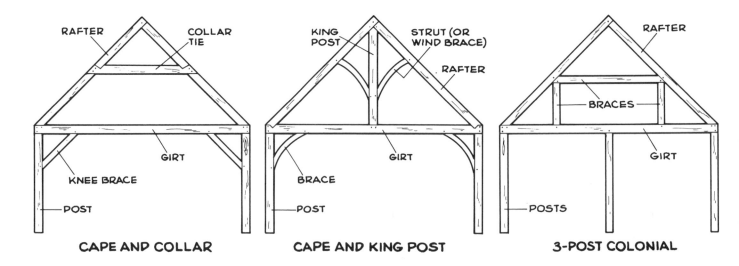

CAPE AND COLLAR   CAPE AND KING POST   3-POST COLONIAL

a second floor that is a bit larger than the first, with weather protection for first-floor openings tucked under the overhang. This design always strikes me as missing a structural counterbalance to the overhang, as though the house would tip over without it.

Saltbox. The two-story Cape plus shed is usually assembled with the shed rafter hooked to the bent before raising. Shed girts are traditionally offset either

above or below the main connecting girts of the central frame.

Saltbox with overhang. This is the two-story Cape with overhang plus the missing structural counterbalance. This bent design has great energy, superb balance, six braces to set per bent, tons of interior space, a specifically forward-looking orientation—to me, stunning lines permitting all kinds of floor plans.

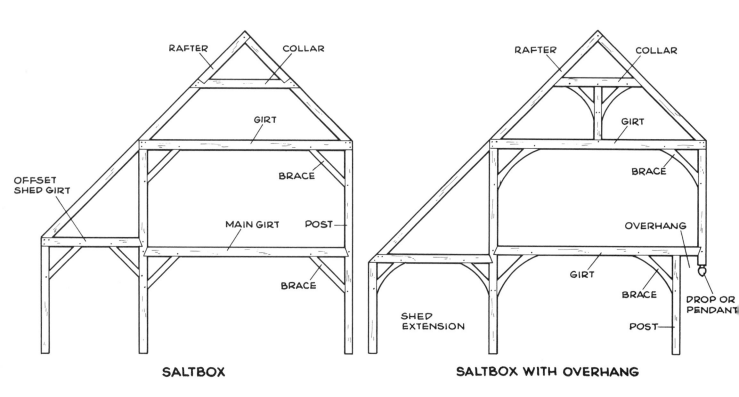

SALTBOX   SALTBOX WITH OVERHANG

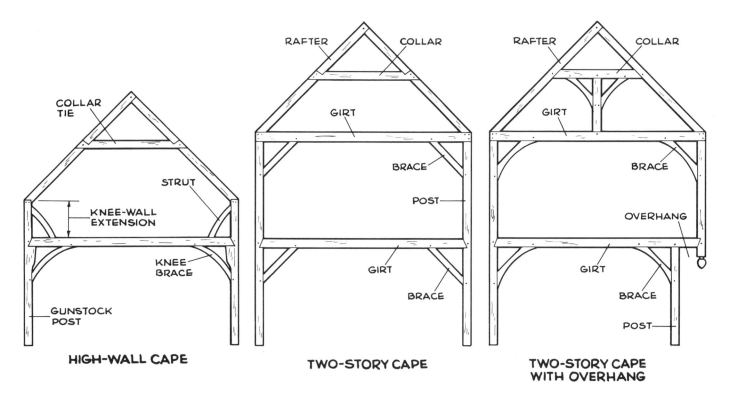

HIGH-WALL CAPE

COLLAR TIE

STRUT

KNEE-WALL EXTENSION

KNEE BRACE

GUNSTOCK POST

TWO-STORY CAPE

RAFTER

COLLAR

GIRT

BRACE

POST

GIRT

BRACE

TWO-STORY CAPE WITH OVERHANG

RAFTER

COLLAR

GIRT

BRACE

OVERHANG

GIRT

BRACE

POST

Gunstock or flared posts are used to increase bearing under multimember joints and to provide more than a narrow shelf for load-bearing girts and plates.

Gambrel. The large gambrel bent is usually comprised of four posts and six braces on the first floor alone—a lot of joinery. A large second-story area is spanned beneath the second girt. On the first floor, the long connecting girt, unbroken by a two-story post, generally includes a scarf joint—an elegant piece of carpentry used to join timbers in the same plane without external scabs.

REAR PLATE

RAFTER

END GIRT

7"

5"

10"

12"

1 1/2"

REAR CORNER POST

6'-10"

PINE WAINSCOT

9"

END GIRT

7 1/2"

GIRT

11 1/2"

22"

11"

8 1/2"

6"

5/8"

7 1/2"

GIRT

5"

4 1/2"

9 1/2"

8"

6"

7"

GUNSTOCK POSTS

RAFTER

GIRT

BRACE

QUEEN POST

GIRT

BRACE

SCARF JOINT

POST

GAMBREL

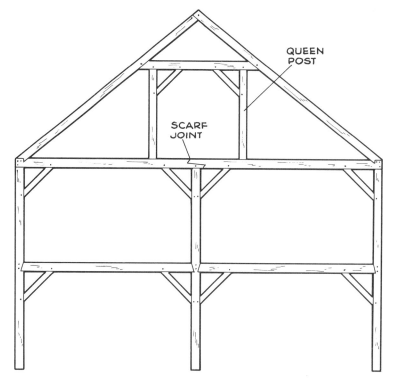

QUEEN POST

SCARF JOINT

TWO-STORY COLONIAL

Two-story colonial. This massive bent calls for two independent levels of three posts and four braces. Aside from great interior space, this bent has a wonderful collection of joinery: shouldered mortise and tenons between first-floor girts and the center post; a scarf joint and tenon between second girts and the center post, and more.

Two-story colonial plus shed. This massive bent calls for two full stories of three posts and four braces each plus a rafter extension for a shed. This is one of the largest bents (you could include a shed on both sides of the bent) that is practical. It is immense. While the most basic Cape has eight timbers and 11 joints, this one has some 26 timbers and roughly 40 joints.

One important point applies to every bent design. Always test-fit joints, compressing them with clamps and come-alongs, then making required adjustments before drilling and pegging.

You can see that there are so many possibilities, including an array of barn frames laid out to form large, open, central halls with side sheds and potential lofts. Other specialties shown in this section include designs that use curvilinear joinery like the hammer beam truss. Some may not produce as much usable floor space for the time and material as traditional Capes, colonials, and saltboxes: That's why these three are classic residential designs. But some of the bent plans are so invigorating, so inspiring, that it seems worth practically any effort to include such special pieces of work as part of your everyday landscape.

JOINERY

You can deal with building plans, layouts, frame design, and site plans (all important considerations) up to a point. When the building process stops being theoretical and starts being practical, you have to lay down your pencil and begin grappling with the timbers. But there's a catch.

Even though architects and engineers spend years in school learning the elements of design and construction, and even though this training should allow them to come up with a better plan than you can devise, it ain't necessarily so. Some of the math and load calculations may stump you, but the lines of the house and the plan and operation of the space inside are subjects every owner-builder, and virtually every homeowner, can tackle. You will be using the space, after all. And, in a way, you know best here.

Carpentry skills, on the other hand, take time to develop. There is no reason you should know about carpentry instinctively. There is no substitute for time on the job. Think of this analogy. You can study the technicalities of music, all the sharps and flats, and you can experiment with tunes and songs and harmonies. So your song doesn't hit the charts. But it sounds nice to you, just the way your smudgy blueprints don't hit the pages of *Architectural Record*, even though they make great sense for your household. But let's say you like your song so much you want to pick it out on a 12-string guitar. Well, your brain may know what the guitar should sound like, but without a lot of training and practice, your fingers can't make it happen.

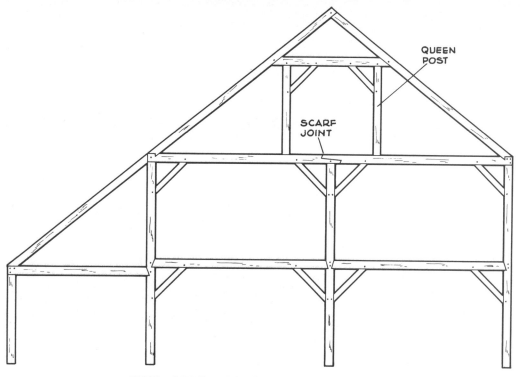

QUEEN POST

SCARF JOINT

TWO-STORY COLONIAL PLUS SHED

Hats off to Ed Levin for the hammer-beam-truss timber frame below. This would make quite a living room, although it is a blacksmith's shop in Canaan, New Hampshire. There is a lot to appreciate here, including curvilinear joinery, let-in diagonal roof struts, and handsome chamfering on the beam's horizontal butt ends. (Photos by Richard Starr.)

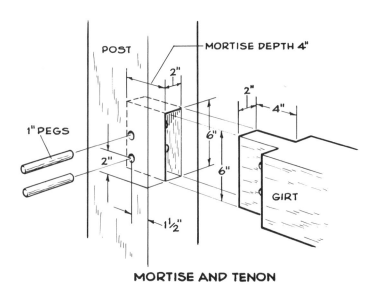

POST

MORTISE DEPTH 4"

2"

1" PEGS

2"

2"

4"

6"

6"

1½"

GIRT

MORTISE AND TENON

Owner-builder schools will help, as long as they offer hands-on classes where you actually build something. Many courses concentrate almost totally on planning and design. And it helps to have the sequence of events laid out in a good how-to guide: where and when you cut, drill, and chisel. It helps, but there is more to it

If you check into owner-builder schools, ask about class size, course projects (do you actually build something?), and actual hours of hands-on instruction. You can learn a lot about tools and techniques in a week or two or three. Getting your hands to move the tools the way your brain now knows they should move is another matter. At some owner-builder schools you may also learn that no matter how thoroughly you study joinery design and span and load tables, you have to eyeball your work, use your common sense and your good judgment.

than that. Try a mortise-and-tenon joint, but don't be too surprised when the first one is not that tight. Try it again. I've cut many frame and even furniture joints that weren't tight enough. But almost always there is a way to run a trim saw kerf through the joint or to pull another of the neat tricks of the trade to improve things. The tricks will help, too. But no book, or guide, or course, or trick can build the frame. A carpenter, albeit a fairly inexperienced owner-builder carpenter, has to do that.

I'm sure you have read all too often that buying top-quality tools and keeping the cutting edges razor sharp is half the battle. Ho hum. This sounds as if it must be good advice. But it's only easy advice. You do not have to buy top-of-the-line tools. And while dull tools will impede a skilled hand, it does not follow that sharp tools will assist an unskilled hand. If you're used to driving an old VW bug and someone drops you behind the wheel of a race-set Porsche turbo, there's no reason to expect that you will suddenly start driving like an experienced Formula I road racer. You can only hope to learn the potential of the machine before you rack it up. Use good tools, sharp tools, heavy-duty and durable tools, not dime-store, do-it-yourself facsimiles. Most important, supply power to guide the tools, but let the tools do the work.

Before any joints are cut, you must eyeball timbers to set crowns up and to orient highlights and defects to your advantage. Work with timbers on secure,

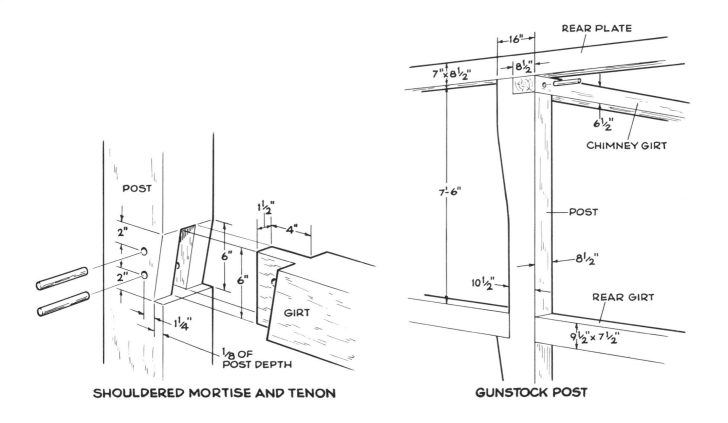

SHOULDERED MORTISE AND TENON

GUNSTOCK POST

heavily braced sawhorses or benches. Long 8×8s weigh a ton. Drop one on your foot and something will go crunch. Square one end of the timber (even if it comes "square" from the mill), orient the timber to the frame (sometimes, to avoid confusion, it pays to lay the timber in line with its future position in the frame), and take measurements from the square end. Finally, establish an easily recognizable distinction between measurements of the exposed timber (the part you'll see when the frame is joined together) and overall dimensions, including, for example, tenons at each end. I generally use Xs over timber ends that will be formed to make joints, and Os on any scrap beyond the joint. Os you can lose. Xs you need.

Mortise and tenon. This basic joint represents the clear difference between conventional and timber framing, where a potential butt joint becomes a wood-within-wood joint. The tenon of one timber fits into the mortise pocket of another timber and is pinned in place with wooden pegs. But it is not a bearing joint. On a large 6×6-inch girt, for example, the tenon might be 4 inches deep and 2 inches thick, held by a pair of 1-inch-diameter pegs—not exactly the amount of material you would place to prevent a shear failure (when a horizontal girt, for example, just lets go). So the basic mortise and tenon is used for light-load girts between posts, for collar ties between rafters, and for connections between posts and sills.

Shouldered mortise and tenon. On horizontal to vertical connections where a timber carries design loads (say 40 pounds psf live load plus materials), shouldering the mortise and tenon adds bearing value to the joint. You prepare a basic mortise and tenon with one alteration, angling the recessed wood sections on each side of the tenon (the shoulder) and cutting a corresponding angle into the post. On a 6×6-inch frame, the tenon could be 4 inches high and 2 inches thick, but 4 inches deep only at the top of the beam. As a rule, the bearing shelf created by the angled shoulder is one-eighth of the post depth, about ¾ to 1 inch on a 6-inch post. It is helpful to make a metal or plywood template of the shoulder angle. This makes it possible to produce a mirror image of the girt angle on the post.

One traditional alternative to a shouldered joint is a gunstock post—that is, a vertical timber shaped to flare out as it reaches the girt, like corbeling brickwork at the back of a fireplace. With one of many possible gunstock patterns the bearing cross section of the post (6×6, 8×8, etc.) is left intact, and the bearing shelf for a connecting girt is cut into the flared, gunstock-shaped section. On some early colonial frames, the entire post was slightly tapered to produce enough wood at girt height for a bearing shelf—for example, increasing from 6 inches at the floor to 9 inches at the ceiling, with many variations of tapering, chamfering, and other structural and decorative detailing.

HOW TO MAKE
A SHOULDERED
MORTISE AND TENON

The indestructible, pinned and shouldered mortise-and-tenon joint is the opposite of what Ed Levin calls the "popsickle stick stud" construction on most houses. Levin's choice of cut-mark lines helps to produce clean, elegant joints. Even a shallow cut from knife or awl prevents surface grain from burring up. If those wood feathers appear, you know the saw is off course. (Photos in this series by Richard Starr, courtesy of *Fine Homebuilding*.)

Depth of the shoulder mortise (the small shelf where the entire width of the tenoned timber will bear) is established by the depth-of-cut setting on a circular saw—the exact protrusion of the blade below the shoe. The time-honored carpenter's rule is to space the kerfs equal to or slightly less than the width of the chisel you will use to clean the mortise.

Clear the bulk of the tenon pocket by boring overlapping holes with a bit whose diameter matches exactly the width of the tenon. (Buy a bit that's close, then make the tenon match it.) Forstner bits, with a small starter knife at the outside diameter, are a good investment if you used a jigged hand drill. Center punch holes so the bit just kisses the tenon layout line.

A sharp, fine-toothed circular or hand saw blade establishes the depth and thickness of the tenon. Some builders (with the right oversize equipment) make one cut from the end grain back to the crosscut to remove the waste in one block. Traditionally, you saw cross grain, then use mallet and chisel to pare off manageable, bite-sized slabs *with* the grain.

Good tools help. Sharp tools are essential. But there is absolutely no substitute for sticking your nose in there, "eyeballing" depth and level and consistency of chisel cuts. Sharp tools, like high-performance race cars, are responsive to subtle adjustments. They go where you tell them. But like a race car driver, you have to see where you're going.

By setting the chisel bevel up and into the cut-mark line, you will cut clean burr-free edges (the high-quality details you'll appreciate later when exposed beams are highlighted against plaster walls). Clear saw-kerfed wood from both sides of the mortise into the center to prevent edge tears and splits. Here the back half wood is already cleared.

Cleaning from both sides will leave some wood scrap in the middle. Make a high-class fix with a bench-rabbet plane, which is a specialized jack plane with its blade flush with the edge of the plane body. This lets you clear all the way into corners. The "self-jigging" plane will stop raising wood shavings when the shoulder mortise is clean and level.

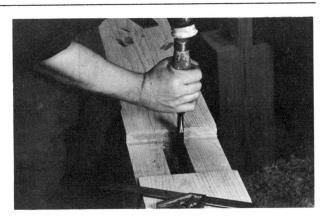

When bored holes are liberally overlapped, hand pressure, or a shot with the heel of your hand, is often enough to clear the chines (little ridges) left between overlapping hole diameters. Work at a slight angle to the shoulder mortise for clean cuts. Be consistent. Treat hidden joints like this one as carefully as those that show.

Corners are easiest to work at the surface of the shoulder mortise where you can see them. That same right angle must run straight down the tenon pocket. Notice the ever so slight cant on the chisel as it digs into the corner. Once past the surface, err by removing a bit too much, not too little, so the tenon will seat without a struggle.

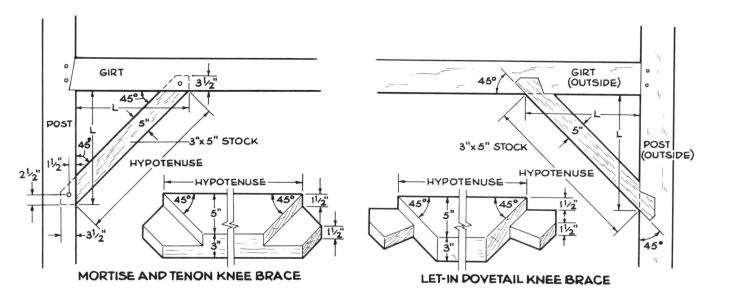

MORTISE AND TENON KNEE BRACE

LET-IN DOVETAIL KNEE BRACE

Braces. On solid timber frames, even where a girt-to-post joint is shouldered, knee braces are used to triangulate and lock the timbers together. These short timbers also reduce the unsupported length of girts. The easiest (and oldest) version is a let-in, dovetail brace—a piece of joinery used in English timber frames 800 years ago. The word "let-in" tells you this system is similar to diagonal corner bracing on some stick-built structures today.

On 12th- and 13th-century English frames, braces were generally longer than those used later. Frequently, they intersected more than two timbers, such as the post and girt at 90°, and were set after major timbers were in place. On a typical 3×5-inch brace, half of each butt end is left at 45° (to fit as the triangle hypotenuse between a post-girt right angle), and half is left in a dovetail pattern (to fit into a corresponding mortise in the post). Any dovetail-like shape can be used as long as it is let in from the side and cannot be withdrawn along the line of the brace. The locking dovetail can be made as a blind mortise (only partway through the depth of the post) or as a full mortise, so the end grain of the dovetail appears on the outside face of the post.

While these joints are elegant, fitted just like pieces of a jigsaw puzzle, they are off-center on the post because they are let in from one side. Consequently, loads coming down the brace will transfer at least some skewing or rotating force to the post. A centered mortise and tenon is theoretically superior to a let-in brace because its load pushes but does not twist. But I don't see an 8×8-inch oak post, 8 feet long, mortised and tenoned between sill and girt rotating from such a minor load, certainly less than the load carried by the girt above the brace.

Whichever method you choose, the most important thing is to remember what the brace is for. It is to keep the major timber in place, to prevent racking, to help resist unexpected, one-time overloads—from wind, for instance. This means that if you cut the brace with a roomy fit, it will be easy to insert but will be more decorative than structural. If you leave anything a little long, looking for the tightest possible fit, and have to beat the poor brace into position, you are likely to distort the main joint and angle that the brace is there to support.

Do you remember being introduced to the X, the unknown, in algebra? You may also remember the detective work of solving an equation for one unknown and the blank stares later in the year when equations had Xs, Ys, and even Zs. The post-to-girt joint is like the X. The same joint with a knee brace is like adding the Y and the Z. If you're a pessimist (and that's about the worst thing an owner-builder can be), you will appreciate that the knee brace gives you two chances (one at each end) to make a mistake. And as in an equation, only one mistake is needed to foul up the entire joint assembly. Braces have to be measured accurately. I'm thinking of the unmarked $\frac{1}{32}$-inch increments between the $\frac{1}{16}$-inch lines on a ruler.

The math analogy is not here by accident. After studying all the mathematical methods of construction (triangulating corners, cutting bird's mouths at rafters according to the umpteenth step off on a framing square, and such) and building houses and additions and furniture and more, I am a firm believer in templates and measuring in place on the job. How's that? Put it this way. Of course I agree that *if* a post and girt do in fact meet at 90°, and *if* you measure equal distances along the girt and down the post, and *if* you

Braces on the large, commercial timber frame shown in these two photos are used between corner posts and girts, between wall posts and wall plates, and between corner posts and end girts.

Softwood frames, as shown in these two photos, are easier to work and more forgiving than cabinet-quality hardwood frames.

use the Pythagorean theorem of $a^2 + b^2 = c^2$ (each leg of the right triangle squared equals the hypotenuse), and *if* you reproduce exactly that c^2 measurement on the knee brace, and *if* you cut 45° angles at the butt ends of the brace, the brace will fit. That's too many *if*s for me. I know numbers don't lie, and it's not that I don't trust them. I don't trust an 8×8 to be

exactly 90° to a post. And if it is at one edge, it may not be in the middle or at the other edge. I don't trust the rivet joints of a box ruler. I don't trust the mill to produce a timber that is precisely 8×8 inches anywhere I choose to measure. I don't trust the shoe of a circular saw to be or stay at precisely 90° to the blade no matter how I push on it. Algebra works oh so well in the class-

You can measure all the pieces, then cut them identically so that each rafter truss duplicates the one next to it. To do this, clamp a collar tie in place, make any needed adjustments, and then mark it on the rafter. Make one complete truss that works, then use the pieces as templates.

As illustrated, 45°-angle braces make a true, timber-pegged frame impossible to tear apart. Notice that the pegged mortise-and-tenon wall plates on this Timberpeg house are built with a bearing lip for floor joists.

BEARING LIP FOR SECOND-FLOOR JOISTS

room. But actual construction can be sabotaged by algebra.

If, after years of experience, you come to value and rely on the mathematical approach, fine. But I think it's a misplaced emphasis for owner-builders and even for relatively inexperienced professionals. There are too many other things to think about and so many skills to learn and master. If you can cut a 4×4 post with two passes of a circular saw so that it is difficult to find where the kerfs meet, if you can cut a truly square butt end (edges and all) straight through the 4×4 with a handsaw, you will be way ahead of measuring to, and relying on, a hypothetical c^2 dimension.

I am not opposed to using a framing square at the 90° joint or to measuring for 3–4–5 triangulation. But I also want to sight down the girt, take floor-to-girt measurements at each end of the timber, take a 4-foot level reading at each end and at the middle of the girt, and check the joint against a template or the girt against elevation lines on a story pole (used with a transit). If all is well, you can get excellent results by clamping the rough-cut brace in place, tapping, adjusting, marking with a fine pencil point, then cutting so the saw kerf just kisses the pencil line. If I trust my work on the main joint, I am happy to mark the brace in place, before I would rely on a template and way before I would rely on a formula. In short, use the math to check your work, not to produce it.

Dovetail half-lap.
A dovetail makes an excellent joint between purlins and rafters, between joists and girts, and between joists and summers. There are alternatives: a simple, square-ended half-lap, and a more complicated housed dovetail, where the flared joint and a small bearing edge on the beam are set into a connecting timber. Mortising—say, a 3×3×4-inch section out of an 8×8-inch girt—doesn't hurt the large timber, even when another joist or purlin is added on the opposite side. But the half-lap may cause trouble in the joist or purlin itself.

Because the joint is so simple, there is a concentration of stress along the inner edge of the lap. This line separates the upper half of the timber, which bears on the girt, and the bottom half, which simply butts against the girt without any support. When one half is at rest and one half is hanging, there is a natural tendency for the two halves to part. You might see limited checking at first, then cracks extending down the joist, then splitting, and then, possibly, failure.

This possibility can be minimized by tapering the joist gradually from the lap to a point 5 or 6 inches away from the girt. When the stress of splitting is spread out along this more gradual separation of the resting and hanging halves, damage is less likely.

Changing the simple half-lap to a dovetail adds lock-

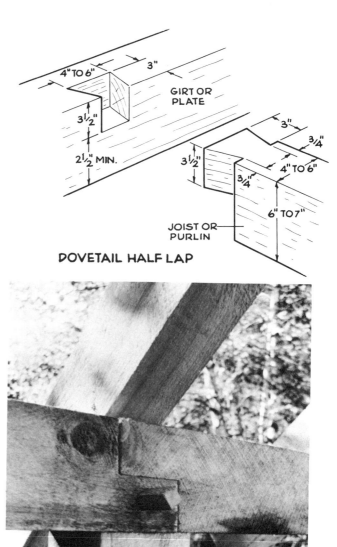

DOVETAIL HALF LAP

The pegged half-lap on the Timberpeg rafter plate above is in no danger of splitting, because the full body of the lapped section plus a good inch of the upper lap timber bear directly on the post. But these softwood joists below are susceptible to checking and splitting, because the lower halves of the joists are unsupported.

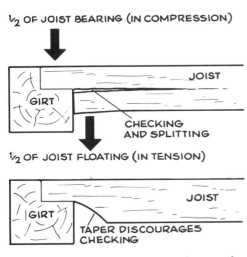

½ OF JOIST BEARING (IN COMPRESSION)

GIRT

JOIST

CHECKING
AND SPLITTING

½ OF JOIST FLOATING (IN TENSION)

GIRT

JOIST

TAPER DISCOURAGES
CHECKING

Tapering joists into mortise pockets in the connecting girt can prevent concentration of stress at the exact point between the forces of compression and tension.

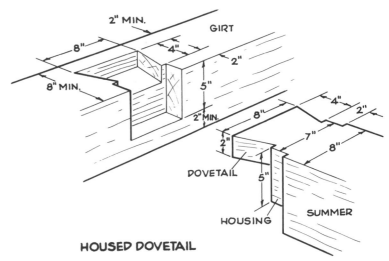

HOUSED DOVETAIL

The housed dovetail requires an additional set of measurements and cuts but will not separate at the line between compression and tension, since a major portion of the beam in section is directly supported by the connecting girt.

ing power to the joint and eliminates the need for pegs to prevent withdrawal. But if the dovetail is 3 inches thick on a 6-inch-deep joist, similar stresses (checking and splitting) can occur. There is a solution for these potential problems with the straight half-lap and dovetail—the housed dovetail.

On a housed dovetail cut into a load-bearing timber like a second-story summer 8 inches wide and 6 inches

deep, the 8-inch-wide, 4-inch-deep, and 2-inch-thick-dovetail extends from a shoulder 7 inches across, 2 inches deep, and 5 inches thick—enough timber to transfer safely all loads from summer to girt. For that matter, a straight half-lap can be housed, too. After the lap cut is made, the corresponding mortise can be moved even an inch into the timber, with a second-stage mortise supporting 1 inch of the entire timber.

Like a mountain climber carrying 175 pounds of body weight on not more than a one-inch ledge, even a marginal post pocket is enough to unload a mortise and tenon. If possible, create these pockets with tapered, two-story (or greater) posts.

Timbers shrink minimally along the grain, but stacks of headers and girders (as shown) in line can cause problems resulting from accumulated compression of wood fibers perpendicular to the load.

POST POCKET

You can see that the array of joints is mind-boggling: bird's-mouth cuts for seating rafters, tongue and fork at sills and at rafters without ridgepoles, elaborate interlocking scarf joints (or simple half-laps with pins), and more. For every situation there is generally a basic joint that relies on wood pegs to prevent withdrawal and also to act against shear. Such a joint is usually not complicated or time-consuming to cut. And there is generally a more elaborate version of the joint that adds structural value by the nature of its design. And there may be a considerably more elaborate version that prevents withdrawal, adds structural value, increases bearing, and more. Naturally, that one will take the most time and trouble.

But the skills of joinery are more important than the exact style and size of individual joints. This skill has six major elements: measuring, cutting, boring, chiseling, adjusting, and fitting.

Measuring.
Orient the timber. Square an end. Work from that end and from initial points without resetting rulers and tapes whenever you can. Adopt a policy on pencil lines and stick to it: Mark the edge of waste material and cut the line off, or mark the cut line of timbers and leave the line on. Don't waffle back and forth. Don't use lumberyard pencils with leads as thick as some saw blades. Use templates on repetitive joints whenever possible. Recheck everything after you are sure you have it right. If you use a 6- or 8-foot folding rule, add a little squiggly line to the 6-foot stepoff mark on the timber. (See drawing and caption, above right.) Finally, if in doubt, if the blade is turning, and if you are not approaching the cut line with absolute confidence about the exact point of contact between the blade and the wood, leave the pencil line on for a thin but additional margin for error.

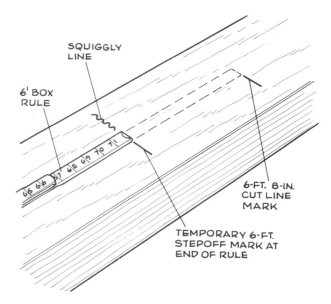

A ruler shorter than the timber you wish to cut may cause a cut short of the intended mark. For example, if you measure to cut off a 6-foot 8-inch timber with a 6-foot ruler, you must make a temporary mark at the 6-foot stepoff (the end of the ruler) and then move the ruler ahead 8 inches to mark the cut line. If the two marks are similar, you may accidentally make the cut line over the wrong mark.

Cutting.
Pick your favorite brand of crosscut and finishing saws. Go to some length to locate a small trim saw. Replace the screws and rivets in the handle if need be. Get the saw sharpened with no set—that means the teeth are directly in line, not splayed to opposite sides. No set produces a minimum kerf. When you set a joint and the shoulder seats first or seats out of square, you can get a near-perfect fit by running the no-set saw through the uneven seam, applying additional pressure to close the joint against the saw as you cut.

The marking and cutting of this timber-efficient, minimal-frame experimental home is done in the shop. Field measurements are replaced by full-scale jigs and templates. Since every "bent" comes out of the same jig, every web of every truss is in line. Similarly, custom timber-frame builders commonly use templates to mark standard joints.

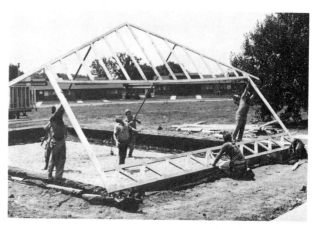

Ed Levin often uses this old Millers Falls coffee grinder boring machine—a kind of portable drill press. Up in the air on the job site a ½-inch power drill and long auger are used to drill and peg connections such as this dovetailed half-lap. (Photos by Richard Starr.)

Boring. Invest in Forstner bits to produce clean holes on visible surfaces. A ragged, split seam really detracts from the elegant, utilitarian appearance of wooden pegs in hardwood timbers. Use colored tape, nail polish, or some other obvious marker on bits to establish boring depths. Cleaning the bottom of a mortise is a lot easier if the four or six holes bored to clean

Photo left: Precise dimensions are needed to build triangulated joints. If the brace must be distorted in order to make both brace and girt fit, the brace begins to lose structural value.

Builders at Native Wood Products use the oversize circular saw (left foreground) for cutoffs on large timbers.

out the heart of the mortise all bottom out at the same level. Although finding the exact tangent between a round hole and the linear border of a mortise (just kissing the mortise outline with the arc of the bored hole) is impressive and pretty, it is unlikely that you will hit this mark six out of six on one mortise. So drill to clean wood and establish depth, and rely on chisels to define the mortise.

Chiseling.

I have a small toolbox lined with an old towel just for chisels: a few old Stanley steel cap butt chisels (old from when the steel in them was really good), a set of long, heavy-duty socket chisels, three oddball sizes from an old carving set, a bunch of Marples cabinet chisels, an extra blade from a finger plane, and more, all wrapped up in the towel. It's great because I can always lay my hands on a size and shape that seems best suited for the work in progress. And even though they are sharp, unless I am in a big hurry I run the one I'll be using over a sharpening stone. With sharp chisels and a varied selection, I use a thin-bladed, wide chisel to make the first cuts through the fibers, along the lines that will show against a connecting beam, setting the flat side of the chisel toward the wood that will remain and the bevel side of the chisel toward the mortise. Don't ask the chisel to do too much. Take away wood in bite-sized pieces. Put down the mallet and clean the walls of the mortise with hand pressure. You don't need a mallet if the chisel is sharp and the bulk of the mortise is already gone. Saw across grain whenever you can, even if the kerfs go only partway through the mortise. Use a small chisel with a low-slope bevel to clean square corners.

Adjusting.

When you cut too much off a tenon or clean a mortise so thoroughly that you wind up taking a fat 1/16 inch too much, you get a loose joint that is difficult to repair. Short of cutting a new one, you might try a piece of veneer cut to the joint pattern and glued in place. But you should be measuring and cutting to be off on the fat side, not the scant side. When measurements fail you, there are ways to make improvements. Sections of square joints can be closed with no-set saw kerfs. You can cheat joints with one clearly visible seam by backcutting—that is, starting a square cut, then applying enough twisting pressure on the saw handle to angle the kerf just slightly under 90°, "cutting back" under the visible seam. Done successfully, this ensures that the tightest, closest-fitting part of the joint is the part that shows. Backcutting requires some finesse. The idea is not so much to remove extra wood inside the 90° cut, but to make sure that the saw kerf does not leave any extra wood outside the right angle that would prevent the top seam from closing. Thinking about backcutting may be enough.

You can also cheat a socket-type joint by slightly chamfering the hard edges of the tenon. Even a clean, square mortise pocket is likely to have a little extra wood and a few stray fibers along the edges—just the way dirt and wax build up in hard-to-reach corners on a kitchen floor. If some of the raised grain in the mortise picks up grain on the tenon, you'll have to beat the joint together. When the corners are clean, the hard tenon edges are slightly chamfered, the dimensions are correct, and the joint should seat but it won't (and that can happen), you might try rubbing the sides of the tenon with a bar of soap or a candle.

Although a boring machine or drill press or power drill can be used to clean most of a mortise, you'll need a mallet and chisel (below) to square up the pocket. Right photo: Chisels should be big enough, heavy enough, and sharp enough to raise shavings of consistent thickness from hand pressure alone.

Last-minute, on-the-spot touchups can be made with saw kerfs, a chisel, or a little pressure. Some joints can be designed with this kind of adjustment in mind. For instance, in this scarf joint, Ed Levin has cut a locking keyway. When tapered wedges are driven against each other from opposite sides of the joint, they press the joint closed in line with the two timbers. (Photo by Richard Starr.)

Fitting. Test-fit all joints before drilling and pinning. You will need a hand-cranked winch (called a come-along), cable, and some form of protection to keep the cable from biting into the wood timbers. There is no better way to close a joint. A mortise and tenon under pressure from the winch and cable may get hung up or skewed out of square, requiring minor persuading with a beetle (a large wooden-headed sledge) on the way in.

Most builders make their own "persuader." You can make big ones or small ones. Call them mallets or beetles or sledges, weighing 5 pounds to 25 pounds. (Bottom photo by Richard Starr.)

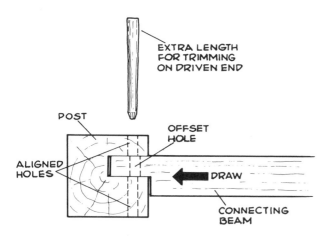

DRAW BORING

You might have fun experimenting with draw-boring, although that last tug, that just-a-little-bit-tighter joint, can be made with a few shots of the beetle or with the cranked-up tension of a come-along.

As a rule, you can get enough torque with clamps and come-alongs to fully seat any joint. When parts of the bent are fitted and you recheck square and dimensions, then you can bore through joints and set pins to hold what you've got. You can buy hardwood dowels, use specially machined bits to cut into a block of hardwood like an apple corer to produce your own dowels, or buy ready-made pegs. At the rate of two per joint, shaving your own white oak or locust pegs is time-consuming and somewhat imprecise. Friction that holds the pegs in place should be as thorough as possible. If you use tapered pegs, the difference in diameter from one end to the other over 8-12 inches

Working with 1-inch hardwood stock on a shaving horse, you can cut down the right-angle corners, then taper roughly round pegs. I get them close by hand, then drive them through a die cut plate mounted on the workbench.

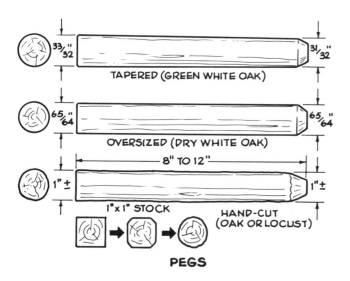

PEGS

should be marginal—$\frac{1}{16}$ inch at most. Green-wood pegs should be a good $\frac{1}{32}$ inch oversize, while dry pegs, which will shrink less, about $\frac{1}{64}$ inch over hole diameter. And leave at least an inch of length extra on pegs so that you will have room to trim, and even a bit more to work with in case the peg starts to split as you drive it but before it is fully seated.

If you are looking for that last little edge, you can experiment with a process called draw-boring. Instead of seating a joint under pressure, then drilling and pinning to maintain the fit, draw-bored holes through the outside timber (a post for example) are aligned, while the hole through the inside timber (the tenon of a girt, for example) is off-centered just a hair, so that as the peg is driven, it will draw the tenon just that hair more tightly into the mortise. Off-center too much, and the peg won't go through.

This system was used to tighten joints before there were come-alongs. It adds considerably to the time spent on each joint, since you have to mark holes, send a pilot hole partially through to mark the exact hole center on the tenon, then withdraw the tenon, cant your drill at the beginning of the bore to shift the tenon hole off center, then drill through the remaining aligned holes in the post, then reassemble the joint, then drive the peg and hope you didn't off-center the hole too much. That's a lot to do, but it makes for good work.

Inevitably there will be a few joints that do not obey the rules. Only experience will tell you when two more shots with a beetle will seat a joint that last $\frac{1}{8}$ of an inch, and when a dozen more shots won't make a bit of difference. It's a subtle but crucial distinction. When a joint is moving, even imperceptibly, there is a certain solid thud as you follow through, as the force of the blow is in fact being soaked up by the timber and being used to edge the joint together. But when you are at a dead end and something is wrong, that solid thud will be missing. You may feel more vibration through the handle of your beetle. Your follow-through may be less complete as the force of the blow hits the dead end and comes back at you. A dead-end blow sounds different, too—more like a *thwap* than a *thud*—as though the force and the sound splatter around the point of impact instead of sinking in.

BENT RAISING

A small bent for a basic Cape is an unwieldy item. You may have moved couches through narrow doorways and refrigerators up flights of stairs, but a bent is different. You simply cannot allow it to fall. If it does, you'll have to refit and possibly rebuild the joints. Also, any bent is big enough and heavy enough to seriously damage whatever or whomever it hits on the way down.

The same kind of curvilinear joinery used on Ed Levin's hammer-beam blacksmith shop (shown earlier) is refined on the scissor-braced balcony in the Manning House. You can't beat this kind of joinery into place with a beetle. (Photo by Richard Starr.)

The Farallones recycled pier beams came together to frame the barn below, which later offered loft space and a galvanized roof over purlins. Quite a student-teacher project.

SAFETY BLOCKING

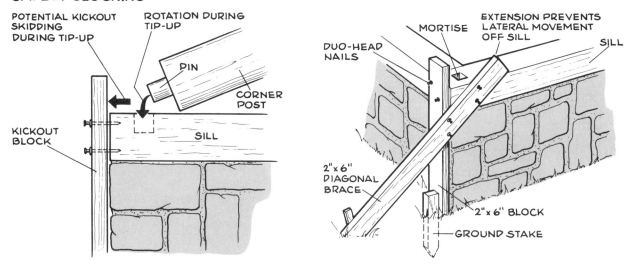

Whether you tip up bents by hand, with block-and-tackle assist, or by crane, take care to protect against kickout. Install bracing and blocking to resist a blow from the bent, not just to hold the bent in place.

Since the first bent is set along the sill at the edge of the deck, and since there are no other bents in place to provide ties and bracing, the first raising is generally the most difficult and potentially the most dangerous. On a basic, four-bent Cape, prepare for raising by laying the first bent over spacers on the deck with the corner post tenons just in front of the corner post mortises. The other bents are stacked on spacers with the girts at the opposite sill. It is natural to be anxious for action with all that planning and labor sitting there ready to go. But you must take the time to ensure that the raising goes smoothly and safely.

Fasten and brace blocking at the outside of the sill, rising a few inches above the sill to prevent the corner posts from skidding off the deck.

Measure a safety line just long enough to become taut between a secure tie-off point on the deck and the center of the girt when the bent is raised to a plumb position. If you have enough bodies, someone can let out slack as the bent goes up. But if a bent goes up and over, one person holding a rope will be in jeopardy. So will the bent. The premeasured rope with a secure tie-off eliminates this dangerous possibility.

Even though the timbers in the bent are set and pinned, snug up the bent with a come-along winch between posts during the raising. On three-post bents with girts joined over the center post, apply heavy-duty scabs (like 2×8s) extending a few feet beyond the joints on both sides of the timbers, positively clamped top and bottom. The joinery is designed to keep the frame in place once it is standing and tied together, not to resist exceptional twisting or an unexpected crack against the deck on the way up.

Have braces and clamps ready. When the bent is set up, run an angled 2×4 brace between the deck and the top of the corner post for a start, secured with a clamp at first. After you have tapped and bumped and nursed the bent into exactly the right position, you can brace this first section of the house solidly enough to serve as a deadman for the other bents (a deadman is a nonhuman structural helper). With 2×4s at each corner post and at the center of the girt (to prevent flexing at midspan), angled down to the deck on one side of the bent and out to the ground on the other side, plus a diagonal brace to resist racking, plus structural scabs clamped over joints, the first girt can anchor a block and tackle to make the rest of the raising easier and safer. Consider 2×6 bracing to make sure the first bent does not move and does not even flex.

If you build piece by piece, practically every new timber in the frame must be securely braced as it is installed. Leave bracing until it impedes progress. (The little boy standing on the beams of this Native Wood Products frame gives you a good idea of headroom along the shed extension rafter.)

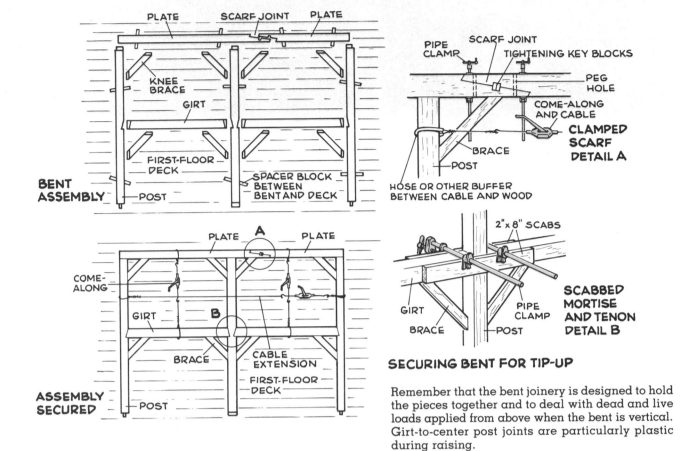

BENT ASSEMBLY

PLATE SCARF JOINT PLATE

KNEE BRACE

GIRT

FIRST-FLOOR DECK

SPACER BLOCK BETWEEN BENT AND DECK

POST

ASSEMBLY SECURED

PLATE A PLATE

COME-ALONG

GIRT B

BRACE

CABLE EXTENSION

FIRST-FLOOR DECK

POST

PIPE CLAMP SCARF JOINT TIGHTENING KEY BLOCKS

PEG HOLE

COME-ALONG AND CABLE

BRACE

POST

HOSE OR OTHER BUFFER BETWEEN CABLE AND WOOD

CLAMPED SCARF DETAIL A

2" x 8" SCABS

GIRT

BRACE

PIPE CLAMP POST

SCABBED MORTISE AND TENON DETAIL B

SECURING BENT FOR TIP-UP

Remember that the bent joinery is designed to hold the pieces together and to deal with dead and live loads applied from above when the bent is vertical. Girt-to-center post joints are particularly plastic during raising.

Photo below: Using initial rafter trusses as a collective deadman to hold block and tackle makes tip-up much easier. Note the series of diagonal braces Ed Levin used down to the wall plate, and an additional rope tie-off opposed to the block and tackle. (Photo by Richard Starr.)

On this Timberpeg job, second-floor joists are decked to provide a working platform for scaffolding. Tongue-and-fork rafters are pegged and braced awaiting the roof deck.

A crane (here barely visible over rafters) is indispensable on large-scale projects, particularly where many timbers must be raised and set two or more stories above grade. Repetitive members should be cut, stacked, and ready to go in order to minimize time on an expensive crane rental.

When the second bent is tipped up, hand-set connecting girts and knee braces between bents. You can use the combination of a come-along and well-chosen shots with a beetle to seat the joints. Clamp them, but don't drill and pin until you go back to the first bent to be sure that all the banging and hauling has not moved the first frame against its braces. Only then should you continue raising bents. Also, it makes sense to add summers between bents as you go, returning later on to set floor joists.

Unless you want to trust an extended forecast of good weather, tack some plywood or scaffold planks over the girts and summers to make a working platform, and start immediately on the rafter trusses. I say trusses because aside from small bents, where you may scab all the joints, including rafters and a collar tie, and tip everything up at once (the higher the bent,

the more space it occupies on the deck and the harder it is to raise safely), the most sensible solution is to set the bents from the deck and the truss sections from above with a crane. I know, you don't happen to have a crane. That's all right. If you can't find one, afford one, or get one to your remote site, you can do the roof by hand.

Without a crane, there is only one option left. What you do is haul the individual rafters, collars, king posts, and whatever else up to your work platform laid on the girts. Then you treat that platform just like the first-floor deck: Assemble the trusses; set braces at the gable-end girts to prevent skidding or kickout; tie on a safety tip-over line; scab and clamp the joints on large trusses to avoid excessive stress during the raising; then tip the trusses up in place in sequence. As you set connecting girts and summers to help stabilize bents on the first-floor deck, add at least one purlin on each side of the ridge (or temporary horizontal braces across the rafters) before moving on.

If you can manage a crane, there are several bonuses: less handling and hauling, increased safety, and the luxury of concentrating on positioning and fitting the timbers instead of wrestling with them. And there are some special considerations. Crane erection can cause unexpected movement in the bent or truss. Instead of a tendency to slide or kick away from you at the feet of the posts, the posts may come back at you as the frame section slips free of the deck and swings on the lift line. Also, you should still scab and clamp center-post connections and add winch lines for security, particularly between first and second girts on two-stage bents. If the frame section flexes on the way up, the joints are unlikely to be as solid or as accurate as you remember them down on the deck.

FRAME FINISHING

Most frames require a little finagling before they are closed in. Despite all your precautions, something somewhere will move. And by the time you have at least 20 major and 16 minor timbers in the frame, it can be very difficult and frustrating to track back through the assembly, checking square and plumb to find the culprit. Sometimes the best you can do is saw-kerf a girt to get it closed just a bit more tightly, for example. Sometimes there is nothing you can do to fix the problem that won't cause reflective problems somewhere else.

This is one reason for waiting on floor joists, roof struts, and at least some of the purlins. You may find that floor joists at opposite ends of a girt-summer span are not the same length. They should be, of course. But if you get variations of 3/8 to 1/2 inch or more, you should think about resetting a few joints to square up the frame. Variations under 1/4 inch are minor inaccuracies you would be wise to live with.

After the frame is assembled, there are a few final jobs to consider before putting on the roof. You can increase the roof rigidity by adding struts (let-in braces in the same plane as the roof purlins) between gable-end rafters, and the purlins nearest the ridge with a full bay between them. Since the roof acts as a single unit when purlins are set, these four additional timbers (set just like let-in braces between posts and girts) can prevent racking across all the rafters. This is a final, high-quality touch that enables you to get a lot of mileage out of a few timbers.

Left: Remnants of the early English cruck are found in braces let into several timbers, creating a truss of sorts. Below: Two let-in struts on each side of the ridge are enough to prevent racking of the rafter and purlin system. Think of these struts as conventional, let-in, corner wind bracing for the roof.

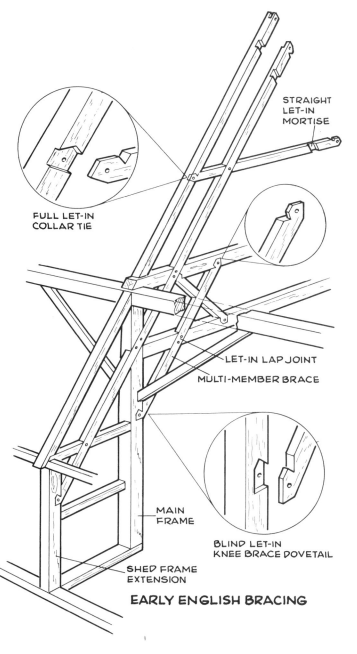

FULL LET-IN
COLLAR TIE

STRAIGHT
LET-IN
MORTISE

LET-IN LAP JOINT

MULTI-MEMBER BRACE

MAIN
FRAME

BLIND LET-IN
KNEE BRACE DOVETAIL

SHED FRAME
EXTENSION

EARLY ENGLISH BRACING

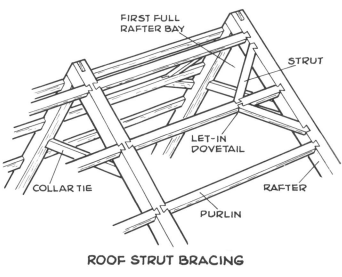

FIRST FULL
RAFTER BAY

STRUT

LET-IN
DOVETAIL

COLLAR TIE

PURLIN

RAFTER

ROOF STRUT BRACING

tacky, collecting anything that happens to float along, then turn gummy for days, impeding finishing work. Cutting the oil with even 20-percent thinner can help. (Most wood treatments will darken the timber frame a little, which can be a pleasing effect against T&G planking. Oiling also tends to accentuate knots and grain anomalies.)

Don't forget the wonderful tradition of tacking a pine bough on the roof ridge at the gable end when you're done; this is the symbol of a successful raising, and of respect for the trees that spawned the frame. Some colonial builders went a step further, treating the frame like a ship and christening the timbers with a bottle of rum.

You won't see the frame this way again—stark, rugged, silhouetted against the sky, a work of carpentry, cabinetry, and sculpture with your sweat and skill in every joint. If friends and neighbors have come to help with the raising, take the rest of the day off, fill their bellies with good food and drink, and swim in the good feelings of your accomplishment. Celebrate! This is the time for it.

FRAME LOADS

You knew this was coming. But I am not going to hit you with endless tables. If you buy from a manufacturer or work with a builder, you won't have to calculate building loads and select timbers to meet architectural and engineering needs. That's one reason I did not risk turning you off timber framing by hitting you with numbers right off the bat. You can still talk to these professionals about let-in dovetails, the different kinds of pegs, the size of summers, and more. You probably should not deal with professionals who regard this information as proprietary, not fit for comment by mere homeowners, much less owner-builders. But as these professionals are responsible for the frame, they must provide code-acceptable specs and plans.

If you are doing all or some of the work, you can get help from some of the trade associations and your local building inspector. You can get a lot of help, but you will have to originate the plan, flawed as it may be, and rely on others to catch and correct any errors.

If you fashion your frame design to duplicate a recorded, traditional design (even if you add a bent or two for additional space), the timber sizes are likely to exceed structural requirements and codes. Of course, you may have a little trouble finding timbers now that were available when those historical plans were drawn up. But this is an interesting way to arrive at a proven design without wading through endless span and load tables. Too often the formulas and meth-

Remember that the timbers in the frame are unprotected at this stage, like trees without bark (literally) until you get the roof on. Even then, the wood will appear a little raw. But with only a few days of extra work you can transform the rough, exterior-grade appearance with sanding and a coat of boiled linseed oil. You'll need a power sander (a belt sander with something like an 80-grit paper is ideal), or the few days will lengthen to weeks of laborious hand sanding.

You might want to experiment with the linseed oil on a not-so-obvious section of the frame. At full strength, the oil is highly protective and lustrous, enhancing wood tone and grain variation. But it can stay

Whether you feel strongly about tradition or not, the bough (here barely visible) nailed onto the ridge signals a completed skeleton worth a long last look before the skin is applied.

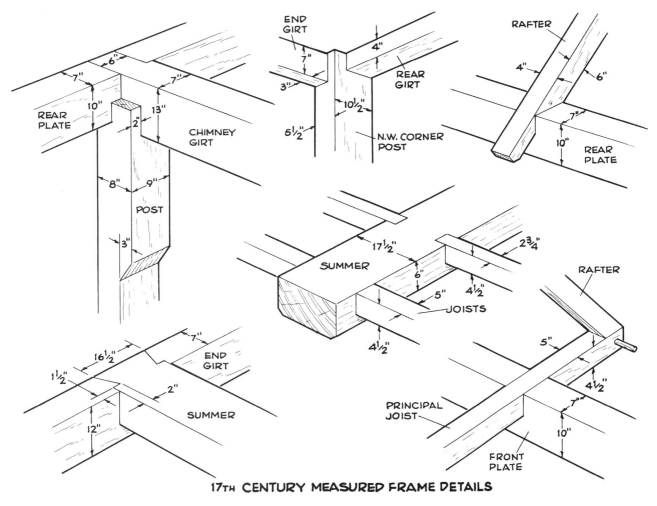

17TH CENTURY MEASURED FRAME DETAILS

One of the most painful accommodations to reality may come when the historic framing plan you intend to follow calls for massive timbers you can't afford (or find) unless you promise the world to the lumber mill.

odical load calculations stop otherwise enthusiastic owner-builders.

It's a shame, but it may be for the best—a kind of self-regulating safety feature. A lack of enthusiasm indicates a lack of familiarity and probably a lack of expertise in an area that is unforgiving, just like the dry textbook math you had in school. Using a facsimile of a documented frame can bridge this troublesome gap and keep you involved firsthand.

One book on timber framing *(The Timber Framing Book* by Elliot and Wallas) sends readers after *Architectural Graphic Standards* (the Ramsey and Sleeper classic) for specifics on loads and spans, which is shabby treatment from a specific how-to book that is devoted to a single subject and purports to tell you "the" way to do one job. *Graphic Standards* has practically no information on large timbers other than Glu-lam beams. The pages on design values for all the different species and grades of stress-rated construction lumber don't even list oak because it is not a conventional, code-restricted, grade-differentiated building material. There are pages on allowable spans for 2×4s, 2×6s, and 2×8s, even a small section on post-and-

beam frames that begin to approach timber frame dimensions (4×12-inch joists and rafters, for instance). But if you want design values on an 8×8-inch girt and an 8×16-inch summer, *Graphic Standards*, as thorough as it is, is a dead end.

The most usable source of design value data specifically for timber frames I've found is Tedd Benson's book, *Building the Timber Frame House*. Load and span data here are presented logically, covering pine and oak. Although I am generally skeptical of dogmatic answers to construction questions and of an absolutely "right" way to cut, fit, and raise (there is no such perfect animal), engineering data has to be "right."

There are alternatives to Benson's system of load and span calculation. But putting together data from wood trade associations with extrapolations from the NDS tables and from measured historical plans, then pulling out safe and sensible design values for your frame, is a real piece of work. Benson's system is dogmatic but sensible and, most important, usable even by first-time owner-builders. It's challenging, but take a look at his thorough presentation before deciding that you can't hack the numbers on your own.

Tracking loads through a frame is more work than finding a clear set of tables, however. You will need both the tables and your common sense. Finding the loads is like following the flow of a river that branches into stream after stream. Think of how much water each stream carries (the size of the timber and amount of the load) and of how some branches lead to wide rivers while others lead to small rivulets (a summer carrying half the floor load uniformly through joists, a girt carrying half the summer).

Start at the roof with dead loads (weight of materials) and live loads (snow and wind), available in *Architectural Graphic Standards*, *Building Construction Handbook*, local building department standards, and many other sources. As you work down through the frame, add on dead loads of rafters and king posts and purlins and such. At floors, add on the extra live load (generally 40 psf, which builds in a healthy safety factor) for people and possessions. The numbers get bigger and bigger, coming together like several rivers at a delta on the masonry footing.

Don't mess with the numbers unless you are sure of the answers. Get professional help if you're not sure. Load and span details don't fall into the category of personal architectural taste. You have to get them right. If you're really in the sticks and building in a no-code area, I urge you to get a second opinion on your calculations, a devil's-advocate opinion to serve in place of the building inspector most owners and builders must satisfy.

Although span-and-load tables may be needed to make engineering sense of the frame, your common sense tells you how loads get from the roof to the sill through a minimal number of exceptionally strong timbers.

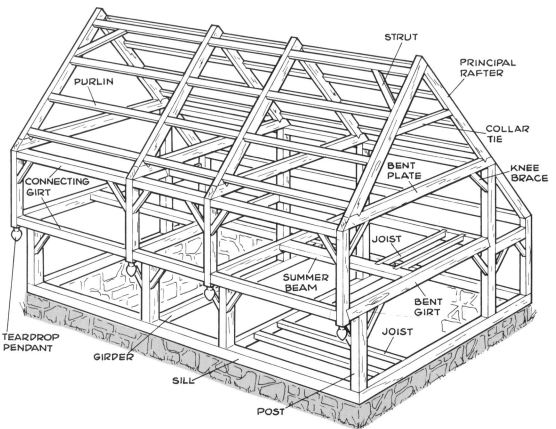

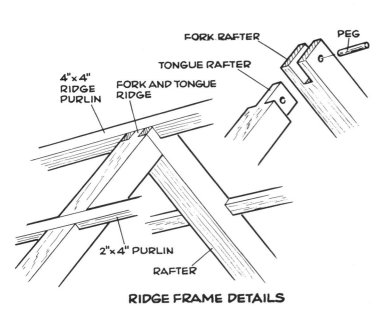

RIDGE FRAME DETAILS

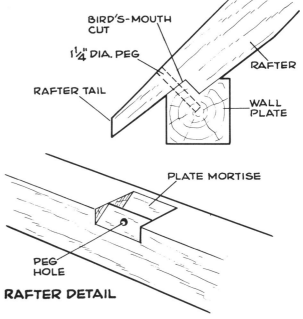

RAFTER DETAIL

Half-lapped or, better yet, tongue-and-fork rafters gather loads through purlins and send them down to the wall plates.

Wall plates collect loads from the rafters and from connecting girts and distribute them to posts. In the same way that you cannot make energy disappear, every load must have a route through the frame.

The rafter applies lateral pressure to the wall plate, which is stopped and transferred along the plate by a bird's-mouth cut and a peg.

Finally, loads reach the first-floor sill through posts and from the joist and summer beam (or girt) frame. On large buildings, concentrated loads under corner posts can compress sill fibers ¼ inch or more if a soft-wood is used.

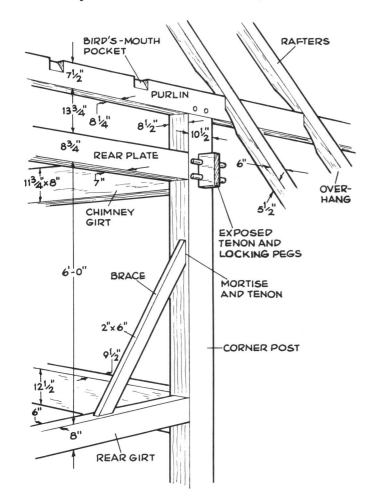

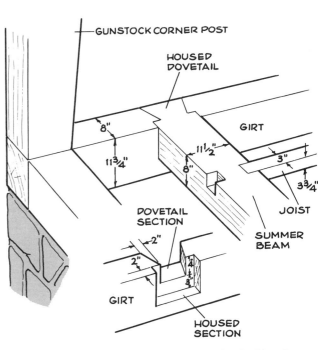

FLOOR FRAME TO GIRTS AND SILLS

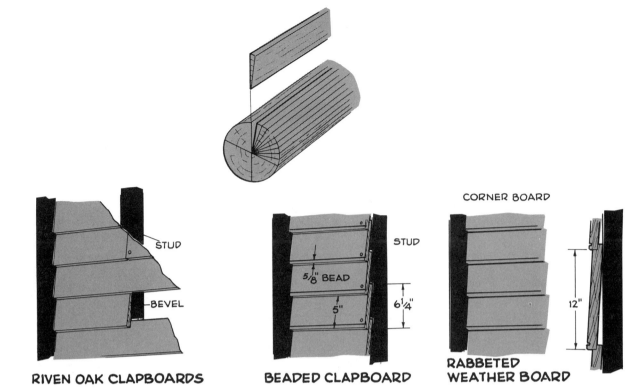

RIVEN OAK CLAPBOARDS   BEADED CLAPBOARD   RABBETED WEATHER BOARD

Tongue-and-groove planking, or board-and-batten siding, or shakes laid over studding or wall girts all can be used, although ground-hugging Capes and saltboxes look exceptional with beveled siding about 4 or 5 inches to the weather.

CLOSING IN

A moment of silence for every carefully crafted dovetail purlin, every shouldered mortise and tenon that will be buried and forgotten—except in your memory.

That done, note that the most scrupulous traditional builders (folks who hunt for antique tools to execute antique plans) part company with history when it comes to the building skin. Hmmm. Tradition seems to be the byword until the wattle and daub starts eroding and winter winds begin whistling through rough clapboards.

With the frame in place, you should feel free to go your own way: to studs, plywood, clapboard, shakes, board and batten, wallboard, lath and plaster, stress skin panels, and other options. The prototypical New England timber frame had a few studs set between the major timbers with a variety of clapboard patterns outside, and lath with plaster inside. The outside-inside building skin system (now with insulation and vapor barrier in between) is used on almost all homes today. But it is one thing to bury a toenailed 2×4 with the organized graffiti of grade stamps in blue and red splotches over the wood, another to hide a white oak, gunstock-tapered corner post. Well, you don't have to do it.

Every 2×4 in a stick-built frame has to be there, not necessarily to carry loads, because you could re-move every other stud and the house would still stand. You need each 2×4 to get the accumulation of stiffness needed in a stick-frame house. Plywood sheathing nailed only at 8-foot centers would be practically worthless. But timber frames are about as rigid as they are going to get without studs or sheathing or siding or roofing. You can fill in between corner and chimney posts with studs, but you don't have to. On timber frames you can select building skins that do not need 16-inch centers.

Wood is not a bad insulator. It's better than glass or steel or aluminum or plastic, but not as good as the 3½ inches of fiberglass or cellulose in the bays between studs. Solid 2×4s make contact between inside and outside wall skins, providing stiffness and, at the same time, a solid bridge for temperature transfer. Sensitive measurements will show that on a cold night the inside wall is a little cooler along 1½-inch-wide strips every 16 inches, right on top of the studs. This phenomenon may produce ghosting—a grayish striping on wallboard caused by these temperature differences. When there are fewer timbers, you can have more continuous insulation, increased vapor barrier protection, and less ghosting.

You could treat an 8×8-inch perimeter frame as the basis of a superinsulated design, even staggering 2×4 studs along inside and outside edges of the frame and winding glass blankets through the maze. It's a neat trick, thermally efficient if a bit stuffy for the inside air, with one very unhappy consequence: The timbers and joinery disappear on the outside and on the inside.

But you have seen one answer to this problem in generations of articles in home how-to magazines about turning the dark, damp, pipe-, wire-, and beam-filled basement into a rec room. You buy a truckload of furring strips and nail one an inch back on each side of the beam face. Then, after you insulate, you nail your wallboard on the furring strips and expose an inch of the beam between individual panels of finished wall. Voila! A massive timber used so that you see the equivalent of a 1 × 8-inch piece of shelving.

So you compromise by using less insulation and recessing the furring strip another inch or so. Now a 6 × 6-inch post wall with a 2-inch recess allows 4 inches of insulation; a 3-inch recess allows 3 inches of insulation; a 4-inch recess exposing most of the timber allows only 2 inches of insulation. First, this is not a happy compromise. Second, as the timbers and foundation settle, your careful seams between wallboard panels and timbers will open. But as long as you are recessing to expose more timber, why not do the obvious?

Why not move the insulation and the wallboard and everything else all the way back and fasten them to the *outside* of the frame? Granted, this may cause some extra detailing at windows and doors, but it is a solid possibility, even though it is a little strange to put the inside skin on the outside of the frame, with the outside skin directly on top of the inside skin. It sounds crazy, but it works. And it leaves that beautiful frame in full view inside the house.

Here's how your all-exterior, inside-outside wall would shape up.

Inside face.
Working outside the frame, you apply the finished interior wall (gypsum board or paneling, for instance). The thinner the material, the more you need support from intermediate studding. Full, 1-inch, T & G planking may not require any additional support, depending on the space between bents, either horizontal or vertical.

Vapor barrier.
Looking at the raw side of wallboard or the unfinished side of paneling, you add foil or plastic. Since you are working outside the frame, there are no interruptions in the barrier.

Insulation.
Again with no breaks, you can use 2 to 4 inches of rigid foam insulating board, the amount depending on local climate. Some manufacturing methods produce values at or above R-6.0 per inch. Concerns about foamboard contributing potentially lethal gases to a house fire can be buried, literally, be-

Timber frames built piece by piece over conventional masonry foundations offer standard operating room for plumbing and electrical lines. (Photo by Native Wood Products.)

tween wallboard and siding (⅝-inch fire-code wallboard for good measure, if you like). And advances with the polyiso boards have reduced this problem in any case.

Exterior siding.
Take your pick here: clapboards, vertical planking, board and batten, you name it.

There is an alternative to this hand-built skin—stress skin panels. These hybrid products are essentially premolded versions of inside panel, insulation, vapor barrier, and exterior sheathing onto which you apply siding. The only negative by-product of eliminating the cavities between wall studs is that you lose the space for wires and pipes.

You could treat the outside finished timber frame as a log cabin, where solid walls force you to be inventive, to run mechanical lines between first-floor joists, and to use concealed receptacles flush with the finished floor. Above floor level you can use raceways or conduit on the surface (hidden with trim, if you want) and cabinet space or purposefully constructed mechanical chases (small, closed-in closets to hide and hold mechanical equipment) to hold pipes, wires, and ducts between floors. You must decide if a completely exposed frame is worth the extra effort. It is a hard choice, because people who really want to see the full frame inside the house may balk if the effect is cluttered with pipes and wires. But there are a few ways to cheat (in the best builder's sense of the word) mechanical access.

ENERGY EFFICIENCY

MECHANICAL ACCESS

After all the work, I'm betting you will give up the view of your timber frame only grudgingly. Anticipating that extra work and planning to route mechanical lines through the house may keep you from trying the all-exterior-building-skin system, here is some help.

Let's take the worst case and say that you want something like an R-24 in the walls (roughly 4 inches of foamboard). Working from the outside over the interior wall and vapor barrier, you can start with a 1-inch-thick board. This insulating layer is unbroken (except for seams) over the entire frame. The next step is to build a cavity into the insulation as you add suc-

In cold climates, insulation must be applied outside the structural skin in order to expose the frame inside the house. Here's a means of keeping interior timbers visible while providing mechanical access and good insulation.

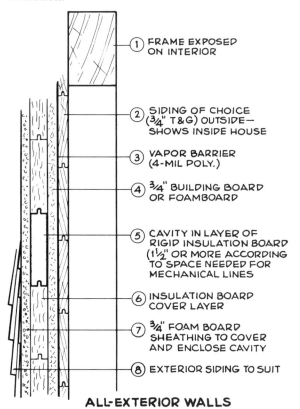

① FRAME EXPOSED ON INTERIOR

② SIDING OF CHOICE (¾" T&G) OUTSIDE— SHOWS INSIDE HOUSE

③ VAPOR BARRIER (4-MIL POLY.)

④ ¾" BUILDING BOARD OR FOAMBOARD

⑤ CAVITY IN LAYER OF RIGID INSULATION BOARD (1½" OR MORE ACCORDING TO SPACE NEEDED FOR MECHANICAL LINES

⑥ INSULATION BOARD COVER LAYER

⑦ ¾" FOAM BOARD SHEATHING TO COVER AND ENCLOSE CAVITY

⑧ EXTERIOR SIDING TO SUIT

ALL-EXTERIOR WALLS

cessive layers. You can create a channel at standard receptacle height, bordered by furring strips or even 2×2-inch studs nailed to the frame like girts on a pole barn. The size of the border (2×2 or 2×4 inches, for instance) determines the air space of the mechanical channel.

In stick-frame structures, electricians generally plan the wiring paths, then drill through anything that gets in the way—well, almost anything. Given the luxury of knowing that all the paths will be hidden by finished floors, walls, and ceilings, these wire patterns are often like crazy quilts. But you can lay out a mini-mechanical chase (insulated to boot) a bit more systematically than that—running branch channels up to the second floor, for instance. This kind of construction (furring on foamboard) deserves to be buried, and after you set required mechanical lines in place, that's exactly what you do.

Again, let's take the worst case, where you need 4 inches of board and a 2-inch cavity for equipment. Simply add a second layer of 2-inch-thick foamboard over the frame cut up to, and flush with, the 2×2-inch cavity border. Then complete the job with a final layer of 1-inch board. This is more time-consuming than using preformed stress-skin panels, but less costly. And foamboard is lightweight, easy to handle, and easy to cut. Some brands are available with tongue-and-groove seams. Also, the board can be set to cover the seam between the frame and foundation and continued down the foundation below grade.

If you take the trouble to finish and insulate the home outside the frame, it is probably worth going to great lengths not to drill through or otherwise mar the structure. If the insulated-cavity system doesn't appeal to you or to the building inspector, and if the idea of a box beam to hide mechanical lines sitting there next to the real thing seems ridiculous, consider conduit painted in sympathy with the wooden frames. Frame joints may fit like furniture, and sanding plus applying linseed oil may help the beams look like furniture, but the frame is a structure first and cabinet-quality carpentry second. It is utilitarian. Simple, functional fittings and fixtures that might look out of place or unfinished in a conventional house often complement a timber frame.

HEATING AND COOLING

I have tried to think of heating and cooling concerns that are peculiar to timber frames, but there are only a few. In a sense, that is good news, indicating that this alternative building system is not so out of the ordinary that conventional heating, waterproofing, finishing, glazing, and other building processes won't work. Here are a few possibilities to get you thinking.

Baseboard. Applied over wallboard, plaster, or planking, hot water "can" (as in tin can) is a necessary eyesore that can disrupt furniture placement because you can't put cabinets and floor-to-ceiling bookcases and such up against the walls. With an all-exterior building skin, the open bays between posts offer a convenient recess for baseboard. You can wind the heating loop up and down between posts, leaving the frame unmarked and unbroken by the baseboard convectors.

Forced hot air. This delivery system can be run below floor joists through an insulated crawl space or cellar. (You can insulate the ducts, too.) A main plenum running along the center of the house with branch ducts to exterior walls is great for the first floor. Registers can be cut between posts, directly on outside walls and in front of fixed glass. This may be all you need in a small Cape. On larger frames with second-story living space (unless it amounts to no more than a small loft), it is difficult to get a controlled amount of heat upstairs. Open lofts help. So do grill vents in the second-floor deck. The problem is how to shut down these natural-draft, gravity-feed features when the upstairs is empty and you want most of the heat produced to warm the first floor. (Grill vents may have operating louvers, but try that on an open loft space!)

Wood stoves. Cape, colonial, and saltbox designs evolved around a central chimney column. Those spaces still work well with wood and coal stoves. But full two-story plans present problems. With closed bays between posts, anything is possible. With exposed bays and no good route for heat and utility systems between floors, think about treating the separate floors as you would with a two-loop, two-thermostat, hot-water heating system. On the first floor, where mechanical access is no problem, you might use hot water or hot air, either oil-, gas-, wood-, or coal-fired. A portion of this heat will move to the second story by purposeful convection and by radiation through beams and planking, enough so you can get by with supplemental heat from a stove or two, fired only when needed. Second-floor stoves are easily vented through the roof with prefab flues. That saves on floor space and costly masonry work.

It is rewarding to heat only with wood all winter long, particularly when you get the wood yourself. I've done it. But wood cutting, gathering, splitting, and hauling, combined with fire laying, damper control, and ash clearing can gang up to be a major presence in your winter life, even if you use long-burning, air-tight stoves.

Multifuel systems. Supplemental wood or coal heat is a kind of multifuel system if it is backed up with oil or gas. And there are multifuel furnaces that present fuel options in the same package. Realize that you may hand-hew timbers for the frame and build with traditional tools, but you are still allowed to have a thermostat in the house. I have never liked to put all my eggs in one basket. I don't do it with work. I wouldn't do it with house design (go ahead, put a big skylight in your log cabin). I wouldn't do it with a heat source either. That's why I like multifuel furnaces. The last few winters I've burned a lot of wood and some oil (about $250 worth), backed up with a greenhouse connected to the main room of the house with two sets of sliding glass doors. I'll put in the time to burn wood, and I am glad for the solar gain. But it's nice to get a break with the oil once in a while, particularly if, as I do, you can harvest some of the endless positive ideas and developments from alternative technology, alternative housebuilding, and even alternative lifestyles without burying yourself away from civilization in the most remote and inaccessible site available.

Superinsulation. You may have gathered that this is not my number-one solution to energy efficiency. And it means you will have to give up most or all of

There are many types of insulating boards: polyiso materials, cellulose and plastic-based panels, foil facings, and vinyl-clad interior facings. Hunting around a little will save the redundancy of applying a finished skin outside the frame, then insulation, and then the finished exterior siding. (Photo by Native Wood Products.)

your frame. But a 6- or 8-inch frame does provide a lot of room. An 8-inch cavity filled with fiberglass nets roughly R-27. You could sheath inside and out with 2-inch foamboard beneath exterior siding and the interior wall to net R-53 walls. At the same time, you better investigate cfm air changes and one of the air-to-air heat exchangers, plus vents in kitchens, baths, utility, and laundry areas to evacuate moisture, plus intakes to support combustion at heat sources.

Fixed glass. On stick-built (2×4 stud) homes, windows have two measurements: one for the window unit and one for the rough opening. They are not the same. The larger rough opening allows space for shimming, so you can plumb and level the unit in the frame, which does not say much about the frame being plumb and level. On rigid timber frames, you don't have to separate the opening from the frame. So you could look for solar gain with an aluminum-ribbed, add-on greenhouse. But how about using the framework of a southern-exposure shed extension? Often, an add-on greenhouse looks like what it is—an afterthought. Not true for a shed on saltbox design. You could cut rabbets inside and outside the frames to hold ⅛-inch plexiglass set in silicone, with trim covering open seams. Double-glazed sliding glass doors can serve between wall posts. Some updated designs do not transplant well to traditional homes. But large, double-glazed panels and doors break down conventional eyeline barriers between house and site and focus attention on the heart of a timber-frame house—the frame itself.

INFORMATION SOURCES

You know that timber frames are durable. Many frames in many styles have survived for hundreds of years, providing a rich resource for owner-builders to-day, in addition to the increasing stream of information that flows when classic materials, methods, and designs are rediscovered and popularized.

TIMBER FRAME BOOKS

Building the Timber Frame House, Tedd Benson (Scribner's). Aptly subtitled, "The Revival of a Forgotten Craft," this how-to book devotes some general text to planning and design, then concentrates on joinery and assembly. The presentation includes a few photographs and a kind of capsule description of, say, a tongue-and-fork joint, then follows through with a planning page complete with measured drawings and step-by-step notes for measuring and cutting. This is not an introductory book. But once you decide to build a timber frame, you'll be grateful for the specific, comprehensive treatment.

The Timber Frame Planning Book, Stewart Elliot (Housesmiths). This interesting book has comprehensive, measured drawings for the frame and every frame component of nine houses and three barns. Including a glossary, sample blueprints, etc., the plans run over 350 outsize pages. The historical plans list the number of mortises, the spaces between mortises, and the sizes of the mortises. It's a limited but instructive look at traditional timber framing.

The Timber Frame Raising, Stewart Elliot (Housesmiths). This well-illustrated book is a diarylike case history of one raising, stressing a community effort. Along with tables and diagrams detailing the center of gravity of bents during raising, the text also covers who showed up with how many bottles of wine, and what the name of their dog was.

HISTORICAL BACKGROUND

Dutch Houses in the Hudson Valley Before 1776, Helen Wilkinson Reynolds (Dover). This book does not teach timber framing, but it does have photographs and plates and descriptions of hundreds of variations on Capes, colonials, gambrels, and other designs.

Early Domestic Architecture of Connecticut, J. Frederick Kelley (Dover). This is one of the unusual historical housing books that concentrates on structure. Detailed, true-to-life drawings illustrate framing plans, bent designs, and joinery. The text is a bit academic, as it takes you through timber frame design and development, joinery, siding, even traditional interior woodwork. A thoughtful, thorough text and excellent drawings make this book an ideal background piece to Benson's detailed how-to book. Another volume along the same lines, *Colonial Architecture of Cape Cod, Nantucket and Martha's Vineyard*, by Alfred E. Poor, is also published by Dover.

Japanese Homes and Their Surroundings, Edward S. Morse (Dover). Traditional Japanese building in general has an affinity with timber framing in this country. This general book covers houses inside and out, entrances, gardens, and more. I would try the library for this one, as much of the information, although interesting, is off the point. But you will get a look at one clean, utilitarian ancestor of timber framing.

The Early American House, Mary Earle Gould (Tuttle). Covering the paraphernalia of household life in America from 1620 to 1850, this book is loaded with pots and pans and the like, also with fireplaces, framing details, cupboard construction, and an interesting chapter on the development and expansion of early one-room houses. Revised back in 1965, it may be hard to find. Few books will give you a full-page picture of a square-hewn, quarter-split, solid timber stairway.

The Owner-Built Home, Ken Kern (Homestead). Even though Kern's books tend to hammer away at waste in construction, inefficient codes, and such, I am comfortable sending you to one of the leading spokesmen for alternative construction. This is a valuable source. Just remember to distinguish between the facts and the proselytizing as Kern pushes his own distinctive point of view.

The Craftsman Buider, Boericke and Shapiro (Fireside). A modest paperback with exceptional color pictures of inventive and unusual homes and construction details. The pictures do the talking.

Country Woodcraft, Drew Langsner (Rodale). This is a book of woodworking projects—things like tool handles, a workbench, spoons, a table, and more, including a small section on mortise-and-tenon joinery. The projects are not about timber framing, but the pictures, drawings, and clear, pared-down, how-to descriptions of the 30 items (all traditionally built with hand tools) offer an introduction to many tools and skills used to make timber frames. No point-of-view preaching here in Langsner's most instructive book.

In Harmony with Nature, Bryère and Inwood (Sterling). A detailed, personal case history of 12 projects is strung together under the theme of the book's

This is a timber-frame replica of the house given enduring fame by Henry David Thoreau in his book *Walden*. The book arose largely from journals Thoreau kept in the house, which he built in 1845 at Walden Pond, Massachusetts. Testimony to Thoreau's wish to "live deliberately," the house measured just 10 × 15 feet. This reproduction was designed and built by Roland Wells Robbins, based on Thoreau's accounts in *Walden* and on Robbins' excavations in 1945.

Courtesy of Robbins, you might use this plan as the basis for building your own Walden replica. Thoreau hewed his frame from "tall, arrowy white pines, still in their youth," and dug his cellar through an unused woodchuck burrow. Robbins sells his book *Discovery at Walden* together with a wall plaque mounted with fragments of brick and plaster from the Walden Pond excavation site. For more information, write Roland Wells Robbins, RFD #2, Lincoln, MA 01773.

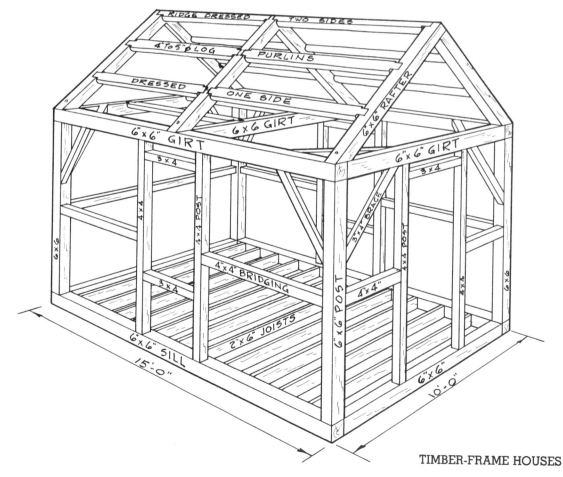

As this Thoreau-house replica shows, the house was essentially a small room with a door, two windows, and a fireplace, here enjoyed by excavator/historian Roland Wells Robbins. One trapdoor opened to the sand floor root cellar. Another trapdoor opened to the garret. Thoreau added a wood shed "made chiefly of the stuff which was left over after building the house."

For a time excavator/historian Robbins sold precut kits of Thoreau's Walden Pond house. In this kit version (no longer available) purlins and roof plankings are exposed and a storage loft has been added. A wood stove here replaces Thoreau's masonry fireplace.

subtitle: "Creative Country Construction." Log, pole, timber-frame, and hybrid designs are covered, and some of the projects are "revisited" in follow-up chapters. This book is a must for rural owner-builders. It's loaded with appropriate, small-scale, usable ideas and designs supported by lively, annotated illustrations. The personal approach used too often by others to rail against the evils of civilization is used instead to explain the realities of alternative lifestyle and building, even to the point of relating how one couple seemed to tire of the day-to-day grind of self-sufficiency.

Fine Homebuilding, The Taunton Press, PO Box 355, Newtown, CT 06470. Covering timber framing and other types of construction, *Fine Homebuilding* consistently provides thorough, thoughtful coverage of homebuilding, stressing traditional designs and hand craftsmanship.

TECHNICAL BACKGROUND

Building Construction Handbook, Frederick S. Merritt (McGraw-Hill). Go to the library for this monstrous reference, which is not lavishly illustrated like *Architectural Graphic Standards*, but has more complete engineering data on large timbers, fasteners, concrete mixes, and more, all without even a hint of how-to.

Design of Wood Structures, Donald E. Breyer (McGraw-Hill). This is another complex, table-loaded, data-stuffed reference more specifically on wood than all kinds of construction materials and building systems. Try the library. Bring a calculator. Don't expect how-to help.

SOUTHERN FOREST PRODUCTS ASSN. (SFPA), PO Box 52468, New Orleans, LA 70152. This association represents mills that produce, among other items, substantial timbers suitable for timber framing. SFPA can steer you to lumber suppliers and provide an array of technical booklets on design values and wood performance in general.

USDA FOREST SERVICE, PO Box 5130, Madison, WI 63705. You might start by writing for Form 81-020, *List of Publications*, which covers the voluminous literature on all kinds of wood use and wood construction. Single copies of many Forest Service reports, on such subjects as racking strength of wood frame walls, are free.

Building the Timber Frame House, Tedd Benson (Scribners). To reiterate, I think the load, span, and other wood performance tables in this book provide a solid and usable technical resource for owner-builders.

ENERGY-RELATED INFORMATION

House Warming, Charlie Wing (Atlantic/Little, Brown). Charlie Wing, who founded Cornerstones, makes a sensible presentation of home heating that starts with basics like the names of house and furnace parts. But the information gets considerably more detailed and practical. You get options, not narrow-minded advice. A lot of interesting and varied illustrations help, but they also take up a lot of room in the 200-page book, leaving a very brief text for the money. Try the library.

Keeping Warm, Peter V. Fossel (Perigee). A thorough text and nice but infrequent illustrations cover heat conservation, including thermal envelopes, wood stoves, solar gain, and interesting chapters on warm clothes and warm bedding. An interesting, modest, basic-level paperback.

More Other Homes and Garbage, Jim Leckie et al., (Sierra Club). Revised in 1981, this weird but wonderfully thorough book has pages of regional mean temperatures and wind velocities, sections on aquaculture, water wheels, waste digesters, solar-assist heat pumps, and the full range of expected alternative energy systems, plus material on heat loss, superinsulation, and other subjects. An excellent buy for some 375 packed, large-format pages.

SCHOOLS

CORNERSTONES, 54 Cumberland St., Brunswick, ME 04011. Founded as an owner-builder school in 1976, Cornerstones presents many courses, including practical, out-of-the-classroom training where you get some splinters—including a timber-framing course where a small frame is cut, joined, and raised under the instruction of Tedd Benson. It would be hard to find a better way to begin your timber-frame project. Write for details of course subjects, times, cost, lodging, etc.

FARALLONES INSTITUTE RURAL CENTER, 15290 Coleman Valley Rd., Occidental, CA 95465. Complementing the sister center (The Integral Urban House), Farallones offers instruction in timber framing as well as solar design, blacksmithing—even a course on gardening for young people. In one of the timber-frame workshops, materials for a 40×50-foot barn project came from a salvage operation at a San Francisco pier and from fir trees cut and milled at the site.

SHELTER INSTITUTE, Bath, ME 04530. Another owner-builder education center teaching efficient, minimal timber housing.

THE OWNER-BUILDER CENTER, 1824 Fourth St., Berkeley, CA 94710. At this writing, the Owner-Builder Center has just opened a branch in New York City. There are many branches across the country, most started by folks who trained at the original center in Berkeley. Regional centers may specialize in regional concerns, offering varied construction training and innovative follow-through programs where you can contract at reasonable rates for periodic professional consultations and troubleshooting services after you leave the Center, begin your project, and encounter problems.

FIRMS

THE BARN PEOPLE, PO Box 4, South Woodstock, VT 05071. This firm doesn't manufacture timber frames; they restore old ones. About $10 nets a detailed portfolio of their inventory—includes photos and measured drawings and explains the process of disassembly, wire brushing, structural restoration (if needed), testing, shipping, and reassembly on your prepared site.

TEDD BENSON, Alstead Center, NH 03602. An authoritative timber-frame teacher, writer, and builder.

DAVID HOWARD INC., PO Box 295, Alstead, NH 03602. This firm custom builds timber frames and imports complete structures from England. Their designs have made the cover of *House Beautiful* and *Better Homes & Gardens*—definitely mainstream acceptance of an alternative building system. David Howard travels British back roads and comes away with some beauties. For instance, the Hunton Court barn is a 101×34-foot structure with unique, half-hip gable ends, an unusual cruck and truss roof frame, all put together about 300 years ago.

DOVETAIL LTD., PO Box 1496, Boulder, CO 80306. Dovetail manufactures completely precut timber frame homes, including prefabricated insulating panels and other finishing options. Blueprints, architectural services, workshops, and books on timber framing by the firm's principal, Stewart Elliot, round out the offering. Some of the company literature may put you off, as it is a little heavy on sales and short on specifics.

FOX MAPLE POST & BEAM, RR 1, Box 583, Snowville Rd., W. Brownfield, ME 04010. In addition to an elegant tool catalog containing the heavy-duty tools that make timber framing a little easier (write away for this one), Fox Maple builds timber frames. They work on custom homes regionally and ship precut owner-builder packages that include blueprints, siding, roofing, insulation, and more, plus the frame. Their most popular frame is a stunning center-chimney saltbox plus shed, with 9-over-6-lite windows and narrow clapboards with 4 inches to the weather.

ED LEVIN, Canaan, NH 03741. An experienced and, within traditional lines, imaginative and innovative timber-frame designer and builder.

NATIVE WOOD PRODUCTS, Drawer Box 469, Brooklyn, CT 06234. This firm provides full custom-building services regionally and ships frame-only packages and more complete homes, including windows, doors, roofing, siding, etc. (You can check regular ads in *Yankee* and *Country Journal*.) The firm also carries hand-forged hardware, planking, and beaded clapboard.

SOLAR NORTHERN, Box 64, Mansfield, PA 16933. This company offers 10 basic stock plans, frame-only packages, and complete, closed-in homes. The frames are rough-sawn Pennsylvania white pine. Composite wall panels use a core of 3¼-inch polyisocyanurate foam to net R-27. Frame packages range from a modest 756 square feet to a large gambrel design of almost 1,800 square feet.

TIMBERPEG, PO Box 1500, Claremont, NH 03743. Timberpeg makes precut timber frames with all materials needed to enclose a weathertight shell. Many stock plans can be altered for special needs and sites. The kits are made from 6 × 6s, 8 × 8s, and 8 × 12s, all with conventional mortise-and-tenon, wood-peg joints.

ILLUSTRATION ACKNOWLEDGMENT

The following contributed illustrations to this part of the book:

CORNERSTONES ENERGY GROUP INC., 21 Stanwood St., Brunswick, ME 04011

DAVID HOWARD INC., PO Box 295, Alstead, NH 03602

FARALLONES INSTITUTE RURAL CENTER, 15290 Coleman Valley Rd., Occidental, CA 95465

FINE HOMEBUILDING, Taunton Press, PO Box 355, Newtown, CT 06470

EDWARD M. LEVIN JOINED TIMBER FRAME STRUCTURES, RFD #2, Canaan, NH 03741 (Most photos by Richard Starr.)

NATIVE WOOD PRODUCTS INC., Drawer Box 469, Brooklyn, CT 06234

TIMBERPEG, Box 1500, Claremont, NH 03743; Box 1007, Elkin, NC 28621

U.S. FOREST SERVICE, 1720 Peachtree Rd. NW, Atlanta, GA 30067

TIMBER EPILOG

Graceful, traditional timber-frame designs and the joinery holding them together are awe-inspiring. Timber-frame longevity is compelling; the simplicity is admirable. Inside an exposed-frame home you get a feeling of quiet strength. It is a soothing environment. But there's a price. Timber frames will take longer to build than conventional homes and probably cost 10-15 percent more as well.

This seems a small price for the benefits if (and it is a legitimate *if*, I think) these housing characteristics deserve such a high priority. Craftsmanship and quality detailing always deserve attention. Going out of your way to build a home to last two, three, even four centuries may be reaching too far. As much as I admire classic lines and elegantly simple woodwork, I can't shake the gently mocking voice that asks if I really believe I know so much that my ideas and handiwork deserve to be here four centuries after I'm gone, particularly since the ideas are not original but a rediscovery of old techniques.

It's something to think about, for a little while, but not something to dwell on. If you use a proven building system and do the best possible job, you can enjoy the work and the results, and let the future take care of itself.

POLE HOUSES

A lot of what I can tell you about pole building comes from time on the job, in the trenches, struggling in the mud to nudge skin-burning penta-sealed 6 × 6 posts onto the centers of their concrete pads, turning my arm to spaghetti after a long day of driving 60-penny case-hardened spikes through 2 × 4 purlins. Free-walking the top chords of 60-foot trusses, trying to get up quickly to set a stabilizing purlin while the 6 × 6s pinned to each end of the truss wobble against their braces—that helps to put all the textbook tables and formulas in perspective.

Practice is important in the field, but also in your head and on paper. It's amazing what you can do with studs and joists and rafters in your imagination—just about anything. Scratching those ideas out on paper, even the most primitive doodling, is more realistic, and very worthwhile. I'm hooked on floor plans and section views, and have probably doodled away hundreds of pounds of yellow pads and graph paper. A few of those ideas have made it off the scratch pad and into a blueprint. But more has come from all the doodles I tossed into the wood stove: practice with outrigger rafters, and cantilevered decks, and the staggering potential for exciting living space you can lace through a rugged grid of building poles.

Photos by Pole House Kits of California.

Pole building vs. other systems. Pole building may not be entirely new to you. But it is still considered an unorthodox building system. Structurally, it is more complicated than log building, earth-sheltered building, and other methods that, at first glance, may appear much more unconventional than pole building. They're not.

Log building, in a sense, is very much like playing with those wonderful children's toys, Lincoln Logs, but playing on a grand scale. Earth-sheltered building is very much like conventional block-on-block or solid concrete construction except that you backfill everything. Sure, that's oversimplified. But those building systems are simple, in the basic, logical, structural sense. The materials on top rest on the materials directly below them. Keep stacking up materials, whether concrete or logs or cordwood or stone or adobe block, and eventually you get a wall. If you make the stack vertical and build it on something solid, you're in business. These systems work on the most basic assumptions we make about why buildings stand up—assumptions we've made about alignment and balance and weight and gravity since we were children playing with blocks.

It may help to think of pole framing as a modern adaptation of log building. But instead of felling and limbing trees and then stacking them horizontally, pole construction uses chemically improved trees replanted, literally, in a solid, systematic way. Pole building does improve on nature. You could try to hang a house from a natural grid of trees, as you would find them in the forest. That would be natural in the strictest sense of the word—a charming idea, yet with obvious drawbacks, such as wind and termites and squirrels, to name a few.

Pole building, unlike some log and timber-frame construction, is not an attempt to re-create the past, even though pole building has a long past. It is not an attempt to return to the farm or the rural homestead, even though that is where pole building first flourished. So much of the information on timber framing, for example, stresses the use of authentic tools, materials, construction methods, and designs that you may wonder if the idea is to build a home or a museum. But each pole frame offers the chance to incorporate suitable materials and methods (even if some of them are modern) and the chance to mold a flexible system to your needs and to the requirements of the site.

Pole building is a little like balloon framing, more like bridgebuilding, and a lot like structural steel work. And those are not familiar residential systems. But it is a true building system, compared to logical stacking of material. A pole structure doesn't rest on the ground with a full foundation like almost every other building. It sticks into the ground at selective locations. It doesn't make a lot of visual sense in the first stages of construction, the way a concrete foundation lets you preview the floor plan. In initial stages, pole building looks as though some crazy maritime construction company got the wrong address and is doing a heck of a job building a waterfront pier. But poles are probably the most versatile building system, certainly one of the most economical, and one of the easiest for owner-builders to use.

A pole-frame house is an obvious choice if you are fortunate enough to have a site on a beach, on a lakefront, on the backwaters of a marsh, or on a steep hillside overlooking a green valley. A pole house also looks pretty good hanging ruggedly on its legs just above a flat grassy field.

You may have strolled along a boardwalk unaware of the structural pole maze below the deck. Set by barge-mounted pile drivers, these treated poles (below left) provide the framework for boardwalks, piers, and slips in the port of New Orleans. The finished slips (below right), are covered with CCA salt-treated decking.

The pole system. In pole building, you have to put away the Lincoln Log instincts and start remembering the Erector Set. With log and stone you build solid, fortress-type walls. With Erector Sets and poles you build open-air space frames, filling in the pieces later on. You can't simply start at the bottom and head up. In pole building, you must synthesize the structure in your mind before you begin. You must see deeper than the floor plan, farther than a perspective view. You must examine the mechanical operation of the structure. In pole building, that structure is the design; the heart of the building that controls how the building looks and works, more so than in other construction systems. That's where you should turn your attention first—to the engineering system. Let the architecture follow.

This is an unfamiliar approach to many owner-builders. We are used to starting with the architecture, the social system, a kitchen with enough room for a table, a coat closet near the front door. And then, after the way we want to live is laid out on paper, we look for ways to support these ideas. Oh, you want to see through the 200-gallon fish tank into the dining-alcove flower arrangement while you're at the sink doing the dishes? No problem. We'll just insert this 1,000-pound I-beam.

But to start from scratch with pole building, you should start on paper, lots of it. Draw yourself some box frames, then rectangles, then triangles. Join some of them together. Then stick some of the boxes partway into other boxes. Draw some floor platforms into the boxes and a platform two-thirds in one of the boxes with one-third out—a cantilever. Don't worry yet about windows and doors and furnishings.

Stick with the linear system of vertical poles, and start to discover just how many different shapes and sizes you can hang from them. You will begin to see the possibilities of hanging floors and decks and balconies at different levels, the possibility of designs with open, free-flowing interior spaces, building economically on a steeply sloped site, and more.

As you'll see on upcoming pages, pole building is an understandable, livable, human combination of modern building technology and one of the most basic and proven building materials—trees straight from the forest.

To develop pole house plans, I often start with a plan, or top, view of the grid pattern, anywhere from 8- to 16-foot on center and then roughly sketch-in partitions defining living spaces. I like to work on an elevation sketch simultaneously, projecting cantilevers, overhangs, floor levels, and rafter angles. A thick pad of graph paper helps too. I position the poles and then experiment with the endless possibilities for free-flowing interiors and exteriors that I can hang on the grid.

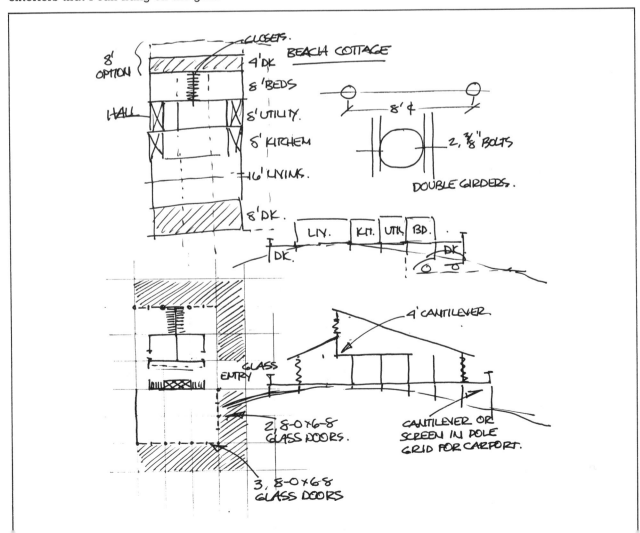

EARLY POLE FRAME HOMES

The Japanese Isé Shrine (left) and a lake dweller's hut in northern Europe are both classic examples of the origins of pole building. The shrine is elegant and refined; the hut is raw-boned. Both use "replanted" trees in a systematic way to form a structure.

Even though pole building can be traced back to the Stone Age, many architectural and history books don't even give it a mention. Poles have supported primitive thatched-roof cabanas in Peru, wind-resistant huts on islands in the South Pacific, communal long-houses of the Iroquois in North America, and Isé shrines in Japan. Hand-driven poles were used to provide soil consolidation under some of the earliest structures in Venice. Poles with primitive preservative treatments were used by the Romans.

Pole building, in one form or another, seems to have arisen independently in many parts of the world. It is the nearest thing to a universal building system; useful as a way to elevate primitive huts above flood plains and hungry animals; useful as a way to secure award-winning modern homes against earthquake damage.

This is just the tip of the iceberg, but enough so that you should understand my amazement when I could not find the briefest mention of pole building indexed in these books: William Dudley Hunt's 600-page, excruciatingly detailed *Encyclopedia of American Architecture*; Eugene Eccli's *Low-Cost, Energy-Efficient Shelter* (a title tailor-made for pole building); J. E. Gordon's intriguing *Structures, or Why Things Don't Fall Down* (and pole structures are among the last to go from wind, fire, or earthquake); even the Penguin *Dictionary of Building* and *Dictionary of Architecture*.

Trust me; the list of no-shows on pole building is unexpectedly extensive.

For me, this is like asking for a Coke with a hot dog and having the waitress say, "What's a Coke?" I'm not sure how pole building escaped the spotlight, but I have a few hunches, based on the more recent history of pole building, particularly in this country.

Can you picture a Clydesdale horse—you know, the breed that pulls the Budweiser beer wagon in TV commercials? They are big, powerful, rugged horses, built for work. They are not the breed you would picture in racing silks, under fine English saddles, with garlands of roses around their necks, posed next to pearl-necklaced ladies in the winner's circle. Oh, no. These heavy-duty clodhoppers grew up on the farm, and that's where they belong. And so it was with poles.

Trees from the forest are big and beefy, built to carry loads, built for rugged construction. But there is a drawback. Take the bark off those trees, leave them out in the weather, expose them to termites, wet rot, and fungus, and they won't last very long. The Romans got more life out of timbers by soaking them in vinegar. In 1716, one Dr. Crook of South Carolina took a patent on an oil-based wood preservative. In the 19th century, a process for manufacturing creosote was patented. And in 1865, the first pole pressure-treating plant was built in Massachusetts. These developments completely changed the life expectancy and possible uses for pole timbers.

These innovations in construction technology coincided with America's industrial and mechanized expansion to the West. There was a feverish need for durable railroad ties, timbers for railroad bridges, poles for electric and telegraph lines. And what could fill the need better than the chemically improved poles—creosoted, pressure-treated.

When pentachlorophenol (penta) was introduced in the 1930s, pole life was extended even further. Later, the U.S. Forest Service published results of tests indicating no decay on poles sunk into the ground for 50 years. So poles could be used for fence posts, feed-storage barns, even stables, where pressure-treating could protect the wood from a potent combination of hay and feed and animal waste and mud and just about anything else.

Pressure-treated poles were the working timbers for the working farm and for the utility companies. By 1960, there were over a million pressure-treated poles in service. (Can you picture them, all together in a clump instead of strung out along the highways?)

And then some agricultural periodicals that regularly included information on poles for fences, poles for stables, and poles for barns began to include information on poles for houses. The idea, at first, was to keep the poles on the farm, but to use them in the farmhouse, too. But pressure-treating had advanced too far, and the building system was too versatile to remain a rural workhorse. The World-War-I song about keeping them down on the farm after they've seen Par-ee sums it up nicely.

Modern residential pole bulding flourished first on the West Coast, particularly California. The Marx-Hyatt house, built in 1957 in Atherton, California, is generally credited as the first famous residential pole building. Its success helped to overcome resistance from building inspectors and mortgage bankers. And, in historical context, you can understand their resistance. Some salesmanship was required to convince pinstriped bankers that this barn-building system would make a desirable, saleable, even foreclosable and resaleable home.

Steel, which has replaced wood as the primary material in many structural parts, is now used increasingly as siding in modern barn buildings. This modern pole barn (yes, square-sawn poles are still called poles and used like poles in pole buildings) built by Farm Builders, Inc., in Remington, Indiana, uses treated poles, web trusses, and steel siding. The company is building structures for private residences in the same way: steel-clad, pole-wall, truss-roof homes.

In 1957, a panoramic shot of the Marx-Hyatt living room made the cover of *Western Building*. The headline: "Pole-Type Residence Lowers Building Cost" (from $36,000 to $31,000, as it was estimated by the architect).

American Builder ran a five-page story complete with exploded-view drawings showing how 3×12 girders overlapped at the poles and were secured with massive ⅞-inch bolts.

The San Francisco *Examiner* ran two pages of photographs with its story that same year subtitled, "Chemicals Team up with Economical Construction in a Woodside Home." The article surprised many builders and carpenters, I'm sure, by saying that the Douglas fir poles "can be sunk directly into the ground without fear of rot, decay, or termite infestation." It raved about the structural value and economy, but was a bit reserved about the appearance. "The poles in the Hyatt house are used in the rough and are an unusual shade of soft green." Reserved but tactful.

The article closed with a prophetic message for interested readers. "Greatest drawbacks to creative experimental effort in the battle of cost and quality in home building are rigid building codes and reluctant traditionalists in the home building field."

After three decades, the Marx-Hyatt pole house shows minimal wear and tear. Extensive deck cantilevers, with upswept detailing, show stains from resealing of the deck surface.

Built in 1957, the Marx-Hyatt house in Atherton, California, was designed by Edwin Wadsworth. Mrs. Marx-Hyatt told me, "We had to go to four or five architects before we found someone who understood what we were talking about." Not only did Mr. Wadsworth understand, he liked the innovative building system so much that he later designed a pole house for himself.

Inside, pole checking and the prairie architectural lines popular in the 1950s are the only indicators of the house's age, and they testify to durability.

Double floor girders with massive bolts and washers have checked only moderately. Such girders look wildly overbuilt by today's minimal standards. What a refreshing difference.

What a contrast! The car is classic 1950s: heavy-duty chrome, two-tone, angles, wings, and fins. The slightly swept-wing deck wall on the Marx-Hyatt house just might make you guess '50s, but the clean, economical lines as well as the use of poles and heavy-timber girders to define the lines of the house and hold it up are features ahead of their time.

Some government agencies became interested in pole frames for housing at the same time. In 1962, the U.S. Forest Service built a test pole building on the beach in Pawley's Island, South Carolina. "On the beach" are the key words. It's almost evolutionary, as poles used for piers and docks were brought up on the sand and used for houses. Over 20 years old now, that house, and the 1963 test house in Surfside, South Carolina, which replaced conventional rafters with roof trusses, are both simplified, economical, easy-to-build homes. The window treatments and deck detailing date the exteriors of both houses somewhat. But the minimal, elegantly efficient pole engineering has never gone out of style.

The Forest Service found that pole building was a safe, economical way to build on the beach, near tidal water, near marshland, in regions subject to high winds and hurricanes. But is is interesting to backtrack a few years to see what generated all this interest.

In the late 1950s, the J. H. Baxter Co. and others found that pole building was the key to making unusable real estate suitable for residential construction. Telling you that first is like telling the punch line before the joke. But consider just how momentous that statement is. With pole building, a steep, unstable, vacant lot could become a building site. This is one reason why pole building flourished early along the West Coast, where real estate was, and is, at a premium.

Following the dictum of form follows function on a larger scale, the West Coast pole-building market grew out of a study of housing in the San Francisco Bay area by Stanford University. This report found that almost all buildable lots (at least buildable in conventional terms) had been used. The Bay area had run out of room. That's where Mr. Marx-Hyatt, working for the J.H. Baxter Co., entered the picture.

The idea was simple: If easy-to-build sites were used up, but there was still a demand for housing (the function), a new building system (the form) was needed—one that could bring the steeply sloped, unusable lots back into the real estate market. The Baxter Co. sank a lot of time and money into the idea of residential pole building. The Marx-Hyatt house was the first well-publicized, visible proof of their work. Not the most elegant house, and not on the most spectacular site compared to other pole homes in this section of the book, it does show the basics. It is difficult, in hindsight, to imagine the impact of this unprepossessing home—what it did for real estate agents, mortgage bankers, home buyers, and builders previously priced out of the market. That is what makes the Marx-Hyatt house so special. There is no first log cabin, no precedent-setting timber-frame home. But a lot of what has happened in residential pole building in the last 25 years can be tracked back to one house: the Marx-Hyatt house in Atherton, California.

DESIGN PRINCIPLES

In solid-wall homes, loads from above travel where they fall. Because the wall is uniformly capable of supporting loads, it does not mattter where the loads fall. In stick-built homes, loads are parceled out to framing members every 16 inches on center. In some energy-tight homes, the loads are split apart 24 inches on center. But in pole building, loads are distributed in 8-, 9-, 10-, even 12-foot and greater increments. The essence of a pole frame system is wonderfully direct. It may be understood most readily as a three-part design system.

1. A minimum number of big, beefy timbers, arranged in a grid system, carry all the building loads, from the roof straight down to the subsoil.
2. A moderate number of moderately beefy dimensional timbers hook the system together, establish structural boundaries, and get all the loads over to the poles.
3. A greater number of light-frame dimensional timbers fill in the frame and support the building skin.

Now, is that sensible or what? You use only a little of the big heavy stuff that is expensive to buy, difficult to transport, and time-consuming to put in place. You eliminate most drainage concerns, most of the excavating, almost all of the foundation work in one shot. You string the poles into three-dimensional box frames with dimensional timbers. With the poles in position, these timbers are easy to handle and easy to bolt in place. Now the fun begins.

You can do just about anything you want inside (and outside) these pole and dimensional timber grids. The loads are transferred and accounted for, so you can install 8×8-foot fixed glass or Texture 1–11 siding. Unlike stick-built homes, in which only some interior partitions are non-loadbearing walls, in the pole frame all partitions, interior and exterior, are non-loadbearing. Within whichever grid pattern you select (and the height of the three-dimensional boxes is up to you), this support system gives you complete design freedom. What a treat!

Another attractive characteristic of pole construction is that it allows you to preserve the building site. Today, the first step in construction on so many homes is to "neutralize" the site, frequently with a bulldozer. This sounds almost like a military operation. Sure, you may be left a few trees, although with 2 feet of fresh fill around the trunks or a few dozer scars through the bark, they may not last long. So often the site is kicked and shoved into a shape that will accept conventional masonry footings and foundations. Most housing developments are like boot camp where everyone gets the same short haircut; the objective is standardization. So what if you destroy the groundcover, cut off a few root systems, or fill in a few natural drainage paths?

Pole foundations serve well where conventional foundations couldn't be built, such as in this hollow.

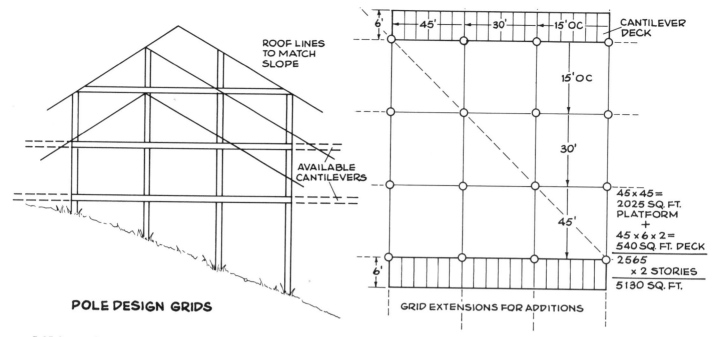

POLE DESIGN GRIDS

ROOF LINES TO MATCH SLOPE

AVAILABLE CANTILEVERS

CANTILEVER DECK

15'OC

30'

45'

45 × 45 = 2025 SQ. FT. PLATFORM
+
45 × 6 × 2 = 540 SQ. FT. DECK
2565
× 2 STORIES
5130 SQ. FT.

GRID EXTENSIONS FOR ADDITIONS

A 15-foot pole grid of 16 poles could net you a small 15 × 15-foot house centered in a massive deck. With cantilevers and another story (but the same pole grid), you can enclose over 5,000 square feet of living space. There are limitless options with cantilevers, roof pitches, and enclosed versus outdoor spaces.

You may wonder if all that standardization was worthwhile when the building is completed, and you start paying bills for topsoil, sod, trees, and shrubs.

On hillsides and other irregular and sloped sites, this approach can do more than financial damage. Neutralize a hilly site, and erosion may cause it to disappear in the next few years. Farmers go to great lengths and expense to prevent erosion because their income depends on the soil. Your home is just as important. If you are lucky enough to find a site where trees and groundcover have taken hold, you would be foolish to destroy this precarious natural balance between vegetation and erosion.

Pole building will leave the site intact. It requires virtually no grading. You don't have to worry about cut-and-fill ratios, dirt disposal, buying bank run for backfilling, or trucking in topsoil. A basic, 16-pole house on a 15-foot grid pattern provides 2,025 square feet of platform space—considerably more if you cantilever dimensional timbers beyond the grid limits. A modest

6-foot cantilevered deck on two sides of the grid, for instance, would add 540 square feet. So far you have 2,565 square feet without a second story (5,130 square feet total). The amazing numbers, though, are at least 720 square feet of excavation for a conventional masonry foundation (a 4-foot-wide perimeter trench with 1 foot of masonry and only 1½ feet of operating room on each side), compared to as little as 16-20 square feet of excavation for the same-size house in a pole frame.

Digging by hand, you might disturb 50 or 60 square feet. Using a mechanical auger (the kind the phone company uses to set poles), you would not disturb more than 18-20 square feet. Even clean holes should be 6-8 inches wider than the pole diameter to allow for plumb alignment and backfilling.

Remember, however, that even if the site is undisturbed, a torrent or water from a leader is not normal rainfall that on-site vegetation can handle. Consider underground drain lines and the possibility of dry wells to retard and redistribute the flow from heavy rains.

DESIGN
DECISIONS

One of the most basic design decisions is whether to use a full- or semi-pole frame. You can use platform poles as a halfway (and, I would have to say, half-hearted) design system. Suppose the building site is steep, unstable, or wet. You can recapture it for conventional construction by installing short poles on the high side of the slope, long poles on the low side. On the high side, 8-foot poles might be embedded 6 feet. On the low side, say 10 feet lower, 18-foot poles might be embedded 6 feet. With this grid system in place, the pole tops are leveled, tied together with dimensional timbers bolted flush with the pole tips, then decked, creating a conventional, level building platform on a steeply sloped, unconventional site.

In this case, you treat the poles as a unique foundation system, even though they could do more for you. The poles give you a flat deck on which you can build a timber frame, a dome, a stick-built home, or whatever you want. But why lop off the poles that can serve above the first-floor joists to hold girts and wall panels, rafters, roofing, decks, and balconies?

Granted, full-height poles can be a handful. I have set 16-foot poles embedded 5 feet solo. It is heavy-duty work. I was able to do it safely by setting a free rotating bolt through the pole tip (the last few inches that would be trimmed later on) and bolting on a couple of 2×4s. After sliding the butt end into the hole, which elevated the pole to begin with, I "walked" the 2×4s forward until the pole was upright.

Working with one other carpenter (someone to handle the butt end), I have set 25-foot poles with a backhoe. The idea is to use a chain with a U-hook to make a self-tightening hitch about two-thirds of the way up the log. Hook the other end of the chain over one of the bucket teeth, and up it goes. With a light touch you can load and unload the pole weight until the butt is perfectly positioned on its concrete pad; then, leaving the bucket in place, you can climb down from the hoe to help with bracing.

For stability and to prevent settling, you can put a concrete pad under the pole or secure it with what's called a necklace—a 12-inch ring of concrete surrounding the backfilled pole, poured just below the frost line. A set of four to six lag screws set to protrude from the pole into the necklace transfers vertical loads to the concrete. But I figure if you can get concrete to the hole in the first place, why not forget the lag screws, forget about possible expansion and contraction between the pole and the necklace, and put the concrete down at the bottom of the hole? This footing is called a punching pad.

If you can't get a ready-mix truck to the site, you

Pole houses do not have to look like pole houses, not unless you look down near the ground or, in this case, the sand. Platform-height poles get homes on the beach or near any potentially wet land up off the ground. The swept-wing roof pole house (below right) is not unlike more ancient Japanese pole buildings.

Full-height poles still get the structure up off the ground. But by setting poles of different heights in different rows of the pole grid, the full-height design anticipates wall plates and rafters. Tall, full-height poles allow for multilevel, open, interior spaces running from the bottom of a sloping site to deck, to sunken living room, to raised living room, to balcony, to skylighted loft.

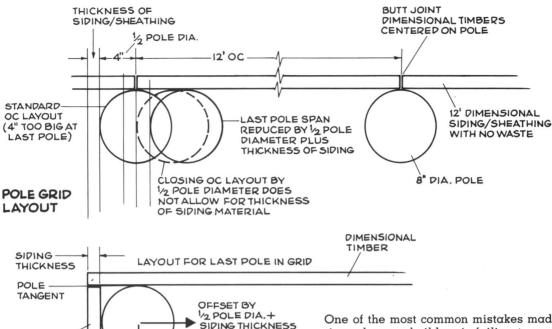

THICKNESS OF SIDING/SHEATHING

½ POLE DIA.

4"

12' OC

BUTT JOINT DIMENSIONAL TIMBERS CENTERED ON POLE

STANDARD OC LAYOUT (4" TOO BIG AT LAST POLE)

LAST POLE SPAN REDUCED BY ½ POLE DIAMETER PLUS THICKNESS OF SIDING

12' DIMENSIONAL SIDING/SHEATHING WITH NO WASTE

POLE GRID LAYOUT

CLOSING OC LAYOUT BY ½ POLE DIAMETER DOES NOT ALLOW FOR THICKNESS OF SIDING MATERIAL

8" DIA. POLE

SIDING THICKNESS

DIMENSIONAL TIMBER

LAYOUT FOR LAST POLE IN GRID

POLE TANGENT

DIMENSIONAL TIMBER

OFFSET BY ½ POLE DIA. + SIDING THICKNESS TO MAKE OC LAYOUT IN GRID

OFFSET BY ½ POLE DIA. ALONE BRINGS SIDING TO POLE TANGENT

One of the most common mistakes made by inexperienced owner-builders is failing to account for material thickness in the structural layout. Modular center-to-center layouts work as well with poles as with conventional construction—and as badly if the end rows are not reduced for the center discrepancy and material thickness.

can use soil-cement (a ratio of one part cement to five parts soil shaken through a 1-inch screen). When the mixture is dampened and tamped in place, it offers an economical alternative nearly as strong as concrete.

In addition to deciding about some of the finer construction options, there are engineering decisions about building in exceptional storm resistance, which poles to use, which preservative to use, and how far to embed the poles in the ground. Three of the most significant design decisions are: (1) what grid pattern to use; (2) whether to build poles within, on the inside, or on the outside of exterior walls; and (3) whether to use trusses or dimensional timber rafters.

The grid system should anticipate the amount of floor space you want, the materials you will use in construction, and the openings you want. Many architects and builders use 10-, 12-, and 16-foot grids. Notice the even numbers. As the building industry is not metric yet (thank heavens), 16- or 24-inch framing bays, extended to 48-inch centers, work efficiently for all dimensional timbers, plywood, paneling, clapboard, wallboard, and more. Square grids can be used to make square houses or rectangular plans. Individual grids may be rectangular as well.

A prototypical pole frame might have a 12-foot grid of 8-inch-diameter poles capable of carrying just about any conventional load with a girder and header system of sandwiched 2×12s (one on each side of the pole, bolted together through the pole). At 16-foot grid

spacing, load calculations may push you into special-order timbers like 3×12s or 2×14s. Floor and roof platforms may be framed on top of the girder system or, with beam hangers, in the same plane as supports. A publication of the American Wood Preserver's Institute suggests economical use of poles 15 feet on center. Their roof design uses double 2×12s on end-wall poles and 3×12s doubled around mid-wall poles.

When you experiment with grid systems on paper, don't make the mistake of drawing poles as dots and walls as single pencil lines. If you do, the sensible paper plan will not make sense in the field, where the dots are 7 or 8 inches across and the lines are about 5 inches thick. To plan efficiently for material use, draw at least a portion of the frame in large scale and figure just where the plywood and T&G decking will fall. And if you are not familiar with framing systems, let me warn you about how the grid has to change at the ends of the building. The phrase "center to center" implies that a 12-foot piece of decking will start at the centerline of one pole and end at the centerline of the next pole. That idea works everywhere except at the outside poles, the poles at the end of the line. Unless you plan to continue the framing indefinitely, you must recapture the on-center offset. You can't make the 16-foot dimensional timber grow a few extra inches, so you must compensate, say, for 4 inches, or half the pole diameter, by setting the last line of poles 11 feet 8 inches, center to center. There's no point in blindly

POLE TO WALL POSITION

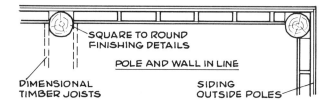

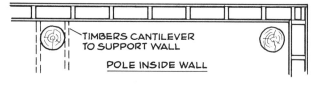

SQUARE TO ROUND
FINISHING DETAILS

POLE AND WALL IN LINE

DIMENSIONAL
TIMBER JOISTS

SIDING
OUTSIDE POLES

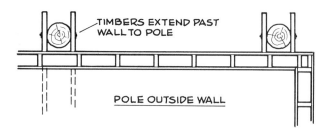

TIMBERS CANTILEVER
TO SUPPORT WALL

POLE INSIDE WALL

TIMBERS EXTEND PAST
WALL TO POLE

POLE OUTSIDE WALL

In conventional construction, it is crucial to align vertically all loadbearing elements. Every time you do not, the structure must be interrupted for headers and jack studs and the like. Poles carry vertical loads, too. But, as shown above, girders attached to the poles can carry walls in line with, inside, or outside the pole line.

This photo below shows the layers of a pole-and-girt barn: exterior ship-lap siding, horizontal 2 × 4 girts set between 2 × 4 spacers to close the end wall, all carried on the pole. From the inside (right), the structure is pleasing but definitely utilitarian. From the outside (below right), the weathered barn siding would look good on any home.

following a plan to the bitter end, particularly if a lot of the lumber you buy won't fit the grid.

The second design question is where to put the poles in relation to the walls. You have three options. On buildings where only one side of the wall is finished—a wood shed, for instance—poles are normally set in the wall line. Girts (2 × 4s on the flat, run horizontally along the poles 24 inches on center) can hold vertical exterior siding. On finished houses, poles can be set within the wall, but with a few problems. Since poles are never perfectly straight and are always tapered, these walls may wobble in and out. Also, seams between poles and exterior siding and glazing and between poles and interior wallboard and trim can be difficult to treat neatly and to weatherize.

To avoid these detailing hang-ups, you can set poles just inside or just outside the walls. Both systems work well. Poles inside enable you to lay a uniform, unbroken building skin on the exterior. But your pole home may appear more like a house on stilts with a lot of smokestacks. Structural poles are graphic inside the house but use up floor space. Sometimes they wind up in places where you want to put a couch or a refrigerator.

Setting poles outside the walls greatly simplifies wall construction. You can build full-length wall frames, complete with openings and headers, and ratchet them up into place. The walls are carried on girders that extend just past the wall to the pole connection. In fact, you can continue the girders past the exterior

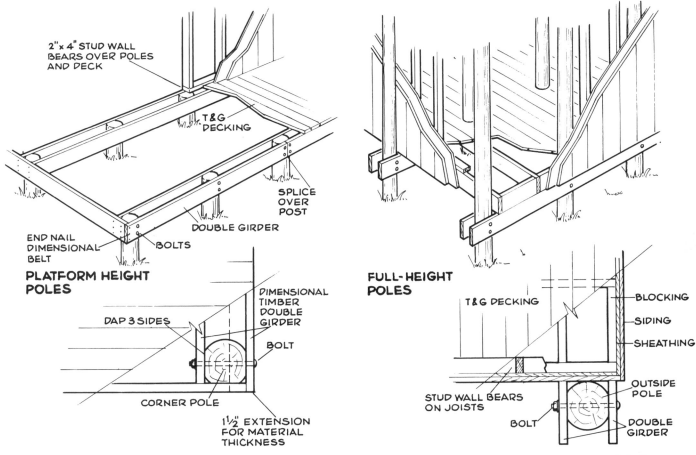

PLATFORM HEIGHT POLES

2"x 4" STUD WALL BEARS OVER POLES AND DECK

T&G DECKING

SPLICE OVER POST

DOUBLE GIRDER

END NAIL DIMENSIONAL BELT

BOLTS

DAP 3 SIDES

DIMENSIONAL TIMBER DOUBLE GIRDER

BOLT

CORNER POLE

1½" EXTENSION FOR MATERIAL THICKNESS

FULL-HEIGHT POLES

T&G DECKING

BLOCKING

SIDING

SHEATHING

STUD WALL BEARS ON JOISTS

BOLT

OUTSIDE POLE

DOUBLE GIRDER

When platform-height poles are enclosed in double girders, the platform can be created with structural decking. In this case, bearing walls should run directly over the lines of poles. On full-height pole structures, the deck and exterior bearing walls can run to the inside or outside girder. When full-height poles are inside or outside the house, leave a little maneuvering room (an inch would be adequate) between pole and finished wall.

Round poles in line with rectilinear 2×4 wall frames create strange, round-to-square butt corners like the one at left between pole and window opening and also below, where headers, wall plates, and collar ties come together in the same plane at the same point. Sometimes, hardware (as shown) is the only answer.

poles for decks, porches, ramps, stairs, and more. This system simplifies interior detailing and keeps heavy-duty bolt connections and other large timber connections outside.

The third design decision is whether to use trusses or conventional timber rafters. In a way, trusses complement pole frames. They cover a lot of ground in one shot and hook up a lot of loads directly to the pole grid. This is particularly true when trusses rest directly on poles and are tied with a system of purlins. I think trusses are great, but only when you have to use them. When you do, for economy, it's usually a good idea to bury them because they don't look very good. That's strike one. Also, the network of angled braces within the truss effectively kills potential loft space for anything but dead, and somewhat inconvenient, storage. That's strike two. The efficiency of truss design relies on getting clear spans with minimal timber use. Often, trusses large enough to clear the pole grid are too large to set by hand. If you haven't had need for mechanical help on the job so far, this structural demand may be strike three for trusses. Finally, once your home starts to show the pleasing details of vertical poles sandwiched between flat-sawn timbers, it is unlikely that you will want to interrupt this system with gusset-plated, gang-nailed, definitely prefab-looking truss details.

On many modern pole barns (and new houses too), economical web trusses span very nicely but leave a lot of dead space (as shown). Even with a low-slope roof, solid rafters tied instead with plywood gusset plates at the ridge, and secured with horizontal collar ties, leave potential storage space. When the rafter pitch is increased, collar ties at wall-plate height can serve as joists for a loft or balcony.

PREPARATION
FOR CONSTRUCTION

Manhandling poles takes a certain bulk, either alone, if you are a large, strong person, or collectively, if you're not. It also takes a special determination. You have to be determined that the pole will move an inch when you wrap your arms around it and heave. In addition to soil tests and other technical preparations for construction, you should prepare yourself for some heavy work.

I spent about six months working on a series of pole buildings with a crew of nearly itinerant pole builders headed by two brothers, Bob and Clifford. Both were religious fundamentalists and traveled with their families in two old Airstreams towed behind an old Power Wagon and a Chevy Travelall. Bob, the younger brother, was probably close to 50. He looked 40 in his hat, 60 without it for lack of hair. He had hands like feet—beefy, calloused, bent, and blotchy colored—and an almost angelic face. He would remove several 60-penny case-hardened, ring-tailed spikes from his mouth to tell someone about the witness he made last Sunday.

Bob would sidle up to a pole like it was something good to eat, wrap his arms around it, lock in with bear-paw hands, and heave silently upward to correct plumb alignment with a little sideways shift before he let go. Bob and Clifford started about seven-thirty in the morning six days a week. They brought in a crane operator to hoist 60-foot clear-span trusses for a main barn, a guy who played recorded crow calls from a loudspeaker on his truck during lunch.

Sometimes, when we stood in the muddy holes to trim braces or first-row penta boards, electricity would bleed through the circular saw handle and through our bodies, which had become grounded in the rain-soaked pole pits. When the buzzing became painful and we complained, Bob would wag his head just a little (what is this younger generation coming to?), jump down in the hole, pull the trigger, and cut away, seemingly unaffected by the electrical buzzing that was, at times, nearly paralyzing to me.

Bob and Clifford were my second exposure to pole building, my first to Skil electric chainsaws, walking 3-inch-wide truss rafters 40 feet in the air, riding down for lunch on the steel crane line, driving 60-penny spikes with 22-ounce ripping hammers, seeing a serious, but not fatal, fall (a guy crawling over purlins a few feet away just seemed to let go, and was on the ground 40 feet down in a heap in an instant). I soaked up enough penta to pickle my innards.

Not all pole building is this eventful, colorful, or bizarre. Although at the time, that seemed to me like just another carpentry job.

SITE SELECTION

When you decide to use pole construction, few conditions at the site can stop you. Pole girders can span gullies and hills. But there are two things to look for: soft soil that cannot support concentrated construction loads, and rocks, big rocks, that prevent pole embedding to adequate depth. You can, by the way, place poles over ledge rock with a minimal amount of stone cutting, using a system of embedded steel pins.

In addition to issues of orientation (see pages 15 and 16 for a discussion of these factors), you have to investigate drainage and soil type. Bedrock, considered a problem by many builders, is actually an advantage. For instance, frost action is eliminated, so the poles won't move unless the entire ledge moves. And when something like that happens, everything goes.

There are several ways to determine what kind of soil you have on the site—specifically, how it bears up under the pressure of building loads. Calculating these soil-bearing pressures sounds like a job for people in white coats examining test tube samples in a lab. Of course, it can be done this way; all very scientific.

One of the more ingenious technical solutions to this question was developed by P. C. Rutledge, a professor at Purdue University. His method is detailed in an excellent, although technical booklet, *Pole Building Design*, published by the American Wood Preservers Institute. The system was designed for on-site use for the Outdoor Advertising Association of America (OAAA), the folks who put up all the advertising billboards on, you guessed it, poles. Rutledge's system calls for a test pull on a 1½-inch diameter indicator

If you let your eyes fall out of focus, the poles and trees here become interchangeable and the house begins to float.

auger, a strain gauge of sorts. The professor's chart then can be used to translate pounds of pull on the gauge into soil classifications and allowable soil stress in pounds per square foot. The resulting chart is impressive but a bit beyond our needs here.

Certainly, if you have unanswered questions about building loads and the ability of the soil to support them, you should seek professional help. Many technical publications, however, also include completely nontechnical descriptions of different soils and their bearing values. The OAAA *Engineering Design Manual*, for example, offers the following description: Good soils are "compact, well-graded sand and gravel; hard clay; well-graded fine and coarse sand; decomposed granite rock and soil. Good soil should be well-drained and in locations where water will not stand." Poor soils are described as "soft clay, clay loam, poorly compacted sand, clays containing a large amount of silt and vegetable matter. Usually, soils of this type are found in low-lying areas where water stands during the wet season."

Most pole-building bulletins, both technical and practical, divide soils into three very basic categories: good, average, and below average. There is nothing overly scientific about that. The FHA *Pole House Construction* bulletin describes the classifications as follows:

1. Below-average soil: soft clay, poorly compacted sand, clays containing large amounts of silt (water stands during the wet season); bearing capacity, 150 pounds per square foot (psf); lateral bearing capacity, 100 psf per linear foot of pole embedment.
2. Average soil: loose gravel, medium clay, or any more compact composition; bearing capacity, 3,000 psf; lateral bearing capacity, 200 psf per foot of pole embedment.
3. Good soil: compact, well-graded sand and gravel, hard clay, or graded fine and coarse sand; bearing capacity, 6,000 psf; lateral bearing capacity, 400 psf per foot of pole embedment.

Note the two values for each soil group: bearing capacity, which is resistance to vertical loads such as the weight of structural timbers traveling straight down the pole, and lateral capacity, which is resistance to horizontal loads (also called tip-over loads) such as wind against the side of the house. Most construction systems, particularly full-height, vertical-pole frames, do a lot more than resist crushing or shearing forces.

When you calculate dead and live loads for the house, and you know what the soil will support, you can get to the heart of the matter: deciding what kind of poles to use, what grid pattern to use, and how deeply to embed the poles in the ground.

PRELIMINARY PLANS

The following tables will help you decide on the right pole system for your site. The tables do not cover every situation, and they do not apply to special cases—for instance, where there is a threat of mudslides. Four of the most common designs are covered.

1. Platform-height poles (long enough to get a level first-floor deck) for level building sites

2. Platform-height poles for sloped sites

3. Full-height poles (long enough to tie to first-floor framing and to structural members up to and including roof rafters) for level sites

4. Full-height poles for sloped sites.

In all cases covered by the tables, pole embedment with either soil or masonry backfilling is assumed to be made with (1) a concrete punching pad (a pad footing below the pole butt), (2) a concrete necklace (the masonry ring placed around the pole just below the frost line), or (3) a full backfilling with soil cement or concrete. See accompanying illustrations.

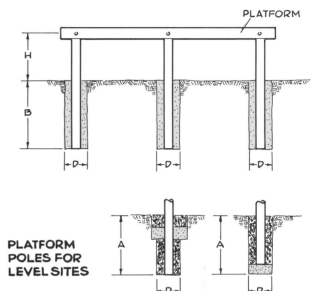

PLATFORM POLES FOR LEVEL SITES

PLATFORM POLES FOR LEVEL SITES
(SLOPE LESS THAN 1:10)

A. Embedment depth with backfill of tamped earth, sand, gravel, or crushed rock
B. Embedment depth with backfill of soil cement or concrete
H. Unsupported height of pole above grade line
D. Diameter of embedment system

H (ft.)	Pole spacing (ft.)	Good soil — Embedment depth (ft.) A	B	D (in.)	Tip size (in.)	Average soil — Embedment depth (ft.) A	B	D (in.)	Tip size (in.)	Below average soil — Embedment depth (ft.) A	B	D (in.)	Tip size (in.)
1½ to 3	8	4.0	4.0	18	5	5.5	4.0	24	5	7.0	5.0	36	5
	10	4.5	4.0	21	5	6.0	4.0	30	5	8.0	5.5	42	5
	12	5.0	4.0	24	5	6.5	4.5	36	5	*	5.5	48	5
3 to 8	8	5.0	4.0	18	6	6.5	4.5	24	6	*	6.0	36	6
	10	5.5	4.0	21	7	7.0	5.0	30	7	*	6.5	42	7
	12	6.0	4.5	24	7	7.5	5.5	36	7	*	7.0	48	7

* Excessively expensive, impractical pole embedment greater than 8 feet.

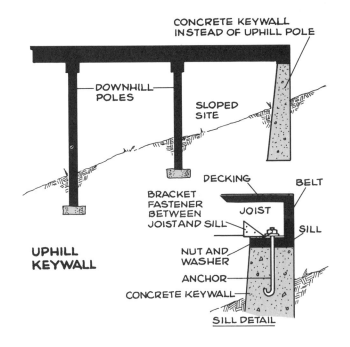

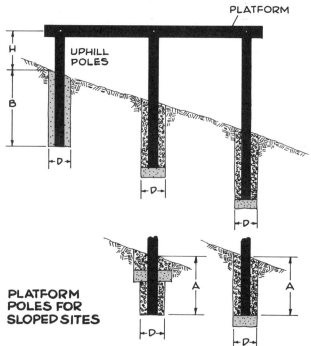

CONCRETE KEYWALL INSTEAD OF UPHILL POLE

DOWNHILL POLES

SLOPED SITE

UPHILL KEYWALL

DECKING

BELT

BRACKET FASTENER BETWEEN JOIST AND SILL

JOIST

SILL

NUT AND WASHER

ANCHOR

CONCRETE KEYWALL

SILL DETAIL

PLATFORM

UPHILL POLES

PLATFORM POLES FOR SLOPED SITES

On sloped sites, as above, the uphill line of poles is critical. When uphill poles cannot be adequately embedded, or where there are special considerations such as a threat of erosion or mudslides, a concrete keywall is used instead of poles. Full-length joists or girders lock the keywall and poles with a rigid diaphragm floor.

The table below lists characteristics of the uphill line of poles that carry vertical loads and resist lateral or sideways loads. The floor connecting the uphill line of poles with the downhill poles should be a rigid diaphragm to transfer the rigidity of short, uphill poles to the rest of the pole grid.

PLATFORM POLES FOR SLOPED SITES; UPHILL POLES ONLY (SLOPE AS GREAT AS 1:1)

A. Embedment depth with backfill of tamped earth, sand, gravel, or crushed rock
B. Embedment depth with backfill of soil cement or concrete
H. Unsupported height of pole above grade line
D. Diameter of embedment system

H (ft.)	Pole spacing (ft.)	Good soil Embedment depth (ft.) A	B	D (in.)	Tip size (in.)	Average soil Embedment depth (ft.) A	B	D (in.)	Tip size (in.)	Below average soil Embedment depth (ft.) A	B	D (in.)	Tip size (in.)
1½ to 3	6	5.5	4.0	18	6	7.5	5.0	18	6	*	7.0	24	6
	8	6.5	4.5	18	7	*	6.0	24	7	*	8.0	36	7
	10	7.0	5.0	21	7	*	6.5	30	7	*	*	*	*
	12	7.5	5.5	24	8	*	7.0	36	8	*	*	*	*
3 to 8	6	7.0	5.0	18	8	*	6.5	18	8	*	8.0	24	8
	8	7.5	5.5	18	9	*	7.0	24	9	*	*	*	*
	10	*	6.0	21	10†	*	7.5	30	10†	*	*	*	*
	12	*	6.5	24	10†	*	*	*	*	*	*	*	*

* Excessively expensive, impractical pole embedment greater than 8 feet.
† Poles where tip diameters may be decreased by 1 inch provided that embedment is increased by 6 inches.

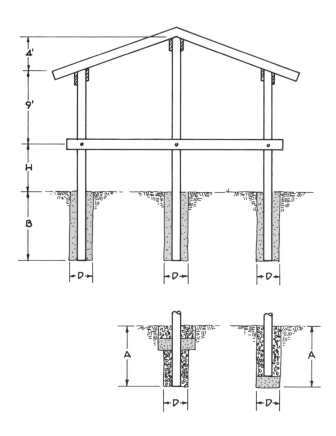

FULL-HEIGHT POLES FOR LEVEL HOUSE SITES
(SLOPE LESS THAN 1:10)

A. Embedment depth with backfill of tamped earth, sand, gravel, or crushed rock
B. Embedment depth with backfill of concrete or soil cement
H. Unsupported height of pole above grade line

H (ft.)	Pole spacing (ft.)	Good soil				Average soil				Below average soil			
		Embedment depth (ft.) A	B	D (in.)	Tip size (in.)	Embedment depth (ft.) A	B	D (in.)	Tip size (in.)	Embedment depth (ft.) A	B	D (in.)	Tip size (in.)
1½ to 3	8	5.0	4.0	18	6	6.5	5.0	24	6	*	6.0	36	6
	10	5.5	4.0	21	7	7.0	5.0	30	7	*	6.5	42	7
	12	6.0	4.5	24	7	7.5	5.5	36	7	*	7.0	48	7
3 to 8	8	6.0	4.0	18	7	7.5	5.5	24	7	*	7.0	36	7
	10	6.0	4.5	21	8	8.0	6.0	30	8	*	7.5	42	8
	12	6.5	5.0	24	8	*	6.0	36	8	*	8.0	48	8

* Excessively expensive, impractical pole embedment greater than 8 feet.
Where a concrete floor slab is used at grade, embedment depths may be reduced to 70 percent of the depths indicated above.

These tables tell you about three pieces of the puzzle: soil strength, pole size, and embedment depth. The tables put the pieces together mathematically so that when you have a lot of strength in the soil, for example, you can reduce pole size, or embedment depth, or both. Bigger may be better, but it may not always be necessary. And before you go out and buy monster poles, you should remember the distinction between vertical and lateral loads on these tables.

DOWNHILL POLES FOR SLOPED, FULL-HEIGHT POLE SITES

	Embedment depth, feet		
Slope of grade	Up to 1:3	Up to 1:2	Up to 1:1
SOIL STRENGTH			
Below average	4.5	6.0	*
Average	4.0	5.0	7.0
Good	4.0	4.0	6.0

* Excessively expensive, impractical pole embedment greater than 8 feet

DOWNHILL POLES FOR SLOPED PLATFORM POLE SITES

	Embedment depth, feet		
Slope of grade	Up to 1:3	Up to 1:2	Up to 1:1
SOIL STRENGTH			
Below average	4.5	6.0	*
Average	4.0	5.0	7.0
Good	4.0	4.0	6.0

* Excessively expensive, impractical pole embedment greater than 8 feet.
Where unsupported height (H) exceeds 20 feet, pole diameters should be at least 11 inches.

FULL-HEIGHT POLES FOR SLOPED HOUSE SITES; UPHILL POLES ONLY (SLOPE AS GREAT AS 1:1)

A. Embedment depth with backfill of tamped earth, sand, gravel, or crushed rock
B. Embedment depth with backfill of soil cement or concrete
H. Unsupported height of pole above grade line
D. Diameter of embedment system

H (ft.)	Pole spacing (ft.)	Good soil				Average soil				Below average soil			
		Embedment depth (ft.) A	B	D (in.)	Tip size (in.)	Embedment depth (ft.) A	B	D (in.)	Tip size (in.)	Embedment depth (ft.) A	B	D (in.)	Tip size (in.)
1½ to 3	6	7.0	5.0	18	8	*	6.5	18	8	*	*	*	*
	8	7.5	5.5	18	9	*	7.0	24	9	*	*	*	*
	10	*	6.0	21	9	*	8.0	30	9	*	*	*	*
	12	*	6.5	24	10†	*	*	*	*	*	*	*	*
3 to 8	6	7.5	5.5	18	8	*	7.0	18	8	*	*	*	*
	8	8.0	6.0	18	9	*	8.0	24	9	*	*	*	*
	10	*	7.0	21	10†	*	*	*	*	*	*	*	*
	12	*	7.0	24	11†	*	*	*	*	*	*	*	*

* Excessively expensive, impractical pole embedment greater than 8 feet.
† Poles where tip diameter may be decreased by 1 inch provided that embedment is increased by 6 inches.

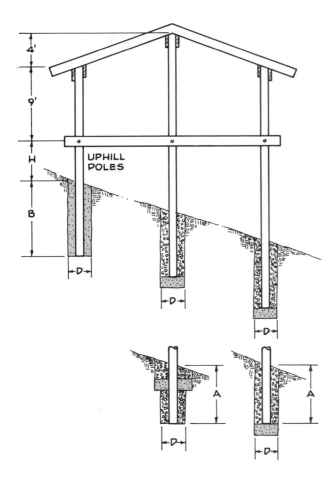

Poles embedded in the ground can do more than carry vertical loads. When the house eaves are 9 feet or less off the ground, 4 feet of embedding is almost always adequate in average and good soils. When the eave height is over 9 feet, the safe rule of thumb is to use 5 feet of embedding. But vertical loads, except in regions with heavy snow loads, may not always be the most critical forces in a pole-house frame.

More than half of the vertical loads on poles are transmitted into the surrounding soil by a process called skin friction. This nifty piece of physics is easy to understand. If you hold a heavy book like an encyclopedia in one hand, your muscles would strain with enough lifting force to counteract the downward force of gravity and the weight of the book. But if someone came along and, instead of putting a lifting hand under the book, pressed the book from the sides with both hands, making a sandwich of the encyclopedia, you would feel your muscles relax. With this lateral pressure and friction on the sides of the book, you might be able to support the decreased load with one finger. It's like stepping into the mud and losing a shoe.

This is why backfilling poles is so important. If you dump loose fill and large rocks around the pole, almost all building loads bear straight through to the butt end of the pole. If you backfill carefully with concrete or soil cement (both somewhat extravagant systems) or firmly compacted soil (compacted in stages as you fill the holes), most of the building loads will disperse into the surrounding soil before reaching the butt end of the pole.

For owner-builders and others on the Gulf Coast and anywhere else where heavy winds and hurricanes can be expected, skin friction is even more important as a force against uplift. Yes, these forces are unusual. We tend to think of structures that fail as collapsing, not blowing away. Watching TV footage of the damage caused by hurricane Alicia around Galveston, Texas, my eye caught the remains of a pole house on the beach. The pole grid was still in place, even and upright, as the wind and rain swirled. But except for sections of the first-floor deck frame, everything else was gone. I'd guess the deck frame was bolted to the poles, and everything else was nailed.

CONSTRUCTION

Both poles are 25 feet high, but the diameter at the tip (top) of the pole determines class size. Once you know that, you can figure about ¼ inch increase in diameter per linear foot to the butt end (bottom) to account for taper.

POLE SIZE BY CLASS

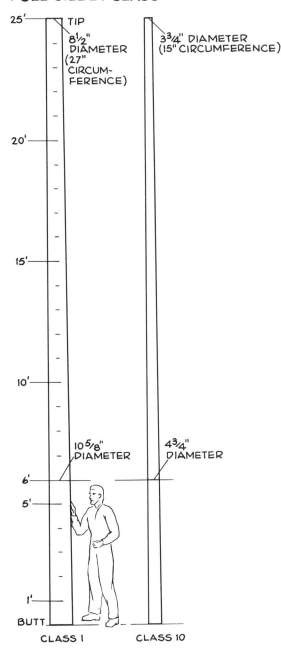

CLASS 1   CLASS 10

POLES

Building poles are classified by a combination of length and circumference. This is hardly surprising, as it reflects the way trees grow. Most short trees have smaller diameters than most tall trees. It's nature's engineering system spelled out in numbers on pole-building tables. Circumference (the length around a pole) is used instead of diameter (the length through the pole) because it can be measured reliably in the field.

Circumference dimensions determine the pole class, which ranges from 1, the largest, down to 10. All circumference measurements determining pole class refer to the length around the tip (top end) of the pole. Once you know that class number, you can reliably account for taper in the pole by adding ¼ inch of circumference per linear foot of pole; 4 feet below the tip would mean roughly 1 inch of additional circumference. If you need to find the diameter of any given point in the pole, divide the circumference at that point by 3.14.

In some special projects, utility poles can be used, although most local building departments will want you to use big Class 1 or Class 2 poles. In some areas you can buy utility poles secondhand at a tempting price, although a thorough and knowledgeable inspection must be made to confirm construction quality. And ask your nose if you will be able to live with the creosote odor. Utility poles generally range from lengths of 25 to 60 feet and are sold in 5-foot increments (building poles are sold in the same increment), considerably greater length than required for almost all residential pole building designs.

Class 1 poles, the largest, have a tip circumference of 27 inches (8½-inch diameter). It is customary also to list circumference at 6 feet from the butt end of the pole, at or about grade. A Class 1 pole, 25 feet long, with a tip circumference of 27 inches (8½-inch diameter), has a 33½-inch circumference (10⅝-inch diameter) 6 feet from the butt end. A pole of the same length (25 feet) in Class 10 has a 12-inch tip circumference (3¾-inch diameter), and a 15-inch circumference (4¾-inch diameter) 6 feet from the butt end. That's quite a difference in size, strength, and handling on the job.

Short poles, 16 feet long or less, are often no larger than Class 7 in shed and other farm designs. Poles in similar designs over 16 feet long are often no larger than Class 4. Still, because pole building is considered unconventional and an alternative to "normal" housing, your plans may not be approved unless you specify "upper class" poles. The smallest pole specified in pole-sizing tables of the Southern Forest Products Association is 10 feet in length, with a 16-inch tip circumference (5-inch diameter) and an 18½-inch circumference (5⅞-inch diameter) 3 feet from the butt end.

The following table shows the American Wood Preservers Institute (AWPI) specifications for Douglas fir and southern pine poles, rated by class size, pole length, tip circumference, and circumference 6 feet from the butt. These two species of wood (both with a fiber stress rating of approximately 8,000 psi) represent about 90 percent of the treated poles sold in the United States every year—enough poles to support over 650,000 existing pole-frame structures.

Poles 60 feet long should cover what you have in mind. Longer poles can be purchased, however. At lengths of 75 feet and more (up to about 125 feet long), almost all poles are Class 1 and Class 2. Just to boggle your mind, a 125-foot long Class 1 building pole, with a 27-inch tip circumference (8½-inch diameter), has a minimum 63½-inch circumference 6 feet from the butt—a mammoth diameter of 20¼ inches or more.

conditions. It also protects against insects and the ubiquitous termite, found in every state except Alaska, including dry-wood termites found in Arizona, California, Florida, and Hawaii.

With thorough, controlled, deep-penetration pressure-treating, pole structures are considered permanent, although some conservative trade and industry literature talks about 50-year lifespans under the most severe conditions. The FHA will give you a mortgage on a pole house. The AWPI states that a house built

Number comparisons may be impressive, but there is nothing like a side-by-side picture. What might look like a standard building pole next to a toothpick is a huge building pole and a small building pole, ideal for platform-height plans on sites with modest embedment requirements.

POLE SIZE VARIATION

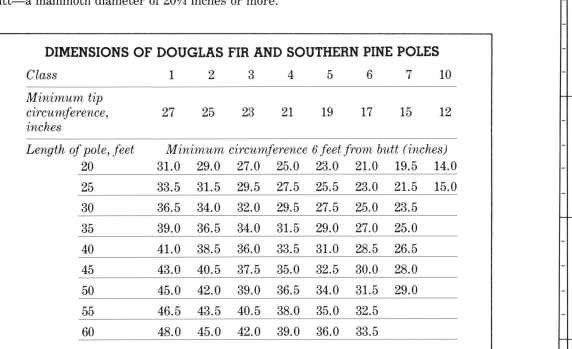

DIMENSIONS OF DOUGLAS FIR AND SOUTHERN PINE POLES

Class	1	2	3	4	5	6	7	10
Minimum tip circumference, inches	27	25	23	21	19	17	15	12
Length of pole, feet	Minimum circumference 6 feet from butt (inches)							
20	31.0	29.0	27.0	25.0	23.0	21.0	19.5	14.0
25	33.5	31.5	29.5	27.5	25.5	23.0	21.5	15.0
30	36.5	34.0	32.0	29.5	27.5	25.0	23.5	
35	39.0	36.5	34.0	31.5	29.0	27.0	25.0	
40	41.0	38.5	36.0	33.5	31.0	28.5	26.5	
45	43.0	40.5	37.5	35.0	32.5	30.0	28.0	
50	45.0	42.0	39.0	36.5	34.0	31.5	29.0	
55	46.5	43.5	40.5	38.0	35.0	32.5		
60	48.0	45.0	42.0	39.0	36.0	33.5		

PRESSURE-TREATING

Pressure-treating is what makes pole building practical. Pressure-treating protects against decay fungi, which can decompose wood under moist and humid

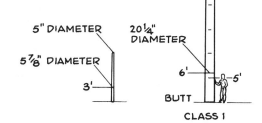

on a frame of poles produced to their standards for size, taper, and preservative treatment "is considered permanent—as permanent as a house on a well-constructed concrete foundation."

So why don't we all run out, buy cans of creosote, penta, chromated copper arsenate (CCA), and ammoniacal copper arsenate (ACA), and splash it on 2×4 decks, and wood shakes, and plywood siding, and every other stick of wood in the house? Recent environmental history, with homes and even entire towns lost to uncontrolled toxic wastes, with residential water supplies fouled by industrial chemicals leaching in subsoils, should deter anyone tempted by the chemical quick fix. Sometimes the much-publicized initial benefits of chemicals used in countless building materials have been overshadowed by the side effects that appear later on. Urea formaldehyde foam insulation (UFFI) is a good example. The dramatic energy savings were attractive. Later on, as more cases of low-grade formaldehyde gas emissions were reported on improperly balanced installations, and more people got sick, some local and state governments, and, finally, the U.S. Consumer Product Safety Commission took action and banned the product. Incredibly, in the fall of 1983 the Justice Department declined to challenge an appeals court ruling, in favor of the formaldehyde industry, that overturned this ban.

The Environmental Protection Agency (EPA) continues to study the possible health hazards of preservative treatments. Before buying poles, check with the EPA in Washington, one of their regional offices, or with your state agriculture department to be sure the treatment process you're buying along with the poles conforms to current standards. Unfortunately, some of the sweetheart deals exposed in the EPA superfund scandal of 1982 and 1983 show that current standards, rules, and regulations may not always be based solely on your health interests.

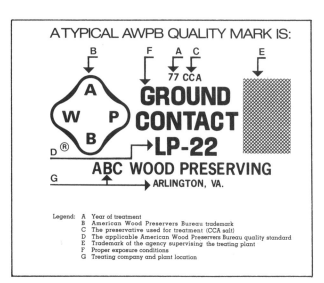

A TYPICAL AWPB QUALITY MARK IS:

GROUND CONTACT LP-22
ABC WOOD PRESERVING
ARLINGTON, VA.
77 CCA

Legend: A   Year of treatment
         B   American Wood Preservers Bureau trademark
         C   The preservative used for treatment (CCA salt)
         D   The applicable American Wood Preservers Bureau quality standard
         E   Trademark of the agency supervising the treating plant
         F   Proper exposure conditions
         G   Treating company and plant location

Interior poles can carry second-floor or, in this case, loft girders and separate functional spaces inside the house without walls. Imagine a warm summer evening sitting around the table, trying to enjoy the taste and smell of your food if this pole were a creosoted telephone pole. Pass the onions and garlic.

As a further safety measure, minimize contact and exposure and do not treat your own poles. I hesitate to plant the idea in your head, but some folks do try it. There are pole-treatment plants in most states, many on the East and West coasts and in southeastern states. And, of course, poles, like any other product, can be trucked from one place to another. Pressure-treated poles are widely available. There are definitive treatment standards, assay tests to confirm the level of chemical retention in the poles, and at least four active industry associations loaded with technical and practical information for builders and buyers on standards, grades, and more.

Most publications and how-to books advise against do-it-yourself pole treatment. For one thing, you cannot get the chemical penetration achieved in a commercial, pressurized cylinder. But one of the best-known and most thorough books on pole building, Ken Kern's *Owner-Built Pole Frame Home*, is surprisingly flawed with pages of do-it-yourself advice on embedding two 55-gallon drums welded together for a "processing center," soaking some species of pole for 8 days (on a small, 12-pole house that's over 3 months on the job just for soaking), even rigging a heat source under

See wood preservative note on page 360.

POLE HOUSES   **186**

See wood preservative note on page 360.

the oil drums to fire up the 100 or so gallons of creosote or penta mix to 180° F.

As long as you are confident about the engineering details, I am happy to have you dig the holes by hand, pour the concrete pads, embed the poles, backfill, bolt on the girders—to do just about any job in the construction sequence. There is enough hard work without trying to save a few dollars by ordering raw poles, cooking up 100-gallon batches of chemical preservatives, and dipping the unwieldy poles, which can weigh hundreds of pounds, into the brew.

When you buy building poles, you may find three available types of preservative treatment: creosote, pentachlorophenol (penta), and waterborne salts. The salt-system poles are normally dry to the touch, paintable, and relatively odorless. Penta may be painted, but you should tell the pole manufacturer in advance that you intend to paint. Sometimes the surface of penta poles is too oily. When penta is carried in a light petroleum solvent, you get an oily smell until the petroleum vaporizes. Penta carried in heavy oil retains both the odor and the oily surface, and cannot be painted. Creosote, commonly used on farm and utility poles, produces a familiar, smoky-smelling, oily finish that does not hold paint. While ideal for platform poles cut off at the first-floor deck, creosoted poles inside the house walls may make the interior air a little stiff. For a preview, get up close to a big telephone pole on a hot summer day and take a whiff.

The table at the bottom of this page lists preservatives and their uses as specified by the Southern Forest Products Association (SFPA).

To simplify matters, table below, right, indicates how specifications for preservative retention in the wood vary for the four recommended treatments in the three wood species commonly used for building poles.

Here's one final reminder. Industry literature, which you might think would tend to minimize the question of toxicity, does not. "Hazardous" and "toxic" are words that appear on a regular basis. Exposing untreated wood on the job—for example, by drilling through poles

CHEMICAL RETENTION IN COMMON POLE SPECIES

	Chemical Retention (lb./cu. ft.)		
Treatment	Southern Pine	Red Pine	Douglas Fir
Creosote	9.0	13.5	12.0
Pentachlorophenol	0.45	0.68	0.60
Waterborne ACA	0.6	0.6	0.6
Waterborne CCA	0.6	0.6	0.6

CHEMICAL RETENTION BY ASSAY OF TREATED WOOD
(lb./cu. ft.)

Oilborne preservatives	Utility poles	Building poles	Trade names of common preservatives
Creosote	7.5	9.0	
Creosote-coal tar	7.5	NR	
Creosote-petroleum	7.5	NR	
Pentachlorophenol	0.38	0.45	
Waterborne preservatives			
Acid copper chromate (ACC)	NR	NR	*Celcure**
Ammoniacal copper arsenate (ACA)	0.6	0.6	*Chemonite**
Chromated copper arsenate (CCA)	0.6	0.6	*Greensalt, Boliden CCA*, Koppers CCA-B, Osmose K-33*, Chrome-Ar-Cu CAC*, Langwood*, Wolman CCA*, Wolmanac CCA**
Chromated zinc chloride (CZC)	NR	NR	
Fluorochrome arsenate phenol (FCAP)	NR	NR	*Osmosalts*, Tanalith, Wolman Salts FCAP*, Wolman Salts FMP**

Note: NR indicates treatments not recommended by the SFPA.
*Proprietary names registered with the U.S. Patent Office.

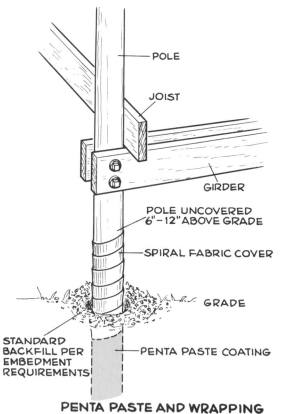

POLE

JOIST

GIRDER

POLE UNCOVERED
6"–12" ABOVE GRADE

SPIRAL FABRIC COVER

GRADE

STANDARD
BACKFILL PER
EMBEDMENT
REQUIREMENTS

PENTA PASTE COATING

PENTA PASTE AND WRAPPING

See wood preservative note on page 360.

The penta paste system at grade does offer extra protection, because wood is most likely to deteriorate (from water and insects) just below, at, and just above the soil line. Except for very special circumstances (and always with the utmost caution), I would skip the paste treatment and stick with pressure-treated poles.

Surface treatments, even "penetrating" stains and sealers, will take a beating on a rough, exposed pole-and-girder frame beneath the house. Theoretically, if you renewed the surface treatment regularly every fall, that would suffice, but that's no bargain compared to maintenance-free pressure-treated poles, which offer better protection in any case.

for bolt holes—may require you to apply small amounts of concentrated preservative on the job site by hand, i.e., not in a pressurizing cylinder. Do so with the utmost caution, wearing protective clothing, and only after you have familiarized yourself with the chemical you are using. The wood and wood-preserving industry associations, which are listed in the "Information Sources" at the end of this section, and the EPA in Washington can provide you with any further details you require.

A particularly caustic penta paste (trade names for the substance include Osmoplastic and Pol-Nu) has demonstrated good results over raw and pressure-treated poles as a ground-line treatment. As most decay and insect attacks occur within the first foot of topsoil, the idea is to apply the paste liberally in those areas—say, 18 inches below and at least 6 inches above grade—then cover it with a spiral-bound waterproof building paper, all before the final stage of backfilling. This relatively inexpensive but chemically powerful penta paste may add 15 to 25 years to the life of your building poles. Again, apply only with the utmost caution.

EMBEDDING

Sure, you can dig the holes by hand. But if you can afford it, hire a mechanical auger first, a backhoe second, and buy a shovel third. A power earth auger bores the most efficient hole and disturbs the least amount of soil. But your building layout has to be right on the button. Because a backhoe is articulated much like a human arm, it needs a little maneuvering room. It moves more dirt than an auger, leaving a larger hole that you can work in, and a series of holes with room in each one to accommodate layout discrepancies, so you can set poles to an absolutely straight string line. Hand-digging with pick and shovel, or even with a posthole digger, is relatively easy in clear soil for the first few feet. After that, it gets harder to keep the walls of the hole vertical and to get the dirt up from the bottom of the hole. Digging in rocky soil is always a struggle. And digging deep holes where you must lift dirt to shoulder height and above is very taxing, even if you are accustomed to digging holes.

As a carpenter first and general contractor second, my prejudice in favor of wood makes me associate digging holes with drainage work and gardening. Maybe I am just eager to get on to the part of the job I like best, the framing. But concrete and foundation work have always struck me as odd combinations of engineering technicalities (in concrete, absolute technicalities) and your basic, grunt-and-groan labor. At the other end of the job, the sheathing, siding, roofing, and all the closing-in work always struck me as matter of fact; a little like doing an oil painting by the numbers. The heart of the structure, the essential ingredient for every successful finishing and closing-in operation, and the most interesting part of the job (even with its own set of engineering technicalities), is the framing. If this rings a bell with you, you'll love pole building because most of the job (after the "gardening") is pure framing work.

If you can't locate a pole-building contractor (there are some firms that specialize only in pole erection and then let someone else finish the house), try a commercial sign contractor. Many of them have trucks equipped with mechanical augers, just like phone-company trucks.

Whoever does the work, and however neatly the holes are dug, always leave at least 6 to 8 inches of extra room for backfilling. And no matter what the tables in this or any other book specify about pole-embedment depth, make sure you get below the frost line. Although national maps showing average and even maximum frost-penetration depths are rampant in how-to books, I've left them out on purpose because I'd rather have you rely on local information—from builders, building inspectors, architects, engineers, and farmers. The maps look good, but the thickness of an ink line representing an additional foot of required depth at map scale may be several counties wide in real scale. And frost depth is too critical to trust to the thickness of an ink line.

FOOTINGS

To lay a concrete pad at the foot of each pole hole, use this rough formula to establish pad size: Thickness should be equal to at least one-half the pole diameter. Then use the preceding design tables covering flat and sloped sites with platform and full-height poles to find the required pad diameter, marked "D" on the tables. This combination should, with proper backfilling, prevent undue settling.

Some pole builders set 12-inch lengths of rebar into the pole butts (4 inches in, 8 inches protruding) and position the spiked pole in wet concrete. This does give you a chance to coax the pole into proper alignment. But with too much coaxing, and with the rough handling that can be expected when you try to get the better part of a tree to stand up in a small hole in the ground, the pole may wind up settling through the wet concrete pad instead of resting on top of it.

A more reasonable approach is to pour a line of concrete pads, then set rebars or conventional J-bolts (anchor bolts used in full concrete foundations) into the concrete-pole pads. Aligning these small pieces of

A simple concrete footing, called a punching pad, poured, without forming, below the frost line, in the bottom of a hole and with undisturbed soil at the base, should be all you need for vertical support. Compacted backfill applies lateral pressure that helps to carry loads and acts against withdrawal.

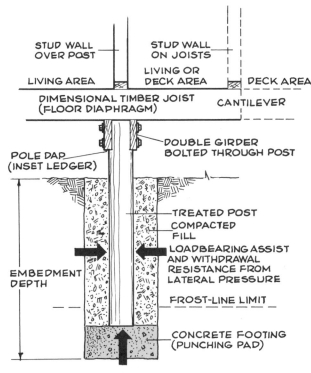

POLE FOUNDATION PLAN

12-POLE LAYOUT

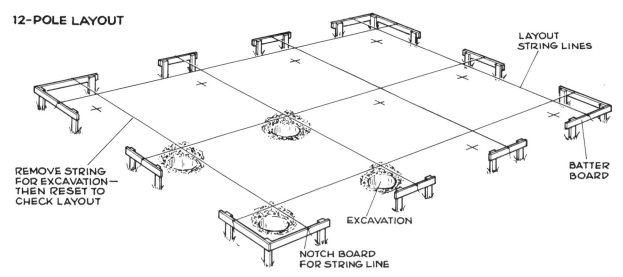

LAYOUT STRING LINES

REMOVE STRING FOR EXCAVATION—THEN RESET TO CHECK LAYOUT

BATTER BOARD

EXCAVATION

NOTCH BOARD FOR STRING LINE

You can lay out a pole frame with the same tools and techniques used on conventional construction. But instead of concentrating on four corners of a perimeter foundation, you must set up batter boards so that layout lines will intersect every pole in the grid. Build them solidly; then you can remove string lines during excavation and have confidence in the layout points when you restring the grid.

Modest-sized poles can be wrestled into place by hand—well, maybe by four or six hands. Owner-builders, particularly on remote sites, have rigged gin poles (as shown), temporary tripods, even lift-bed pickups to get a little mechanical advantage. Poles are heavy. If you have taken some trouble to make excavation depth uniform, stake a 2 × 4 on the flat so the pole will ride into the hole smoothly without dumping a foot of loose fill in first as it crumbles the soil at the edge of the hole.

Within reason you do not have to be concerned about elevation as the poles are set. But vertical alignment is crucial. Use 2 × 4s, not 1-inch braces, and 2 × 4 ground stakes. Clamp everything first. Check and recheck plumb. Check alignment against a string line. Tap, push, shove, and nudge with the poles tightly clamped. Then, finally, nail the braces in place.

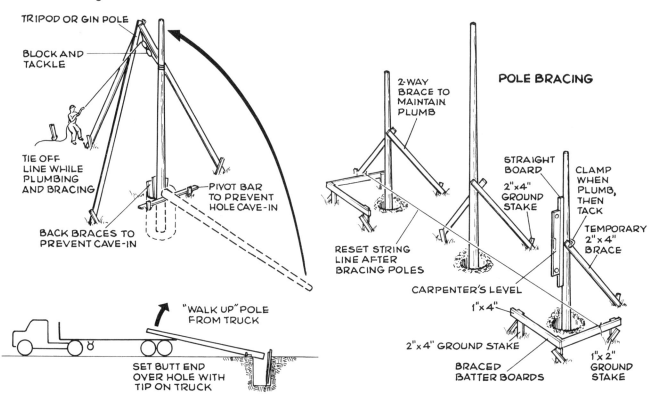

TRIPOD OR GIN POLE

BLOCK AND TACKLE

TIE OFF LINE WHILE PLUMBING AND BRACING

BACK BRACES TO PREVENT CAVE-IN

PIVOT BAR TO PREVENT HOLE CAVE-IN

"WALK UP" POLE FROM TRUCK

SET BUTT END OVER HOLE WITH TIP ON TRUCK

POLE BRACING

2-WAY BRACE TO MAINTAIN PLUMB

STRAIGHT BOARD

2"x4" GROUND STAKE

CLAMP WHEN PLUMB, THEN TACK

TEMPORARY 2" x 4" BRACE

RESET STRING LINE AFTER BRACING POLES

CARPENTER'S LEVEL

1"x 4"

2"x 4" GROUND STAKE

BRACED BATTER BOARDS

1"x 2" GROUND STAKE

hardware is a lot easier when they do not have 20-foot, 350-pound poles attached to them. When the concrete is set up, you can bore pilot holes in the pole ends, flood the holes with preservative, and lower the poles onto the steel rods.

I must say that most of the pole buildings I've worked on have been built with poles on concrete without any steel hooking them together. A pole with a 12-inch diameter at the butt end, for example, requiring a pad 6 inches thick and 16 to 18 inches in diameter, can safely carry some 25,000 pounds of building load. This setup at each pole in a small, 12-pole house would provide a total capacity for building loads of 300,000 pounds—about what you would need for a home with armor plating on the roof.

My experience is that conscientious backfilling (and proper pole design for strength, of course) is adequate provision against lateral, or sideways, forces. Obviously, a steel pin cannot help to support vertical loads. With the simple pole-on-pad design, you don't begin the pole's life with a 4-inch hole in its base, and you don't run the risk of bending over rebars or break-

ing the bar from the pad as you wrestle the pole into place.

For extra embedding strength without a punching pad below the pole, you can use a concrete necklace, poured 12 inches thick in all cases, to the diameter "D" suggested in the pole design tables, and encasing 4 or 5 lag bolts protruding from the pole, all just below the frost line. With this necklace system, you set and plumb the poles and backfill to a foot or so below the expected frost line, then widen out the hole, pour the necklace, and complete the backfilling after the concrete is cured.

Another embedding option is to encase the entire pole below grade in concrete. This is relatively expensive and requires a lot of time and material for bracing against the force of a continuous pour from a ready-mix truck. A less-expensive wrinkle on the all-masonry backfill method is to use a five-to-one soil cement, which requires a lot of mixing on the job site, or dense pea gravel, or even clean sand. If available locally, clean sand, flooded with water and backfilled in stages, can result in very firm, 100-percent compaction when the surrounding soil offers good drainage.

It is risky to use one braced pole as a deadman that draws or pushes another pole into plumb. Some of the drawing and pushing is likely to wind up in the first plumb pole no matter how well it is braced. Leave bracing in place as you connect poles with girders and plates. Continue to check plumb as you build. Remove the braces only when it is impossible to continue building with them in place.

GRID READY FOR CONNECTION

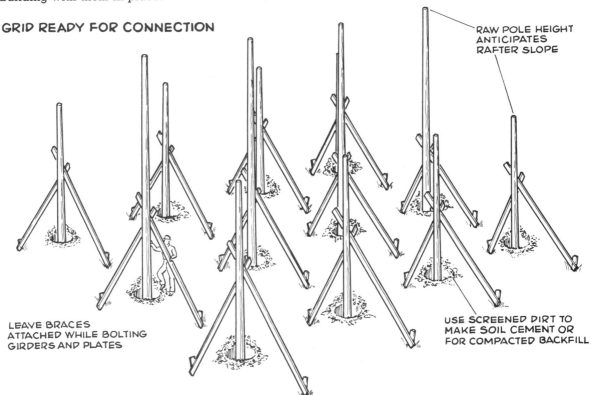

RAW POLE HEIGHT ANTICIPATES RAFTER SLOPE

LEAVE BRACES ATTACHED WHILE BOLTING GIRDERS AND PLATES

USE SCREENED DIRT TO MAKE SOIL CEMENT OR FOR COMPACTED BACKFILL

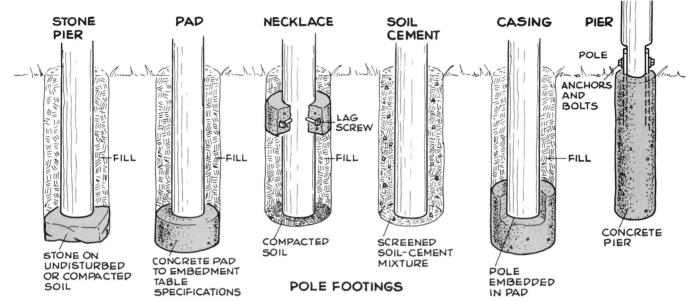

STONE PIER	PAD	NECKLACE	SOIL CEMENT	CASING	PIER

POLE

ANCHORS AND BOLTS

LAG SCREW

FILL — FILL — FILL — FILL

STONE ON UNDISTURBED OR COMPACTED SOIL

CONCRETE PAD TO EMBEDMENT TABLE SPECIFICATIONS

COMPACTED SOIL

SCREENED SOIL-CEMENT MIXTURE

POLE EMBEDDED IN PAD

CONCRETE PIER

POLE FOOTINGS

Six types of pole foundation systems: (1) a wide, flat, sound stone; (2) a concrete punching pad; (3) a necklace (set below the frost line) with compacted fill below; (4) a full embedment of soil cement compacted and set in stages; (5) a concrete casing acting as a punching pad with additional withdrawal resistance; and (6) a poured or precast concrete pier set on sound, compacted soil and reaching to or slightly above grade, where strap hardware connects pier to pole.

Unless you look closely, different foundation systems have no effect on appearance. On the beach house below and at right, poles sit on concrete piers at grade. The pole bases are dapped (chamfered) to provide bearing for strap hardware. On the beach house shown on the next page, poles are jetted down into the sand.

Unless you are building the only pole structure ever seen in your area (and that is a remote possibility), temper all the technical information with firsthand reports from local builders, your county extension agent, and other housing pros—practical reports on soil condition, and successful pole installations.

And remember not to waste time precisely leveling the holes and the concrete pads (or other embedding materials). Small differences in elevation from hole to hole don't show up until you get all the way to the pole tips, which, in almost all pole designs, are trimmed flush with the rafters. This acceptable lack of precision on concrete and foundation elevations is a real bonus of pole building and gets you out of the holes in a hurry.

It is a luxury to concentrate on squaring up the pole grid without the added consideration of elevation. Order poles in increments (plus at least a foot of waste) to account for roof pitch, and use a water level or a practiced eye to maintain uniform, if not identical, excavation depth at each pole hole.

TERMITE SHIELDS

With all the information about chemicals and pressure-treating, this may seem like a superfluous issue. It would be, too, if every stick of wood in the house was treated as thoroughly as the poles. But that would cost you quite a bundle, as you can imagine: Wolmanized

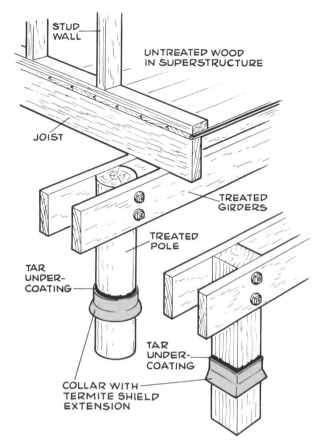

STUD WALL

UNTREATED WOOD IN SUPERSTRUCTURE

JOIST

TREATED GIRDERS

TREATED POLE

TAR UNDER-COATING

TAR UNDER-COATING

COLLAR WITH TERMITE SHIELD EXTENSION

Pole homes are not immune to termites just because they are built on pressure-treated poles. Termites have been around for thousands of years, and any wee-brained creature that has managed that is smart enough to crawl up a pole to an endless free lunch.

kitchen cabinets, penta-sealed chair legs. Termites may not munch their way up treated poles to get to that nice raw T&G decking. They may use the poles as a highway, the way they use concrete foundations in conventional homes, as a route from the nest and necessary ground moisture to the food source—your home.

On round poles, effective termite shields should have a 2-inch rim extending down toward the ground at roughly a 45° angle, secured to the poles in beds of tar. Even minuscule openings (like 1/32 of an inch) permit access for the insects. Regular coatings of termite repellants approved by the EPA for above-ground use will help. Soil poisoning with chlordane and other powerful treatments also is a tempting measure. Don't do it. There are just too many possible problems with water source contamination, children and pets contacting the soil around the poles, and more.

Pole buildings in the extreme southern, southeastern, and southwestern areas of the United States must have very thorough protection against flying termites, including pressure-treated timbers on all exterior surfaces and a combination of screening and caulking to prevent access to framing bays, crawl spaces, and any other edible parts of the house.

The poles below and at the bottom left of the next page were set by Pole House Kits of California. The poles rest on piers and are locked in with strap hardware. Two-way braces are double-staked to the ground and pinned where they intersect. Note that many diagonal braces are still in place as the poles are dapped (mortised) for double girders and locked into a platform.

BRACING

Properly embedded poles are strong enough to stand on their own. You can think of the depth of embedment as compensating for the root system on a tree in the forest. And those trees stand very well without brackets and concrete and steel pins and soil cement. But on less-than-ideal sites (and there are more and more of them compared to excellent natural sites), bracing between poles can reduce embedding requirements. If you hit rocky soil, you can use timber knee braces or diagonal threaded rod between poles to reduce embedding requirements to the minimum, which is 4 feet deep. If you hit bedrock, a 3-inch-deep pocket cut into the ledge will suffice.

Knee braces, generally made of 2×6 treated timbers, can be bolted to run between the pole and the double floor girders, as a hypotenuse completing the triangle. Threaded rod (⅝ to ¾ inch in diameter), set with beveled washers to compensate for the diagonal angle, also provides extra stiffness and resistance to lateral forces such as wind loads on the house.

Shear walls, continuous floor diaphragms, concrete keywalls, and other specialized types of construction can be used alone or in combination to overcome embedding problems. But when these methods are required, you should seek professional help with the design, even if you are an experienced builder. Wall studs, interior partitions, and generally any framing that is used to fill in spaces between poles cannot be

On modern pole barns, the first 2×6, T&G plank (above) is called a kickboard or splashboard and establishes the elevation for the building. Generally, the first one or two planks are pressure-treated material, even if the girt frame is covered with siding. Below, braces are still in place on this horse barn built by Brescia Pole Builders in Montgomery, New York.

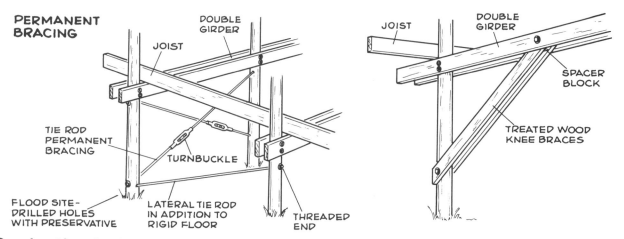

PERMANENT BRACING

DOUBLE GIRDER

JOIST

DOUBLE GIRDER

JOIST

SPACER BLOCK

TIE ROD PERMANENT BRACING

TURNBUCKLE

TREATED WOOD KNEE BRACES

FLOOD SITE-DRILLED HOLES WITH PRESERVATIVE

LATERAL TIE ROD IN ADDITION TO RIGID FLOOR

THREADED END

On pole grids with substantial unconnected spans, above-grade diagonal wind bracing adds greatly to frame stiffness. Bolted, treated timbers work well, although most architects specify threaded rod with turnbuckles. In the photo below, you can see the steel lines if you look for them, but the vertical lines of the poles are not challenged by bulky diagonals and cross-hatches, as is the case if you use wood bracing.

counted on to hold the poles together as a unit against loads. (I am reminded again of the pole house on the Galveston waterfront with all the stick framing blown away by hurricane forces.) Continuous floor girders bolted along an entire line of poles, with wall plates and ridgepoles connecting continuous lines of poles at right angles to the floor girders, do tie the pole grid into a single, strong unit.

GIRDER AND WALL PLATE CONNECTIONS

Tenpenny nails, the standard residential fastener, are enough to hold stick frames together. And conscientious stick-frame builders will throw sixteen-penny nails into built-up girders. But on pole structures where you want to connect a 2 × 12 to each side of an 8-inch diameter pole (a structural sandwich 11 inches thick), using conventional nails is like using toothpicks.

You want pole fasteners not only to resist shear, or the downward loads that would snap the nail in two, leaving half in the pole and half in the 2 × 12, but also to resist withdrawal, or the sideways loads that would pull the 2 × 12 and the nails away from the pole. Also, you want a fastening system that can accommodate both the round surface of a building pole and the flat surface of a dimensional timber like a 2 × 12—an unusual match not found in ordinary structures.

Shear resistance for nails and bolts of equal diameters is the same. In other words, it takes the same force to break two equal pieces of steel no matter what

The inelegantly nailed wall plate on this pole barn (six sort-of-staggered and sort-of-spaced nails) will be structurally faced with barn siding—still no substitute for bolts. Notice the gang-nail plate to the left of the pole holding truss members together.

shape they have. And you can buy huge spikes, like 10 and 12 inches long, with ⅜-inch and greater diameters. Even big nails, however, cannot match the withdrawal resistance of bolts, washers, and nuts. If your poles or dimensional timbers are relatively green (unseasoned), nails can lose half their holding power when the wood dries. Threaded, ringed, and spiral-pattern nails can help, but not enough.

For holding power, good-better-best means common galvanized nails (hot-dipped if you can find them), increased holding with annular-thread common nails, and even more holding with spiral thread nails. Naturally, more holding power costs more money.

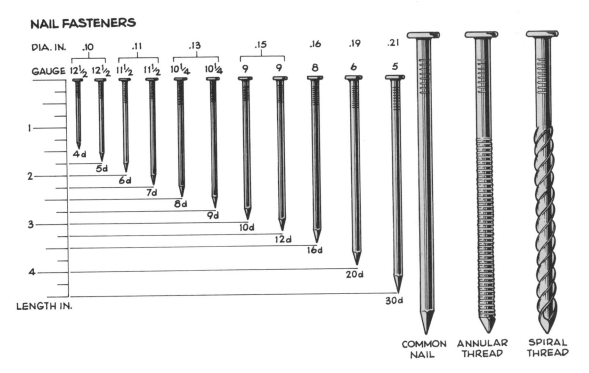

NAIL FASTENERS

COMMON NAIL   ANNULAR THREAD   SPIRAL THREAD

HARDWARE FASTENERS

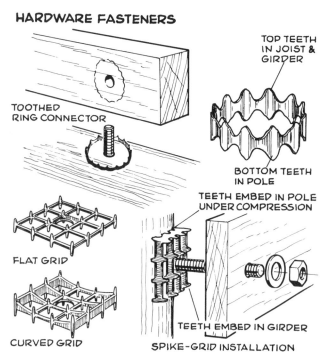

TOOTHED RING CONNECTOR

TOP TEETH IN JOIST & GIRDER

BOTTOM TEETH IN POLE

FLAT GRID

CURVED GRID

TEETH EMBED IN POLE UNDER COMPRESSION

TEETH EMBED IN GIRDER

SPIKE-GRID INSTALLATION

I have built many pole buildings, some of my own design, some designed by engineers, and I have seldom used connectors other than ⅜- or ½-inch diameter bolts. On one very large barn I did use split ring connectors (shown above) on truss sections, which were delivered to the site in halves because even half the truss span was an extra-long trailer load. Spiked grids do offer exceptional holding and some vertical support as well, but on most designs they are needed only for one or two exceptional structural connections.

Bolts are the obvious answer for connections between poles and floor girders, and between poles and rafter pairs or the wall plates the rafters will rest on. The only drawback (aside from the cost of bolts) is the size of the bolt holes in the timbers. Two ¾-inch-diameter holes in a 2 × 12, for example, do create problems. Like a hole in the ice on a pond, these perforations attract stress. Does a piece of 2 × 4 decking split along the end between the two common nails that hold it in place? No. It splits in a line along one or both of the nail holes.

There is some specialized hardware, the kind you won't find in the local hardware store, designed to overcome these problems. Toothed-ring connectors and spiked-grid connectors both isolate the bolt hole from shear stress. Stress lines attracted to the bolt hole are deflected by the toothed ring and absorbed by the spiked grid, so that the load cannot concentrate at the hole. Neither connection system requires special installation tools. The teeth in both systems are forced into the surrounding wood as the dimensional timber is drawn against the pole under the pressure of a washer and nut.

For excellent resistance to shear and withdrawal, a single, 2½-inch spiked connector has the holding power of about 5½-inch-diameter bolts and some 15 sixteen-penny nails. And they would make a mess out of any dimensional timber. Teco, Simpson, and some of the other fastener manufacturers offer many specialized connectors, including heavy-duty joist hangers and strapping to resist wind lifting, that are an ideal choice instead of nails.

One of the most suitable fasteners is a curved spike grid. The curved side makes an improved bond against a round pole, while the flat side embeds in a flat dimensional timber. A bolt is fitted through a ring in the center of the grid. To get full penetration of the teeth on each side of the grid, it is good practice to use a high-strength threaded rod and a multiple set of washers with increasing diameters to squeeze the timber-and-pole sandwich together. You might try an auto-parts store that handles foreign cars and get yourself a few lengths of metric, fine-thread rod. The fine threads transfer a little more torque applied from an impact wrench or simple box end wrench, and thus minimize thread stripping.

A single ¾-inch bolt used with a curved spike grid can carry about 3,800 pounds; a 1-inch bolt, about 4,100

Pole builders who have access to an ironworks can take advantage of all kinds of specialty connection hardware generally not available from firms such as TECO and KC Metals, which manufacture joist hangers and the like for conventional stick-built buildings. Shown at left is a curved bracket allowing bolt connections outside the pole. Above is a T-connector to join headers above a post. And above right is an arrow connector to support a valley rafter when there is no more room for bearing on top of the pole.

Dapping, or cutting a flat bearing surface into a pole where it meets a dimensional timber, adds structural value to the joint without significantly weakening the pole. Usually reserved for pole-to-floor girder connections, it makes sense on rafters, too, as shown below.

pounds. As good as these grids are, I have never built pole frames with single-bolt connections. With large boards like 2 × 12s and special-order 14-inch stock, solitary connection points just leave too much of the board flapping in the wind.

Dapping the connection between dimensional timbers and poles can take a lot of the load off bolt connections and, as an added bonus, pin timbers into the pole to resist uplift forces. Dapping is like making a shallow mortise in the side of the pole—a bearing pocket. A dap for a 2 × 12 could be 11½ inches vertically (up and down the pole) and ¾ of an inch deep, roughly half the thickness of the 2-inch nominal (1½-inch actual) timber. A cut of this depth on a roughly 8-inch-diameter building pole leaves a bearing surface between 4 and 5 inches across, which should be flooded with preservative. Two 2 × 12s, each dapped about ¾ of an inch into the sides of a pole, pinned with two ½-inch- or ⅝-inch-diameter bolts set about 2 inches in from each edge of the 2 × 12, with nuts on top of heavy-duty, 1½-inch-diameter washers—that's a structural sandwich that won't shear, won't pull away from the pole, won't sail away in a storm, and will provide a very sturdy, loadbearing deck frame. (See upcoming pages.)

HOW TO DAP AND PIN
TIMBERS TO A POLE

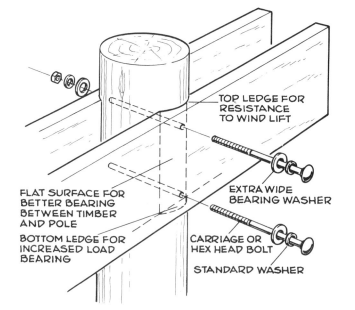

TOP LEDGE FOR RESISTANCE TO WIND LIFT

FLAT SURFACE FOR BETTER BEARING BETWEEN TIMBER AND POLE

EXTRA WIDE BEARING WASHER

BOTTOM LEDGE FOR INCREASED LOAD BEARING

CARRIAGE OR HEX HEAD BOLT

STANDARD WASHER

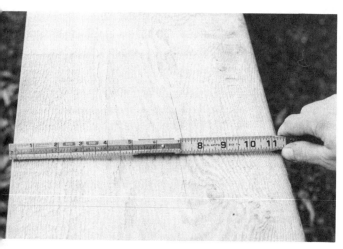

STEP 1. Verify the depth of joists, rafters, and girders before laying out a dap cut or bolt centers. This one is 11⅜ inches, a bit short of the expected 11½ inches.

STEP 2. The ½- to ¾-inch depth of the dap pocket is controlled by the circular-saw shoe. Test-cut a piece of scrap lumber, and tighten the saw depth.

STEP 3. Double-clamp a 2×4 cleat onto the pole to support the girder or joist temporarily. Minor elevation changes are made by tapping the clamped block.

STEP 4. With girder elevation established, securely clamp joists or girders onto the blocks and against the pole, not only to locate and drill bolt holes, but also to mark dap cuts. If in doubt about needed pole height or girder length, leave any excess. Trim last, not first. Note the 2×4 block spacer, under the girders, preventing deflection.

STEP 5. Sight across a ruler to measure the pole diameter, and mark center lines on both girders (joists). Center in the pole mass, not including protrusions.

STEP 6. Transfer center lines down both girders. If clamps impede marking, move them one at a time, and do not disturb the temporary elevation blocks.

STEP 7. On a 2 × 12 girder, drill bolt holes about 2 inches in from each edge to keep the girder flat. You may need a bit extension or a long-stem auger to bore all the way through the backside.

STEP 8 (SMALL POLES). Two ⅜-inch-diameter carriage bolts will secure a dapped girder. I always spend the few extra cents for a second, larger washer so the bolt head won't bite into the wood grain.

STEP 8 (LARGE POLES). On large-diameter poles, you may not be able to obtain bolts long enough for the job. In this case, drive long lengths of threaded rod through and cut to suit.

STEP 9. With clamps and the spacer block still in place, mark the top and bottom of both girders against the pole. These are limit marks for the dap cuts.

STEP 10. After removing clamps and girders, use a circular saw (as shown) or a chainsaw to cut just up to the girder limit marks for the dap cut. And yes, it is a little difficult to keep the circular saw shoe square to the pole.

STEP 11. Use a level or simply sight along a piece of scrap lumber set just at the edges of the top and bottom dap cuts. These guidelines will help keep the dap edge uniform.

STEP 12. Circular saw or chainsaw? Take your pick. The main goals are to make a series of horizontal cuts

while keeping the blade square to the pole and to maintain a consistent depth of cut on successive kerfs.

STEP 13. After cutting evenly spaced kerfs by eye, clean up the dap with a hatchet. The dap can be left slightly ragged because it is concealed.

STEP 14. After flooding the area with preservative with a paint brush, tap the girder in place, then drive in the bolts and lock up. Avoid splashing preservative onto your skin.

FLOOR FRAMING

With the pole grid in place, all braced and backfilled, you will begin to get the first real sense of your pole house—the three-dimensional space of the structure and how that space lays with the land. This is an exciting part of the job, when you will be tempted to rush ahead with floor girders and joists and wall plates, explaining to anyone who will listen how you'll be cantilevering a series of 2×12s on one set of poles for a 6-foot deck outside the kitchen, how you'll be dapping the tops of poles on one side of the roof so that pairs of rafters will slope out 3 feet past the sliding glass doors.

Sometimes, when you are most anxious to proceed, you have the most to gain by pausing. There are so many framing options, so many ways to use the strength and design flexibility of the pole grid. National Design Specification (NDS) tables on upcoming pages indicate that a southern pine 2×12 joist can clear span 21 feet 6 inches and, at 16 inches on center, still provide a 30-pounds-per-square-foot live load capacity with deflection stringently limited to $\frac{1}{360}$ of the span. That's a big piece of a strong wood species, and it probably offers more span than you'll need.

Suppose you have a 12-foot-on-center pole grid, and the poles are roughly 8 inches in diameter at the elevation of the floor diaphragm. If you sandwich 2×10 or 2×12 girders bolted through the poles, there will be 8 inches between the dimensional timbers, an area that can carry copper pipe, Bx, Romex, ductwork, drain lines, and more. The 12-foot span between poles

is then reduced as follows: by 4 inches on each side of the span, representing half the pole diameter (8 inches total); by 1½ inches on each side of the span, representing the inside dimensional timber of each structural sandwich (3 inches total). The 12-foot-on-center grid connected with sandwiched floor girders now requires floor joist lumber to clear span only 11 feet 1 inch.

According to the NDS tables, this span requirement can be met with southern pine 2×8 joists, 24 inches on center, providing a 40-pounds-per-square-foot live-load capacity. With a pole grid 10 feet on center (a nice

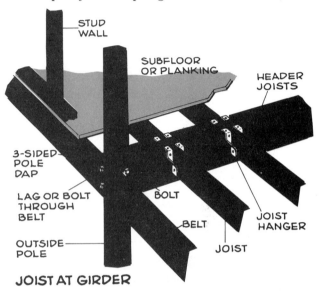

JOIST AT GIRDER

Using joist-hanging hardware, it is possible to lay pole-to-pole girders and the floor joist system in the same plane. Joists can be headed off at the pole or cantilevered beyond it.

Once girders are pinned, the floor-joist system resting on top of them can carry the exterior wall to or well beyond the pole and girder line. Joists extending past the poles in one direction can carry enclosed space, while girders cantilevered past the poles at right angles to the joists can carry decking and outdoor space.

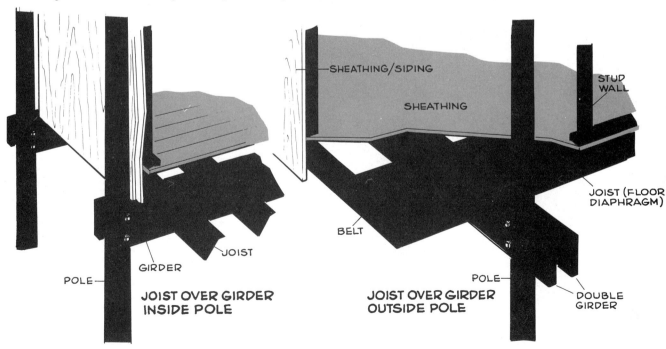

JOIST OVER GIRDER INSIDE POLE

JOIST OVER GIRDER OUTSIDE POLE

choice anticipating 8-foot sliding glass doors), the girder sandwich span reduction of 11 inches from 10-foot centers to 9 foot 1 inch, means you could use southern pine 2×6s, 16 inches on center for the floor frame.

Don't let all these numbers intimidate you. I have included a few span tables later in this section so that you can look at some of the options in detail and also see how changing any one of several factors (wood species, on-center length, timber size, live-load requirement, etc.) gives you different design options.

You might decide on double (4-inch nominal) joists suspended in the same plane as the girder system with beam hangers, covered with 1-inch structural plywood decking. You might decide to get cantilevers with the girders, or with the floor frame resting on top of the girders, or with both levels of framing.

The pole house under construction at right uses two 4-inch dimensional girders to tie the pole lines together, while the finished pole house below uses exposed 4-inch dimensional joists over hidden 2 × 12 girders. Notice how this plan accommodated one natural pole (the twisted tree) with a recess in the cantilevered deck and a hole in the roof overhang.

The NDS tables are extensive, covering different woods with differing strengths, used in a wide variety of situations. You don't have to become involved with these technicalities. You can leave them to a pole-kit manufacturer, pole-building company, architect, engineer, plan-service company, maybe even a building inspector. But, I figure the more you know, the better off you are. And if you are taking the idea of being an owner-builder from scratch, generating your own design, at some point your design will undergo professional scrutiny. If your plans call for 2×6s where an architect or engineer would specify 2×12s, most building department officials will tell you to take a walk—straight to an architect or engineer. But if you specify 2×10s for 2×12s, or take 6 inches more than allowed for joist spans, many building department officials will work with you.

The accompanying tables should help you make your specifications very close to accepted standards, if not exact. The first table includes construction-grade lumber of several framing-wood species. The characteristics of the wood are generally called design values, which are really specific performance ratings for different jobs done by the wood under different circumstances. Two of the most important, as far as spans are concerned, are: extreme fiber stress in bending, or the resistance to stretching tension in the fibers along the bottom edge of a beam loaded from above; and modulus of elasticity, technically, the ratio of stress (force per unit area) to strain (deformation per unit length)—in plain English, the timber's resistance to downward bending, called deflection, when loaded from above.

It is possible to buy lumber in different grades that produce different design values. On the negative side, this makes wood-frame design even more complicated. On the positive side, this gives you even more flexibility, depending on what species and what grades of timbers you have available. Maybe the lumberyard you deal with wants an arm and a leg for southern pine, but has common and select structural grades of another species that perform just about as well.

You have to know basic information about the wood you use for floor frames, rafters, and other parts of the structure before you can make any sense out of NDS span tables. Once you know the fiber stress in bending and modulus-of-elasticity values for the wood species and grade, the lumber size, and the framing module (normally 16 or 24 inches on center), the NDS tables indicate the maximum allowable span.

The two tables on the next two pages cover floor joist spans, first, for light-duty spaces such as sleeping areas and attic floors, where the live load is designed to be 30 pounds per square foot, and second, timbers of the same size for living areas with a design load of 40 pounds per square foot. Both tables include 2×6s, 2×8s, 2×10s, and 2×12s. The live load is stringently limited by deflection to no more than $\frac{1}{360}$ of the span. Each column of figures in these tables (like the NDS tables) reflects a different modulus-of-elasticity rating. The span limits listed below each category are coupled with required fiber stress in bending values.

Bending (fiber stress in bending) and deflection (modulus of elasticity) are distinct characteristics. There is no constant mathematical relationship between the two that applies to all wood timbers. Douglas

FRAMING LUMBER DESIGN VALUES
(CONSTRUCTION GRADE)

Wood species	Fiber stress in bending single/rep. (psi)	Modulus of elasticity (psi)	Compression perpendicular to grain (psi)	Compression parallel to grain (psi)	Horizontal shear (psi)	Green-dry shrinkage (%)
Western cedar	775–875	900,000	265	850	75	0.8
Douglas fir	1,000–1,150	1,100,000	385	1,150	95	1.7
Hem-Fir	825–975	1,200,000	245	925	75	1.2
Eastern hemlock	900–1,050	1,000,000	360	950	85	1.0
Western hemlock	925–1,050	1,300,000	280	1,050	90	1.4
Eastern white pine	700–800	1,000,000	220	750	70	0.8
Southern pine	1,000–1,150	1,400,000	405	1,100	100	1.6
Calif. redwood	825–950	900,000	270	925	80	0.9
Eastern spruce	775–875	1,200,000	265	775	70	1.1
Sitka spruce	800–925	1,200,000	280	825	75	1.4

In this pole barn at right, 2 × 10s cantilevered past two 2 × 12 wall plates on the interior pole line form a hay loft with easy access to the horse stalls below. On the elegant, hillside pole house (next page), 4-inch beams are distinctively notched, as are the major girders under the deck and over the entryway. The touch I like best is the free-standing pole extension running up through the deck next to the hot tub.

fir in one grade may have a deflection value that permits a 16-foot span for a 2 × 10 joist but a bending value that permits only a 15-foot-6-inch span. For California redwood or another wood species and grade, the deflection limit may be more severe than the bending limit. To plan safe framing details, to meet local building codes and keep the inspector happy, always work with the most limiting factor, deflection or bending. One other note: Despite all the energy-saving, forest-saving, ecologically sound advice about minimizing

timber sizes, using 1-inch instead of standard 2-inch dimensional timbers, using plywood web joists, plywood laminated beams, "engineered" frames, and all the rest of it, I still want to err on the side of overbuilding, not underbuilding.

I have renovated and torn down enough frame structures to know that even rotted, poorly nailed, termite-infested timbers have more strength than seems possible. Somehow, when all the little sticks get together into a frame, they gain extra strength, greater

FLOOR JOISTS: 30 LBS./SQ. FT. LIVE LOAD

Joist (in.)	Spacing (in.)	Modulus of elasticity (in psi)						
		900,000	1,000,000	1,100,000	1,200,000	1,300,000	1,400,000	1,500,000
2 × 6	16	8–10/830	9–2/890	9–6/950	9–9/1000	10–0/1060	10–3/1110	10–6/1160
2 × 6	24	7–9/950	8–0/1020	8–3/1080	8–6/1150	8–9/1210	8–11/1270	9–2/1330
2 × 8	16	11–8/830	12–1/890	12–6/950	12–10/1000	13–2/1060	13–6/1110	13–10/1160
2 × 8	24	10–2/950	10–7/1020	10–11/1080	11–3/1150	11–6/1210	11–10/1270	12–1/1330
2 × 10	16	14–11/770	15–5/890	15–11/950	16–5/1000	16–10/1060	17–3/1110	17–8/1160
2 × 10	24	13–0/950	13–6/1020	13–11/1080	14–4/1150	14–8/1210	15–1/1270	15–5/1330
2 × 12	16	18–1/830	18–9/890	19–4/950	19–11/1000	20–6/1060	21–0/1110	21–6/1160
2 × 12	24	15–10/950	16–5/1020	16–11/1080	17–5/1150	17–11/1210	18–4/1210	18–9/1330

Note: No span should or will exceed 1/360 deflection at these values shown in the table. For example, after the span figure 8–10 (8 feet 10 inches) is the limit, shown in italics, for fiber stress in bending (Fb) values in case you use a species that satisfies the deflection limits with a good E value, but has a poor Fb value.

than the strength of the individual parts. The modern buzzword for this phenomenon is synergism. Management types use the word a lot these days—to get more production out of a smaller work force, for example.

You might not want these plywood laminated beams on any exposed section of the pole frame, but wood engineering (including the plywood joists above the beam) can carry a lot of weight across a long span. These beams are manufactured by Truss Joist.

FLOOR JOISTS: 40 LBS./SQ. FT. LIVE LOAD

Joist (in.)	Spacing (in.)	Modulus of elasticity (in psi)						
		900,000	1,000,000	1,100,000	1,200,000	1,300,000	1,400,000	1,500,000
2 × 6	16	8–0/860	8–4/920	8–7/980	8–10/1040	9–1/1090	9–4/1150	9–6/1200
2 × 6	24	7–0/980	7–3/1050	7–6/1120	7–9/1190	7–11/1250	8–2/1310	8–4/1380
2 × 8	16	10–7/850	11–0/920	11–4/980	11–8/1040	12–0/1090	12–3/1150	12–7/1200
2 × 8	24	9–3/980	9–7/1050	9–11/1120	10–2/1190	10–6/1250	10–9/1310	11–0/1380
2 × 10	16	13–6/850	14–0/920	14–6/980	14–11/1040	15–3/1090	15–8/1150	16–0/1200
2 × 10	24	11–10/980	12–3/1050	12–8/1120	13–0/1190	13–4/1250	13–8/1310	14–0/1380
2 × 12	16	16–5/860	17–0/920	17–7/980	18–1/1040	18–7/1090	19–1/1150	19–6/1200
2 × 12	24	14–4/980	14–11/1050	15–4/1120	15–10/1190	16–3/1250	16–8/1310	17–0/1380

Note: No span should or will exceed 1/360 deflection at these values shown in the table. For example, after the span figure 8–10 (8 feet 10 inches) is the limit, shown in italics, for fiber stress in bending (Fb) values in case you use a species that satisfies the deflection limits with a good E value, but has a poor Fb value.

Safety factors built into design tables, the cautious scrutiny of building inspectors, mortgage bankers, and insurance agents, and other factors do tend to make modern residential structures somewhat overbuilt. I'm still happy to spend the extra time and money at this end of the job to ensure strength and durability and to save time and money on maintenance and repairs as the structure ages.

With a grid of poles in place, you can establish the floor line at any elevation. For protection against water and insects, 18–24 inches of clearance from wood to soil is a sensible minimum. But you can allow more room for such concerns as floods or tidal action, access to mechanical systems, storage of a boat or car.

One of the most satisfying floor details I've worked on used 2×12 girder sandwiches, 24 feet long. Roughly 16 feet of the girder spans was pinned to the 8-foot pole grid, and 8 feet of the girders was cantilevered outside the grid over a sharply sloped grade. The cantilevered porch was about 8×24 feet, about 1 foot off the ground at one end and 8 feet off the ground at the other end. I put solid bridging 2 feet on center inside the four girder pairs, left the outside girders 1½ inches longer than the rest, and headed the cantilever with a 2×12 belt, which pinned the free-floating ends of 2×12s; 2×10 joists, set in beam hangers 24 inches on center, at right angles to the girders, completed the cantilever frame; 8-foot 2×4s from the house wall to the 2×12 belt completed the decking.

Aside from the unexpected, strikingly handsome floating effect, the large outdoor platform was rock-solid. I was taught not to exceed the ⅔-to-⅓ cantilever principle (⅔ pinned, ⅓ free-floating). Some how-to literature makes much of counterbalanced loads, how loads past the point where timbers are bolted to the poles equalize and counteract forces on the other side of the pole. Some suggest 1-inch-wide joists and cantilevered sections as great as the secured span (half pinned, half free-floating). But in addition to learning about principles of carpentry, I was also taught to build for real people, who perceive structures by the way they work and feel rather than by the way they match up with some idealized, minimal-material, textbook design.

Fifty-percent cantilevers may be safe to stand on, but they flex so much they do not feel safe. Noticing the deflection and flexing in a floor is like noticing the shimmy on a car that needs an alignment. You may know that the car is safe, that you may be adding a little wear and tear to the tires, but that the car will turn when you turn the wheel and stop when you put on the brakes.

Once you notice the flex of a floor, the absence of that rock-solid feeling may make you ill at ease every time you walk on it. If you notice that the risers on a set of steps are particularly high or particularly low, so that you have to make a conscious effort to adjust your stride, the design is inferior, certainly unsettling,

Platform-height poles offer a solution for conventional house designs stuck on unconventional or problem sites. Once you treat the poles like a conventional masonry foundation and span them with a girder, standard joists, belts, bridging, insulation, subfloor, finished floor, and full stud walls can be built over or cantilevered beyond the pole line.

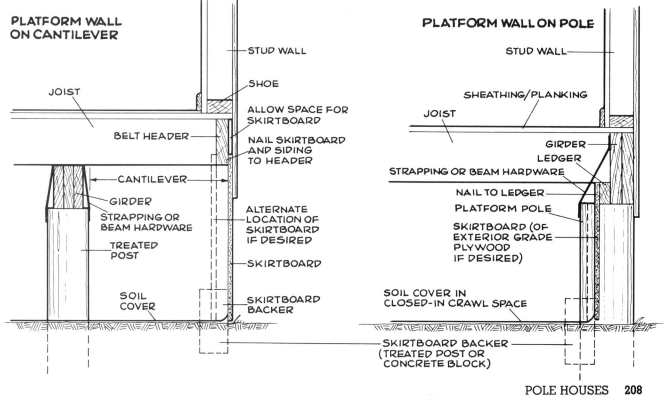

even unsafe. The more you notice the treads and risers, the less likely you are to notice where you're going. As a designer, on your own house or someone else's, that's not a situation you want.

Between the lines here, I am suggesting that you follow design tables, observe building codes, and seek professional help when you need it, but also design and build with your common sense. Use all the technical information you can get to help you build human-scale houses—houses built for people, not technical tables.

WALL PLATES AND RAFTERS

Wall plates work the same way girders work on the first-floor frame. Run either parallel or perpendicular to the first-floor girders, double plates (like sandwiched girders) are common. The outside plate may be set first to establish the rafter elevation, with the second part of the sandwich set after the fact. Like the first-floor frames, roof frames can be made with all kinds of different timbers in many different configurations.

One of the most basic options is to treat the pole frame roof like a conventional roof. Run two ridgepoles (one on each side of the midline of poles in the grid), two wall plates (sandwiching the outside line of poles),

Pole-frame cantilevers offer many options: first, a roomy exterior corner by projecting girder past pole and joist past girder; second, the somewhat more conventional effect of interior flooring reaching out past the walls; and third, a special corner for a hip roof with exterior decking inside and outside the pole line.

then set rafters on top of the ridge and on top of the plates 16 inches on center. That's one way to put a very familiar roof on an unfamiliar frame. Here are a few of the other options:

Purlin on rafter. This is the way most farm buildings are framed. Pairs of rafters bolted through poles run from the high line of poles (the ridge) to the low line of poles (the exterior walls). Purlins, usually 2×6s, are set on edge, 16-24 inches on center, depending on the span between poles, on top of, and at right angles to, the rafters. Any conventional roofing (plywood, tar paper, and shingles, for instance) can be used on top of the purlin system.

Below, cleats can be used to set and level outside wall plates as well as intermediate and ridge plates that tie individual pole lines together. Full-length or lapped rafters tied to the poles across the girders complete the grid. Photo right: Double wall plates can be level to bear against trusses, but they must be offset to match the pitch of solid rafters.

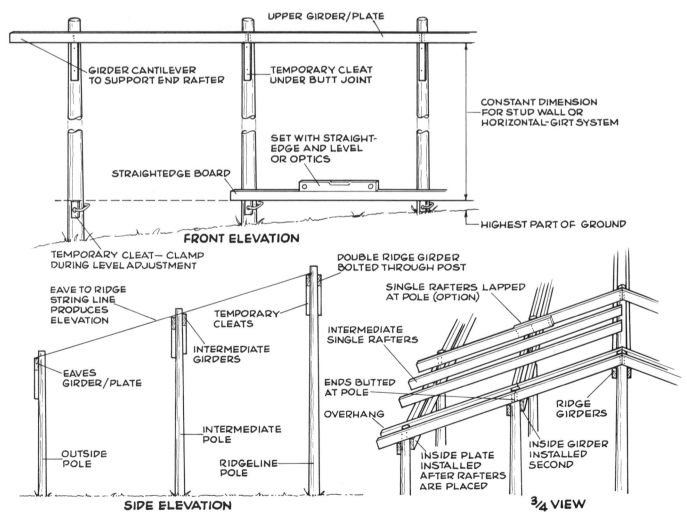

POLE HOUSES **210**

These two pole barns show different roofing options. At left, overlapped 2 × 4 purlins on edge carry plywood. Above, 2 × 4s on the flat hold a stiffer material, corrugated sheet metal, which is the classic barn roof.

Purlin in rafter. This roof frame is similar except that the purlins are set in the same plane as the rafter pairs. Since the rafter sandwich does not leave much room for end-nailing, hangers can be used to pin the purlins in place. This framing makes the roof diaphragm thinner. It also works with any conventional roof covering.

Structural hardware can save a layer of material on the roof as well as on the floor. Depending on roof pitch and load, you can put up an exceptionally economical roof with 8-foot 2 × 6 purlins on edge in brackets 2 feet on center.

Pole and truss. Large interior spaces, free of interior structural supports, can be framed with trusses. Even 60-foot spans can be brought in by truck and assembled on the site. Fully dapped trusses bear directly on the poles bordering the grid and are commonly bridged with a system of purlins 16 or 24 inches on center. Inelegant trusses demand a finished ceiling and create a considerable amount of dead attic space.

PURLINS IN RAFTER PLANE

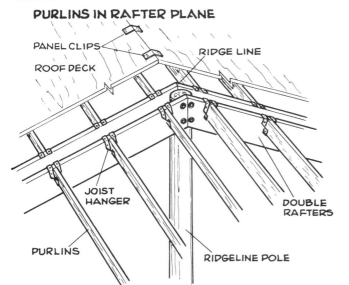

PANEL CLIPS

ROOF DECK

RIDGE LINE

JOIST HANGER

DOUBLE RAFTERS

PURLINS

RIDGELINE POLE

The section drawing of one of Pole House Kits of California's models shows two mini-trusses joining to make one very large modified truss. Inside, you get lofty space and living areas defined by furnishings instead of stud walls, shown in photo.

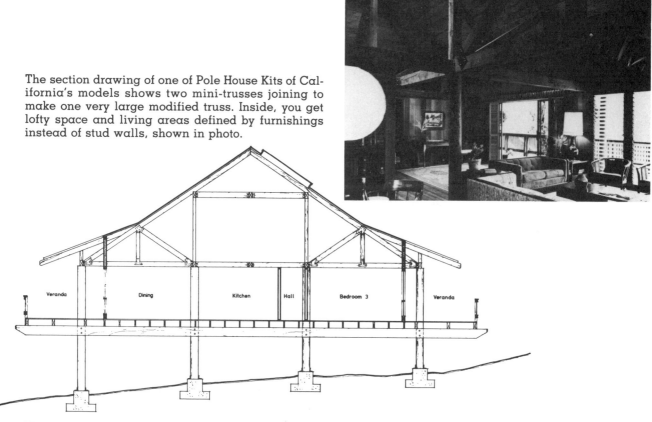

Veranda   Dining   Kitchen   Hall   Bedroom 3   Veranda

This sawtooth truss creates a unique and practical design for a barn. On the uphill side, the truss angles up to the ridge, leaving access to a loft. Downhill, a large overhang has no perimeter post obstruction. Wonder how they got those double plates woven into the truss webs? Only one way: after the trusses were in place, and before the siding went on.

Girder and deck. With a small-enough pole grid, like 8 feet or less on center, it is possible to use sandwiched girders for the floor frame, sandwiched rafters for the roof frame, with no other rafters, joists, bridging, or purlins to fill in the spaces. Structural 1-inch plywood will suffice at 4-foot center supports, but at 5-, 6-, 7-foot and over centers, heavy timber decking is required. Some planking is about as strong as many framing systems—for instance, a 4-inch-thick select structural Douglas fir with a double tongue-and-groove locking system to keep the boards together. When heavy timber decks span from girder to girder, all additional framing is eliminated.

When you get to the point of setting plates and rafters, you will likely be working amidst a pole grid cluttered with 2×4 bracing. As you carry in materials, as the wind blows, somehow some of those carefully plumbed poles will be a bit out of plumb by the time you get ready to lock them into a frame. Reset braces where required. Take your time. Don't drill bolt holes first and check with the level second. When you establish the elevation for wall plates, for example, set the outside plate on a 2×4 cleat securely clamped to the pole. Do the same thing at each end of the girder. Also, clamp the girder to the poles. Then play devil's advocate and check everything with the level (a water level or transit over long spans) all over again. If you have enough clamps, you can leave the wall plate and set to work on the ridge girders. You can safely lock up poles aligned in one direction; you know they are all plumb; you know you have the same dimension between the poles at grade as you have at the plates. Lock up another line—along the ridge, for instance—

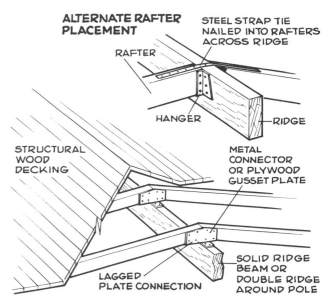

ALTERNATE RAFTER PLACEMENT

STEEL STRAP TIE NAILED INTO RAFTERS ACROSS RIDGE

RAFTER

HANGER

RIDGE

STRUCTURAL WOOD DECKING

METAL CONNECTOR OR PLYWOOD GUSSET PLATE

SOLID RIDGE BEAM OR DOUBLE RIDGE AROUND POLE

LAGGED PLATE CONNECTION

DECKING OVER RAFTERS

Solid 4×8s or 4×10s, 4 feet on center, butted against or joined just over the ridge, form a beautiful roof when decked with T&G planking. Insulation and shingles sit on the deck, leaving a pure wood interior.

Overhangs at the gable end can be made by reducing the depth of the outside rafter so that it will carry cantilevered purlins (shown below) to an outrigger rafter. Or wall plates (below left) can easily be extended one-third of their length to carry one or more rafters outside the end wall.

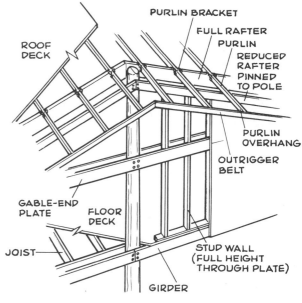

PURLIN BRACKET

FULL RAFTER

PURLIN

REDUCED RAFTER PINNED TO POLE

ROOF DECK

PURLIN OVERHANG

OUTRIGGER BELT

GABLE-END PLATE

FLOOR DECK

JOIST

STUD WALL (FULL HEIGHT THROUGH PLATE)

GIRDER

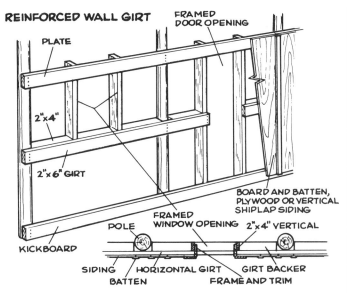

REINFORCED WALL GIRT

PLATE

FRAMED DOOR OPENING

2"x4"

2"x6" GIRT

KICKBOARD

POLE

FRAMED WINDOW OPENING

BOARD AND BATTEN, PLYWOOD OR VERTICAL SHIPLAP SIDING

2"x4" VERTICAL

SIDING

BATTEN

HORIZONTAL GIRT

GIRT BACKER

FRAME AND TRIM

Wall-girt construction can be adapted for residential construction, complete with window and door openings, although the pole-in-wall girt system is commonly reserved for utilitarian structures.

When pole frames are made of 4 × 6 timbers (common for barn frames), there is always room to straighten up a long wall run. On this barn, scrap blocks are set at the corners and connected with a string line. On any pole in between, a 2 × 4 block should just fit inside the string against the pole—an easy reference system.

when you are equally sure of pole spacing and pole plumb. You really have to be careful when you start to lock one line already pinned to a girder to another line. This is a good time to clamp rafters in place, check plumb in both pole lines, and check each set of poles against a string line to make sure they stay in line while other adjustments are made.

Extra-thick decking may be required for 40-pound live loads at large pole spans. The table on page 216 indicates how you can use readily available 2×6 tongue-and-groove decking at least for roofing, if not flooring. When you check your design against these span limits, remember that an 8-foot-center pole grid is not the same as an 8-foot span, not if you build sandwich girders which shorten the span. An 8-foot grid with good-size poles may send you to the section of the table on 7-foot spans. For 2-inch flooring across a span like that with a 1/360 deflection limit and a 40 pounds-per-square-foot live load, you would need a wood species with a modulus of elasticity of 2,725,000, almost twice the value of southern pine and Douglas fir-Larch. In other words, you might have to go for a 3-inch or greater deck material.

The 8-foot pole grid (roughly 7-foot span) on a sloped roof (3 in 12 or greater), where the load requirement is only 20 pounds per square foot (and that's common for all but heavy-snow-load areas), 2-inch southern pine or Douglas fir-Larch decking provides enough strength,

Even after wall plates and girts and rafters and decking and roofing are in place, there are a few touches for extra strength: for instance, the vertical stiffener face (below) nailed through intermediate wall plates and spiked to the rafter, and the corner brace (below right) pinned above a beveled block and face nailed through exterior wall plates.

The unusual plates in the pole house above are lapped over each other and pinned with massive bolts running through four pieces of 4-inch nominal timber plus a pole. Classic post-and-beam framing calls for 4-inch dimensional rafters and exposed 2×6, T&G decking with insulation beneath finished roofing.

Remember that grid poles inside the house need cross connections and ties too, as shown in accompanying photos. Frequently, second-floor girders or joist systems require the tie beams for support. If not, use ties to define space, carry cabinets, and hang shelving.

with the accepted deflection limit of $\frac{1}{240}$ (less stringent for sloped roofing than for flat flooring) and even $\frac{1}{360}$.

Heavy timber decking will cost more than $\frac{1}{2}$-inch plywood. But you will save the extra expense in time and material by eliminating 16-inch center rafters or a system of rafters and purlins. And the best part is what you get underneath the roof deck—a finished ceiling. You may want to seal the decking on the roof before you turn it over and nail it, or spray the decking and rafter spans in place later on. Insulation board and roof coverings will work on top of this assembly, which looks good and works well with pole construction.

SPANS FOR T & G DECKING
(2-INCH NOMINAL)

Span (ft.)	Live load (lb./sq.ft.)	Deflection limit	Fiber stress (psi)	Modulus of elasticity (psi)	Span (ft.)	Live load (lb./sq.ft.)	Deflection limit	Fiber stress (psi)	Modulus of elasticity (psi)
Roofs	20	$\frac{1}{240}$ $\frac{1}{360}$	160	170,000 256,000		20	$\frac{1}{240}$ $\frac{1}{360}$	300	442,000 660,000
4.0	30	$\frac{1}{240}$ $\frac{1}{360}$	210	256,000 384,000	5.5	30	$\frac{1}{240}$ $\frac{1}{360}$	400	662,000 998,000
	40	$\frac{1}{240}$ $\frac{1}{360}$	270	340,000 512,000		40	$\frac{1}{240}$ $\frac{1}{360}$	500	884,000 1,330,000
	20	$\frac{1}{240}$ $\frac{1}{360}$	200	242,000 305,000		20	$\frac{1}{240}$ $\frac{1}{360}$	360	575,000 862,000
4.5	30	$\frac{1}{240}$ $\frac{1}{360}$	270	363,000 405,000	6.0	30	$\frac{1}{240}$ $\frac{1}{360}$	480	862,000 1,295,000
	40	$\frac{1}{240}$ $\frac{1}{360}$	350	484,000 725,000		40	$\frac{1}{240}$ $\frac{1}{360}$	600	1,150,000 1,730,000
	20	$\frac{1}{240}$ $\frac{1}{360}$	250	332,000 500,000		20	$\frac{1}{240}$ $\frac{1}{360}$	420	595,000 892,000
5.0	30	$\frac{1}{240}$ $\frac{1}{360}$	330	495,000 742,000	6.5	30	$\frac{1}{240}$ $\frac{1}{360}$	560	892,000 1,340,000
	40	$\frac{1}{240}$ $\frac{1}{360}$	420	660,000 1,000,000		40	$\frac{1}{240}$ $\frac{1}{360}$	700	1,190,000 1,730,000

Span (ft.)	Live load (lb./sq.ft.)	Deflection limit	Fiber stress (psi)	Modulus of elasticity (psi)
7.0	20	$1/240$ $1/360$	490	910,000 1,360,000
	30	$1/240$ $1/360$	650	1,370,000 2,000,000
	40	$1/240$ $1/360$	810	1,820,000 2,725,000
7.5	20	$1/240$ $1/360$	560	1,125,000 1,685,000
	30	$1/240$ $1/360$	750	1,685,000 2,530,000
	40	$1/240$ $1/360$	930	2,250,000 3,380,000
8.0	20	$1/240$ $1/360$	640	1,360,000 2,040,000
	30	$1/240$ $1/360$	850	2,040,000 3,060,000

Note: Spans are based on 10 pounds per square foot dead load with provision for a 300-pound concentrated load on a 12-inch-wide deck.

Span (ft.)	Live load (lb./sq.ft.)	Deflection limit	Fiber stress (psi)	Modulus of elasticity (psi)
Floors				
4.0			840	1,000,000
4.5	40	$1/360$	950	1,300,000
5.0			1060	1,600,000

Inside the Kent House, below, a busy interior somehow manages to be serene, an interior where every piece of wood is there for a good structural reason. Even if you have no experience with building, I bet your eye can follow imagined loads around this frame. This has the fascination of a house with two staircases: every structural element is pinned in at least two directions, so there is always one way in and another way out.

The Kent House, designed by architect Edward J. Seibert of Sarasota, Florida, was built on a remote island off the coast. It is fair to think of it as a modern, engineered version of ancient Solomon Islands huts. Here you can compare the finished Kent House and Mr. Seibert's first presentation to his client. Storm panels that pivoted up and down in the sketch swing laterally in the house. Also, the clerestory on the finished house mirrors the hip roof of the first floor and vents on four sides.

FRAMING AND MATERIAL ALTERNATIVES

With all the talk about options and alternatives, I thought you would enjoy learning about a specific pole house that successfully pulls together many pole-building options. Edward J. Seibert, the architect, is not an "alternatives only" designer. But he does allow that although the big commercial projects take up most of the time and pay most of the bills, it is enjoyable, a break from the routine, even pure fun to do a house like the one you see here, the Kent house. Here are a few of the distinctive features. You don't see any cars in the pictures, or roads, or telephone poles, because the site is a remote island off the Florida coast accessible only by boat. Mr. Seibert, who works out of Sarasota, says the house is a vacation cottage, almost a retreat. There are no exterior walls or windows on the first floor. There is no furnace. The only enclosed room is the bathroom. The movable siding and the roofing are metal, painted green. The floor is 3-inch-thick T&G cedar. The "hat" on the hip roof is a ventilation chimney sided with operating clerestory windows to control the draft.

In the vernacular of architects, Mr. Seibert notes that, "the big thing about pole houses is to articulate the structure, i.e., not cover it up." Tucked into the rough, scrubby landscape, the poles are the controlling element in this house, both inside and outside. The "walls" between poles are made of insect screens, protected from the weather with a liberal overhang on the hip roof. And in wet weather the hip design protects each sidewall equally. (More conventional peaked roofs shelter the eaves but expose the gable ends.) When the owners leave the island, hinged, metal storm shutters are closed to button up the house.

In the mild climate, a wood stove is provided to get the chill off winter evenings. The climate may be temperate, but the wind along the Florida coast is not. For security in hurricane country the 16 treated pine poles in the 30 × 30-foot grid are set like mooring piles 20 feet into the sand. The unsupported pole height above grade is pinned with diagonal timber bracing to help resist lateral wind loads.

You can imagine that sitting in the 30 × 30 room, where interior poles rise starkly through the 3-inch cedar floor, is almost like sitting outside amidst the foliage and the solid trunks of palm trees. Except inside the house you get the vistas without getting wet or bitten.

The rafter system is straightforward, running from double girders along the clerestory down to the plates joining the exterior poles; 2 × 4 sleepers (on the flat) laid across the rafters support galvanized steel, V-lock roof panels. But notice how thoroughly the pole grid is braced and interconnected. It would be very reassuring if you had to sit through a hurricane. Aside from the extreme embedding, the full-height instead of platform-height poles, and the lateral bracing, notice how the perimeter poles are pinned in line to each other with wall plates and tied to one of the interior poles with double braces. These interior poles carry double braces to the adjacent wall as well. And above these stacked braces, clerestory girders connect the interior poles to each other.

There are more conventional, more expansive pole-frame buildings in this book. But this Kent house should give you a lot of ideas, even though it is a very specialized design for a unique site. Your plans may not include walls made of screening. But some of the framing details here apply to any pole design—for instance, the interesting structural relationships where knee wall blocking is set between the sandwiched girders around the clerestory to support the windows and frame the lines of the "hat."

I was excited when I first discovered Charles Eames' "factory" house, built of materials normally reserved for commercial and industrial buildings, all of which were ordered out of catalogs. Somehow, these harsh

In the pole grid for the house shown in accompanying photos, there are some platform-height poles to carry floor frames, and some full-height poles to carry floor, walls, and roofing. This utilitarian "forest" was set by San Antonio Rigid Pole Construction Co., of Brea, California, in Beverly Hills. The house, designed by Nyberg & Bissner of Pasadena, uses 20- to 60-foot poles to enclose 5,154 square feet. Cutoffs of Koppers Cellon-treated poles (right) are used for landscaping, while full poles (next page) run through the spa on one of five levels in the home.

materials—wire web trusses, steel roof decking, and the like—were combined to make a warm, inviting, human-scale house, more appropriate than some of the award-winning, extremely bleak modern monsters with pipe railings, endless glass, and large bare spaces inside. Industrial and commercial products and designs do not always work in residences. But industrial products, for instance, generally offer great strength and durability compared to their single-family residential counterparts. This makes me appreciate the barn roofing on the Kent house, even though it must make quite a racket during a heavy rain. And I appreciate a design in which the roofing, the frame, and the connections are left out in the open, not buried as with other buildings.

The Kent house is a comfortable combination of simplicity and complexity. The plan is simple, but the engineering is complex. The house is wide open, but the pole grid is locked up, like a chain without weak links. In a sense, the interior poles are very much like trees in the forest, with branches extending to other trees, providing a protective umbrella of foliage (the roof) like a rain forest. Except that this model of the forest is executed in treated timbers and girders and decking and nuts and bolts.

The wide-open room, a basic shelter design not unlike a primitive hut, is a deceptively simple idea carried out in an updated, hurricane-resistant pole frame. Which is more important? That the house is nice looking and has workable floor plan, or that the house will still be there after the storm? These two ideas are not mutually exclusive. Both are worthy goals. And no matter how this chicken-or-egg question works out, the pole frame is too important, too good-looking in its own way, to be buried.

The poles and the grid connections in the Kent house, and in most pole houses, are the heart of the home. Personally, you may not be the kind of person who wears his heart on his sleeve. Architecturally, it would be nice if your pole house did.

ENERGY EFFICIENCY

MECHANICAL ACCESS

As you read about structural architecture and buildings with their bones bared, it might have occurred to you that the living room ceiling would be draped in Bx electrical cable and the kitchen would be decorated in copper pipes. I'm partial to bright copper, all steel-wooled and lacquered. But don't worry. If you build a pole house, you don't have to see any of the mechanical systems if you don't want to.

Obviously, 16- or 24-inch center floor, wall, and roof framing has ample room for plumbing and electrical lines. But let's take the worst case: T&G decking over girders bolted to the pole grid. Dimensional timbers sandwiched around poles leave a lot of room inside the sandwich. Two 2 × 12s pinned to an 8-inch pole leaves enough room for an 8-inch-diameter hot-air duct, or a soil pipe, or drain lines, or recessed light fixtures.

In houses where mechanical work is done after the fact (converting a small bedroom to an extra bath, for example), box beams, which are just fake girders made of something like 1-inch pine, are used to carry and cover up pipes and wires. You can treat sandwich girders the same way, if you have to. One-inch pine can be used to cover the bottom of the pole-girder box beam, hiding any equipment inside.

As you install girder and rafter pairs and select the bays that will hold pipes and wiring, you can add nailers (something small like ½ × ½-inch molding) about 1 inch up from what will be the bottom, inside edge of the girders. If you rip your own pine and rip it a shade fat, it will fit up against the nailers, held in place by the pressure of the timbers. A few long finishing nails will make sure it stays in place and make the pine easy to remove if you have to go after a pipe leak, for example. Setting the nailers 1 inch up from the bottom

Photo right: With solid-timber rafters, the only way to get electricity into the hanging fixtures is to run conduit up a rafter.

Below: Using double joists and rafters instead of 4-inch beams provides room for all kinds of equipment such as forced-hot-air ducts, plumbing supply and waste lines, lighting, and electrical lines.

MECHANICAL ACCESS IN DOUBLE BEAMS

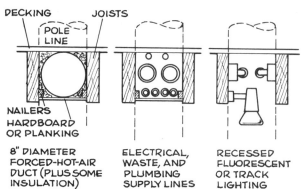

DECKING   JOISTS
POLE
LINE
NAILERS
HARDBOARD
OR PLANKING

8" DIAMETER
FORCED-HOT-AIR
DUCT (PLUS SOME
INSULATION)

ELECTRICAL,
WASTE, AND
PLUMBING
SUPPLY LINES

RECESSED
FLUORESCENT
OR TRACK
LIGHTING

of the girders leaves a neat ¼-inch reveal (recess) between the girders and the pine, which measures ¾-inch actual.

Granted, your second-floor joists or roof rafters may not run in a line where you want a light fixture. That's a problem. You have to build the structure, keeping in mind the layout and the building loads, not individual light fixtures. Despite careful planning, a hot-water pipe, a hot-air duct, a water-supply pipe, or an electrical line may fall outside easily concealed girders, wall frames, kitchen cabinets, and other hiding places.

Instead of showing a little piece of copper here and a tail end of Bx there, think of concentrating these extraneous mechanical tidbits in one package. For example, if you don't have floor or ceiling access to get pipes and wires past an exterior wall made of poles and sliding glass doors, you might run a combination light box and mechanical chase along the header above the doors. This can be as simple as a wooden T-shaped frame with lighting below and pipes and wires above, both concealed by the head of the T facing the room. Or you might run a large-diameter PVC drainpipe along the header, painting it brown to blend with the headers and poles. It's a little unorthodox but will probably meet with code approval. And in the same way that control joints collect cosmetic stress cracking in concrete, plastic pipe, box beams, and similar pieces of construction can serve to collect mechanical equipment and get it where you need it.

You are more likely to have trouble getting plumbing lines in and out of the house, particularly when the first-floor frame is elevated well above the ground. Electrical service lines running from pole (that's utility company pole) to house are not affected because of the pole (that's your pole house) framing. Lattice panels or some other skirt construction between floor girders and the ground offer a reasonable cosmetic solution for piping. But if you live in an area with seasonal freezing, a lattice won't prevent the water inlet from bursting.

In many pole homes that are full or modified "dog-trot" designs, with a lot of area for sheltered storage (or a dog) beneath the floor frame, a central mechanical core is constructed. In full-height dog trots, this core might contain an entrance to the house, stairs to the living quarters one flight up, storage space, and a small utility room. If the sheltered area beneath the floor frame amounts to no more than an open crawl space, you can excavate a small entryway for mechanical systems only. An elaborate mechanical pit would sit on a concrete slab, have poured-concrete or block walls, and an entry hatch in the first floor. In just enough space for elbow room and a pipe wrench, the pit could be partially underground, equipped with a sump pump, a cleanout T for the soil pipe, a water-supply inlet and main cutoff valve, an electrical line for a light, and a waterproof outlet for a thermostatically controlled heat tape wrapped around the water-inlet pipe.

I have found these tapes indispensable in conven-

I am reluctant to include the drawing (below left) because the conventional masonry construction, even the slightly more sympathetic AWWF (all-weather wood foundation), has nothing to do with poles. But such construction can get you a mini-mechanical room. Of course, you could hang a weatherized floor from intermediate poles to accomplish the same goal. Below: The smallest mechanical room is a hotbox; a highly insulated mechanical compartment from the first floor to below the frost line where plumbing lines enter.

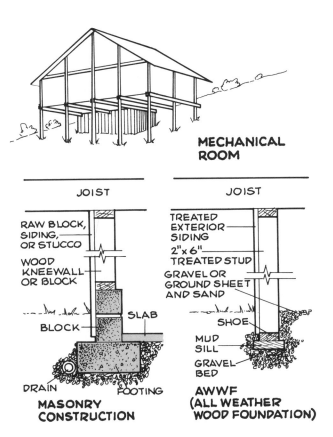

MECHANICAL ROOM

JOIST

RAW BLOCK, SIDING, OR STUCCO

WOOD KNEEWALL OR BLOCK

BLOCK

SLAB

DRAIN

FOOTING

MASONRY CONSTRUCTION

JOIST

TREATED EXTERIOR SIDING

2"x6" TREATED STUD

GRAVEL OR GROUND SHEET AND SAND

SHOE

MUD SILL

GRAVEL BED

AWWF (ALL WEATHER WOOD FOUNDATION)

MECHANICAL ACCESS "HOT BOX"

WASTE LINE VENT

WATER SUPPLY LINE

FINISHED FLOOR

JOIST

GRADE

RIGID FOAM INSULATION

FROST LINE

SOIL PIPE

PRESSURE-TREATED AWWF WALLS

LIGHT

CLEANOUT

DRAIN

GRAVEL BED

JOIST

AWWF OR BLOCK WALL

SWITCH AND RECEPTACLE

HEAT TAPE WITH THERMOSTAT

tional, drafty old houses. If, for some reason, the air temperature in a mechanical pit falls, the tape thermostat starts pulling power in very small amounts through the resistance coils of the tape. This generates enough heat to prevent ice from forming in the line. In a vacation home used only seasonally, include a drain-down valve for hot-water heating and all supply piping past the main valve.

A less complicated but effective mechanical pit can be excavated, finished with a gravel floor and a pressure-treated, all-weather wood foundation system sitting on wooden mudsills. With a lining of insulated wall studs or rigid insulating board, pipes should not freeze (I would add the heat tape anyway), and the wooden foundation wall may be less conspicuous than concrete block.

Basic floor plans that concentrate mechanical needs are always efficient, but particularly in a pole house. Instead of a bathroom in each corner of the building, try to find a central location for back-to-back baths, adjacent to (or above) a utility room or mechanical pit and next to the kitchen. Short plumbing runs are less expensive to install. You are less likely to need box beams and other camouflage construction. You'll have less mechanical maintenance. When you need to do some work on the systems, they will be easier to get at. Also, the hot water will arrive hot if it travels to the next room instead of to the next county.

HEATING AND COOLING

There is no heating or cooling system that is unique to pole-frame houses. They can take conventional gas-, oil-, or electric-fired systems, solar systems, hot-air, hot-water, and all kinds of passive-solar configurations. There are, however, several insulating and ventilating details that are unique to pole buildings.

If the floor of a pole house is a concrete slab (an unlikely but possible choice), it can be insulated with conventional 2-inch rigid foamboard set between the gravel bed and the concrete. If you pour for a ground-level utility room or entry and stairs, allow for some pole movement and flexing by leaving an expansion lip between pole and concrete, sealed with an elastomeric rope or resilient caulking like butyl or silicone.

It is more likely that the first floor of your pole frame will be elevated off the ground and, thus, need protection from temperature and moisture. Normally, the floor of a conventional crawl space is covered with sand, gravel, or a layer of concrete. A sand base covered with 55-pound felt paper or polyethylene sheeting wrapped up the foundation walls will keep ground moisture out of the insulation between floor joists. This

In warm climates, you may be able to tuck mechanical lines under a deck or entry ramp on the high side of a sloped site. If it weren't for the bulky gas meter at the bottom left of the photo, you might not notice the pressure regulator and gas line.

won't work on a pole house, however, even if there is a cosmetic skirting, such as a lattice, around the perimeter of the building.

All the jobs done in the cellar or crawl space of a conventional house must be done at the first-floor frame in a pole house. The bottom of the floor is the first line of defense. A 4-mil-thick polyethylene sheet may resist moisture, as will foil or double-cover felt paper. But such materials are not likely to resist wind whipping under the house, birds, mice, raccoons, and all kinds of other creatures that will find a weak link, a hole, or a rip, and turn it into their own front door to a warm bed.

Probably the most thorough closure of the floor frame is achieved with ¼-inch exterior plywood. There is only one way to put the sheets in place, and that's from under the floor, usually on your back, which is about the worst position in which to measure and hammer. It's murder. Everything you drop that, normally, would fall to the ground, falls in your face instead. Hammering upward, you will discover how much hammer power is due to gravity.

I have seen floor frames closed with either felt paper or polyethylene covered with a narrow-grid, welded-wire screening—like superdense, heavy-duty chicken wire, not like insect screening. This will keep out just about everything but insects.

This is the kind of job you should think about before finalizing your pole-house plans. A floor 2 feet off grade means you'll do this work on your back; if it's 4-5 feet off grade, you'll be able to work from your knees, which is lot easier; 6-7 feet off grade, and you can work while standing; 8-10 feet off grade, and you'll have to work from planks on sawhorses or some other kind of low scaffolding.

What about the worst case—working on your back with minimal operating room? You can make an unbearable job a little better by laying scaffold planks or plywood sheets on the ground in line with bays between timbers, and scooting along them on a mechanic's dolly—the kind used in garages, called a crawler. If you measure and cut insulation and vapor barriers first, then stretch them out in place under the framing bays, it is not all that difficult to work your way down one bay at a time, with the material ready to lift and staple. To secure the plywood sheets you will definitely need at least one deadman (or a live helper)—something like a 2×4, I-shaped brace cut to fit between the ground and the plywood. It can be used to hold one end of the sheet up against the floor timbers while you wrestle with the other end and get a few nails set.

Remember that while the rule of thumb is to put the vapor barrier between the heat source and the insulation (to keep household moisture out of the wall cavity), you will be working with an additional vapor barrier to keep ground moisture out of the floor. If the heat source is warming the air above the first floor, this means the foil facing on the insulation, which serves as a vapor barrier, would be face up to the heated air above the first floor. This doesn't do much for the raw insulation facing toward the ground.

If I had the time and money, I would probably opt for a full plywood soffit. It sure would make the floor frame rigid, pinning the timbers top and bottom. Ideally, I would use ⅜-inch exterior-grade plywood and slop a lot of Woodlife on it to boot. Working under the floor frame, I might go for 1-inch styrene insulation board, a rigid, lightweight, tongue-and-groove panel directly over the timbers. Then I might add the ground vapor barrier, probably 4-mil polyethylene overlapped about 6 inches at every seam. Finally, I would add the plywood, then go up into the light and air and add foil-faced insulation with the foil up, stapling the insulation flanges to the sides of the timbers and keeping the insulation an inch or so above the insulating board to leave an airspace. Then, with plumbing, electrical, and heating or cooling lines in place, I would deck the first floor. This would give a moisture-resistant floor frame with something like an R-27 rating. You can adjust the amount of insulation depending on climate.

If you do get stuck adding insulation from below the floor, remember to button yourself up tightly. Fiberglass is bad enough when you are working above your head. On your back, endless little invisible fibers will find their way down shirtsleeves and collars. Knowing

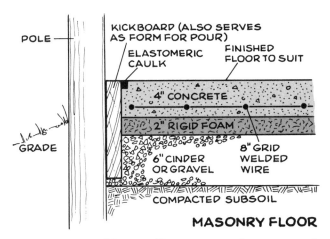

MASONRY FLOOR

Slabs that would normally run up against a screed board can rest against a treated kickboard, which establishes building wall elevation and ties poles together.

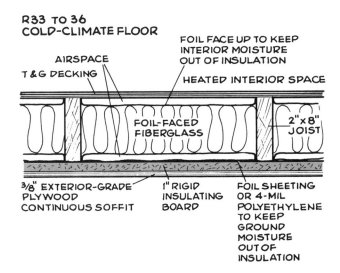

**R33 TO 36
COLD-CLIMATE FLOOR**

Rigid-diaphragm floors on pole houses can be weatherized, depending on climate, with as much insulation and as great an R-value as in any conventional home. Because the floor is exposed, though, batt insulation left open in standard cellars must be protected from ground moisture and damage.

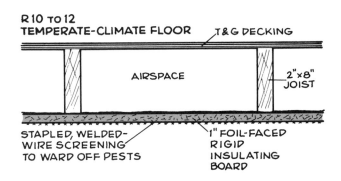

**R 10 TO 12
TEMPERATE-CLIMATE FLOOR**

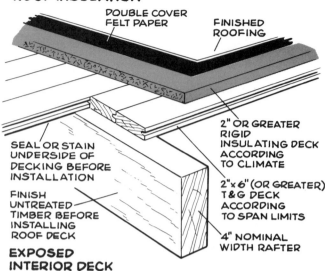

ROOF INSULATION

DOUBLE COVER
FELT PAPER

FINISHED
ROOFING

SEAL OR STAIN
UNDERSIDE OF
DECKING BEFORE
INSTALLATION

FINISH
UNTREATED
TIMBER BEFORE
INSTALLING
ROOF DECK

**EXPOSED
INTERIOR DECK**

2" OR GREATER
RIGID
INSULATING DECK
ACCORDING
TO CLIMATE

2"x 6" (OR GREATER)
T&G DECK
ACCORDING
TO SPAN LIMITS

4" NOMINAL
WIDTH RAFTER

Three drawings at right: Exposed T&G decking limits your choices, as the roof will always be a solid, unventilated mass, no matter how many inches of rigid insulating board you need. Purlin roof systems may be combined with vent space but generally need two decking layers: a finished board to the inside and a structural board under the finished roofing.

you will wind up with a faceful of the stuff, wear plastic goggles, a respirator mask (if you can stand it), two long-sleeved shirts, and double-seal collars and cuffs—i.e., put gloves over the cuff on one shirt and under the cuff on your overshirt. Seal up with rubber bands. I know this sounds paranoid, but unless you are immune to the sometimes intense itching and prickly heat produced by fiberglass, you will appreciate the protection.

Before you crawl out from under the house, check the seams between plywood and poles. Those are the seams that will be used by insects and mice. I would lay a thick bead of elastic caulk on the underside of the plywood where it will meet the pole, just before installation. After nailing, you can go again with another bead on top. It's like a seven layer cake—five-ply sheathing with caulking top and bottom.

And by the way, if I was building in the South, and the ground was mucky, and there was only 2 feet of clearance to the floor frame, and it was about 100° F. with 90-percent humidity during the day, I might well decide to put on a wetsuit, blitz my way through a stack of T&G foam panels, several rolls of welded wire, several boxes of heavy-duty staples, and let it go at that. And I would need a heck of a good reason (and offhand I can't think of one) for leaving so little clearance between the floor frame and the ground.

ROOF SYSTEMS

A pole roof frame with girders and a rafter system can be treated as a conventional roof, filled with pipes, wires, ducts, and insulation. Again, the peculiar case of girder sandwiches and exposed structural decking causes a departure from conventional insulation. And if you are spending the money and possibly enlarging

the girder size or reducing the pole grid to get a finished, heavy-timber deck, the last thing you want to do is slap insulation and wallboard on the ceiling.

Assuming a sound structural deck, exposed to the inside, you can add a purlin network above the decking to create bays for insulation—which is structural overkill—or use layers of rigid foamboard directly on the deck. Depending on the manufacturing process, some insulating boards give an R-value as high as 6.5 per inch. If this is the only insulation you have on the roof, you may need 3 or 4 inches or even more, depending on local climate.

If you pick up a 2×8-foot sheet of foamboard, a stiff wind can snap it in two. You can easily put your fist through it. Yet, on a solid wood deck it has surprising resistance to crushing, at least when loads are spread out. By the time you get several inches of the foamboard stacked up, the roof will feel just a little spongy underfoot. Still, distributed loads can be adequately supported. The problem comes when you concentrate compression loads by nailing down shingles. The foam cannot resist the pressure of a roofing nail head driven by a hammer blow. Overnail, and you will send the nail through the shingle and through the foam, down toward the wood deck. This gives you two options: either you can accommodate the foamboard thickness with exceptionally long, galvanized roofing nails and work carefully to control nail depth; or you can add a plywood roof skin to distribute roofing and nailing loads over the foam, to control nailing depth, and to increase the nail holding power. It's an added expense, but with a thick layer of foam it is one way to get a strong, well-insulated, cold-climate, durable roof.

I have seen a few books that suggest reversing the order of materials—laying rigid foam directly over joists and roof rafters, then structural decking on top of the foam. Avoid this method. Solid nailing will pull

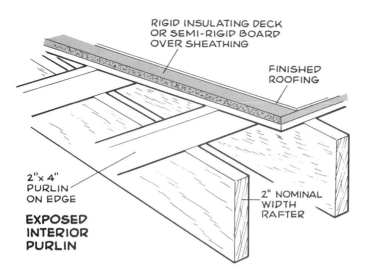

RIGID INSULATING DECK OR SEMI-RIGID BOARD OVER SHEATHING

FINISHED ROOFING

2"x 4" PURLIN ON EDGE

2" NOMINAL WIDTH RAFTER

EXPOSED INTERIOR PURLIN

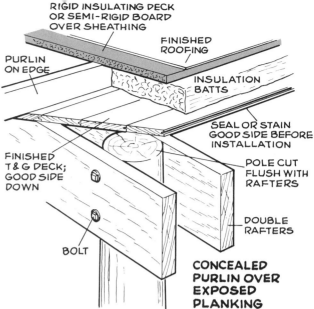

RIGID INSULATING DECK OR SEMI-RIGID BOARD OVER SHEATHING

FINISHED ROOFING

PURLIN ON EDGE

INSULATION BATTS

FINISHED T & G DECK; GOOD SIDE DOWN

SEAL OR STAIN GOOD SIDE BEFORE INSTALLATION

POLE CUT FLUSH WITH RAFTERS

DOUBLE RAFTERS

BOLT

CONCEALED PURLIN OVER EXPOSED PLANKING

Three photos at left: Modern pole barns do not present the most attractive roofing details, but they are cheap and quick to build: lapped 2 × 4 purlins; site-built, low-slope, three-member trusses; and, for large structures (if you have the money and a crane), shop-built, gang-nailed web trusses.

the deck down to the dimensional timber—a 2 × 10 rafter, for example—concentrating the compression along the 1½-inch width of the timber. Foamboard won't stand the strain. You will hit the nail again and again, and the foam will fold up like a sponge.

There are some products that can be laid directly on dimensional timbers, but I hesitate to call them insulating boards. I have used a Homosote, 2-inch, vinyl-wrapped, tongue-and-groove panel system over 4-inch-wide timbers 4 feet on center. But this pressed paperboard does not have a great insulating value. Of course, you could add a layer of foamboard on top of it.

There are a lot of options, and they are interconnected. The pole grid affects the girder span and decking requirements. Fewer poles means bigger girders and heavier timber decking. Rafters and purlins can hold ample insulation. Rafter pairs and exposed decking means insulating above the roof deck. It's a little like playing chess, where it pays to look ahead and examine the options and the consequences of making a particular move before you plunge into a job.

This may be a good place to note that alternative housebuilding systems can have, but do not necessarily require, alternative energy systems. In other words, it's okay to build a pole house and heat it with a 95-percent-efficient pulse-combustion furnace. As a matter of fact, that will be very effective. Pole buildings, and the other types of construction in this book, can also be heated with active and passive solar systems, electricity supplied from your own wind generator, wood stoves, mass-heat air tubes, mass-heat water tubes—take your pick.

Frequently, unconventional building systems are automatically lumped together with unconventional energy systems. The idea seems to be that if you are interested in one you will be interested in the other. Maybe, but not necessarily.

I have heated with wood for years, and, for a time, cooked and got domestic hot water from a huge kitchen cookstove. I have run loose-joint stoves and airtights. I have felt the benefits of thermal mass delivered long after the fire has died in a monstrously huge, six-burner cookstove. Heavy timbers, poles included, also store and reradiate heat. An imaginative designer might find a way to combine wood poles in a 12-foot grid with air or water mass storage tubes of a similar diameter on a 4-foot grid intermingled with the poles. Pole frames are flexible enough to work with conventional energy systems, unorthodox energy systems, and all the active and passive alternatives in between.

In contrast to the C-faced plyscore of pole barns, this kit home has a pyramid-frame roof covered with purlin-backed veneer plywood, and a foil/foam roof deck.

If the poles in this home were stained dark, you
wouldn't pick out the stove pipe.

INFORMATION SOURCES

Remember in the beginning of this pole-building part of the book I mentioned that I was surprised to find so little information on pole building, from a historical, practical, or any other point of view? This means that aside from several books, a lot of the sources of information that follow are companies and other groups that are a bit out of the way, not all that used to dealing with owner-builders, and with more than a gee-whiz, passing interest in alternative housing.

Also, pole-building plans, like other out-of-the-ordinary designs in this book will get close attention from your local building department. It's like having an unusual occupation or innovative accounting method cause an IRS audit. Your plans may get flagged because they are unusual, even though they may be as safe and as durable—or even more so—as the plans the inspectors are used to seeing. When this happens to your tax forms, an auditor may feel compelled to justify the flagging by finding hidden taxes or inconsistencies in your forms. In the same way, the building department may call you on bolt diameters, for example, only because they stopped to look, and had to find something.

My advice is not to fight, as long as the suggestions (often, suggestions made as offers you can't refuse) do not dramatically alter the design and cost of your home. Giving in on bolt size or shortening a cantilever may avoid a needless challenge on pole-grid size or girder dimensions.

POLE-BUILDING BOOKS

Barns, Sheds, and Outbuildings, Byron D. Halsted (Stephen Greene Press). This book is a nice way to ease into the subject of pole building and heavy timber construction. You may have to check other sources for technical details, but you will enjoy the plans and designs for all kinds of structures.

Farm Builder's Handbook, R. J. Lyle et al. (Ideals Publishing Co.). A more technically oriented companion book to the one above, this handbook makes assumptions about your basic carpentry and layout skills and goes straight to the details of spans, stress, loads, heating, and cooling requirements.

A building inspector may appreciate some of these strengthening details, even though pole frames are stronger, more resilient, and more resistant to fire, wind, and earthquake than standard construction. When rafters do not bolt through the pole grid, wind plates and strapping can make the roof frame nearly as secure as the rest of the pole home.

WIND AND STORM RESISTING ROOF DETAILS

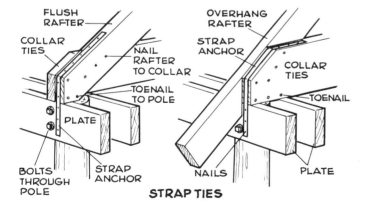

STRAP TIES

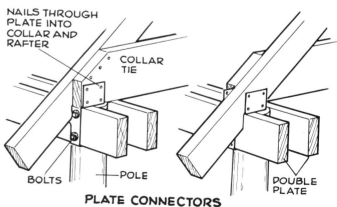

PLATE CONNECTORS

Low-Cost Pole Building Construction, Ralph Wolfe (Garden Way). This large-format, well-illustrated book is an excellent introduction to pole building, combining a modest amount of technical information, many reprints from technical bulletins, a few pole-building plans, and a large picture-caption section showing many types and examples of pole houses.

The Owner-Built Pole Frame House, Ken Kern (Scribner's). This detailed book has very specific solutions to pole-building questions. If you agree with Kern's philosophy and approach to pole design, the book is an invaluable guide. If you are not yet decided about fixed-glass frames with screen vent panels below windows, about sawdust-filled tubes for wall insulation, and other such options, you might find another book with a wider variety of options and—more important—opinions, before you try this one.

Recreational Buildings and Facilities (Agricultural Research Service, U.S. Dept. of Agriculture). USDA Handbook No. 438 contains 75 pages of specific plans for vacation cabins, farm cottages, log cabins, A-frames, barns, greenhouses, and several pole-frame cabins. The plans and drawings are basic and clear.

The Wheeling Farm Building Book, W. A. Bell et al. (Popular Library). This small paperback, which is produced by the Wheeling Company and contains a center section of product advertisements, is not particularly well written or illustrated. Yet, it is a blunt, straightforward account of practical design and construction. The book is filled with tables on shapes, sizes, nailing specs, and more.

TECHNICAL BACKGROUND

Construction Poles and Posts, American Society of Agricultural Engineers, 2950 Niles Rd., St. Joseph, MI 49085. The ASAE has a series of short technical bulletins detailing pole-design values, preservative retentions, and other data.

Construction Poles Design and Treatments, Southern Forest Products Assn., P.O. Box 52468, New Orleans, LA 70152. This short bulletin lists pole dimensions, soil-bearing tables, formulas for figuring allowable soil-bearing pressure, and similar data.

FHA Pole House Construction, American Wood Preservers Institute, 1651 Old Meadow Rd., McLean, VA 22101. In 32 pages this booklet presents an incredible amount of design and construction data, mostly in condensed form. Illustrations are primitive, but all the options are included. Take a look.

How to Build Storm Resistant Structures, Southern Forest Products Assn., P.O. Box 52468, New Orleans, LA 70152. This short, well-illustrated booklet covers all kinds of strengthening details, from foundation to sill, from rafter to plate. Pole dapping, spiked grid connectors, and many other pole-connection details are shown.

Mechanical Properties of Wood, and *Physical Properties of Wood*, Forest Products Laboratory, USDA Forest Service, Washington, DC 20250. These two reprints from the USDA *Wood Handbook* are thorough, well-illustrated treatments loaded with tables of material weights, design values of timbers, moisture contents, decay-resistance factors, and more.

Pressure-treated Southern Pine, Southern Forest Products Assn., P.O. Box 52468, New Orleans, LA 70152. Offers a look at grades, grade stamps, design uses, preservative treatments and retentions of one of the most common building pole wood species.

Pole Building Design, American Wood Preservers Institute, 1651 Old Meadow Rd., McLean, VA 22102. This booklet takes some effort to read. It is filled with engineering formulas and nifty tables called nomographs that are like building formulas all graphically displayed so that you can plug in your own set of numbers. Some of the information, like the formulas for stress on different truss members, went right over my head, but a lot of information on soil values, bolt sizes, and embedment depths is invaluable. Try this one.

Principles for Protecting Wood Buildings from Decay, USDA Forest Service, Washington, DC 20250. This booklet covers woods in general, not just poles.

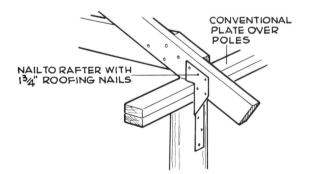

CONVENTIONAL PLATE OVER POLES

NAIL TO RAFTER WITH 1¾" ROOFING NAILS

Photographs and drawings document the areas, even specific fungi, that do the damage. The technical text tells how to prevent the problems.

TOOLS AND HARDWARE

KC METALS, 1960 Hartog Dr., San Jose, CA 94577. One of the companies offering a wide selection of structural connectors.

SIMPSON CO., 1450 Doolittle Dr., P.O. Box 1568, San Leandro, CA 94577. Simpson's catalog includes all kinds of connectors, straps, bridging brackets, and more.

TECO, 5530 Wisconsin Ave., Chevy Chase, MD 20815. In addition to endless structural fasteners, TECO also offers plan books with framing details.

POLE FIRMS

CROWN ZELLERBACH, P.O. Box 449, Mobile, AL 36601. A major pole producer with outlets in Louisiana, Michigan, and Oklahoma.

INTERNATIONAL PAPER CO., P.O. Box 231, De Ridder, LA 70634. Another large supplier with outlets in Iowa, Michigan, Missouri, and Washington.

J.H. BAXTER & CO., 1700 S. El Camino Real, San Mateo, CA 94402. Baxter, as noted in the previous chapters, is the firm that was in on the Marx-Hyatt house and the infancy of residential pole building in California.

KOPPERS CO., INC., Pittsburgh, PA 15219. A very large pole supplier-treater with some 25 outlets in the U. S. and Canada. An excellent source of information on pole treating and pole grades.

WADSWORTH LUMBER CO., P.O. Box 98, Bunnell, FL 32010. Another large timber company with many outlets in southern states.

ASSOCIATIONS

All the groups listed below have practical and technical information on pole building, pole materials, and pole treatment. Most offer so much literature that I suggest you write for a publications list first, telling the agency what your plans are, briefly. The American Wood Preservers Institute, for example, sends out a 15-item bibliography with publications that would keep you reading for weeks.

Directory of Wood Industry Associations, Staff Members, and Headquarters Offices, National Forest Products Assn., 1619 Massachusetts Ave. NW, Washington, DC 20036. This little directory compiled by the NFPA lists names, addresses, and phone numbers of many personnel at the agencies and associations listed below, at both national and regional offices.

AMERICAN FOREST INSTITUTE, 1619 Massachusetts Ave. NW, Washington, DC 20036

AMERICAN INSTITUTE OF TIMBER CONSTRUCTION, 333 W. Hampden Ave., Englewood, CO 80110

AMERICAN SOCIETY OF AGRICULTURAL ENGINEERS, 2950 Niles Rd., St. Joseph, MI 49085

AMERICAN WOOD PRESERVERS INSTITUTE, 1651 Old Meadow Rd., McLean, VA 22101

CANADIEN WOOD COUNCIL, 85 Albert St., Ottawa, Ontario, Canada K1P 6A4

NATIONAL FOREST PRODUCTS ASSN., 1619 Massachusetts Ave. NW, Washington, DC 20036

SOCIETY OF AMERICAN WOOD PRESERVERS, 401 Wilson Blvd., Suite 205, Arlington, VA 22209

SOUTHERN FOREST PRODUCTS ASSN., P.O. Box 52468, New Orleans, LA 70152

SOUTHERN PRESSURE TREATERS ASSN., 2920 Knight St., Shreveport, LA 71105

WESTERN WOOD PRODUCTS ASSN., 1500 Yeon Bldg., Portland, OR 97204

POLE HOUSE KITS

There seems to be a cycle as alternative building systems are rediscovered. With log construction, for instance, there were individual handcrafters, then companies (now dozens of them) that standardized a particular log-building system, and then companies that went a step further and produced precut log homes in kit form.

The pole cycle is not as far along. I know of only one firmly established pole-kit firm, Pole House Kits of California (see address below). I know, if you live on the East Coast you just groaned, thinking about the shipping costs. Gordon Steen, the company president and designer, started building custom pole homes in Hawaii. His custom work has been synthesized into many plans, including two townhouse designs and three designs for a guesthouse with garage. Standard kit packages of the full homes range from a cost of $50,000 for about 2,000 square feet with 900 square feet on a covered veranda, to about $95,000 for almost 4,000 square feet with 1,000 square feet on the veranda.

The Pole House Kits catalog, which will cost you about $7, is an elegant presentation of pole housing. The price seems intended to discourage browsers. But some of the standard and optional kit features are special—such as Japanese handcrafted glazed roofing tiles, oak plank flooring, 14-inch-diameter Douglas fir poles, heart redwood plank siding and trim.

These catalog pages from Pole House Kits of California show one of a dozen stock pole plans. This 16-pole grid encloses about 3,500 square feet—2,200 inside and 1,300 outside on a covered border veranda.

THE GRANDE PAVILION

Total Living Area   3,472 sq. ft.
Interior   2,176 sq. ft.
Veranda 1,296 sq. ft.

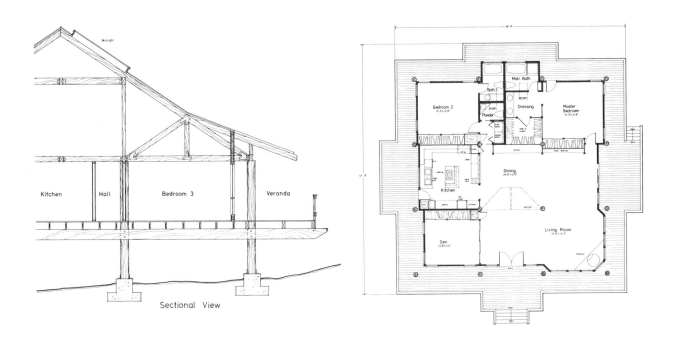

Sectional View

Other companies may push the pole cycle further and make pole homes more accessible for buyers and builders who do not want to start from scratch. But pole frames are probably the easiest for owners to handle themselves. Pole-frame carpentry does call for brawn but not for many of the fine finishing skills that many owner-builders never get the chance to develop.

ILLUSTRATION ACKNOWLEDGMENT

The following contributed illustrations to this part of the book:

J.H. BAXTER CO., 1700 S. El Camino Real, San Mateo, CA 94402

BRESCIA POLE BUILDERS, PO Box 278, Montgomery, NY 12549

FARM BUILDERS INC., RR2, Box 190, Remington, IN 47977

KOPPERS CO., Pittsburgh, PA 15219

POLE HOUSE KITS OF CALIFORNIA, 220 Newport Center Drive, No. 10, Design Plaza, Newport Beach, CA 92660

EDWARD J. SEIBERT AIA, 81 Cocoanut Ave., Sarasota, FA 33577

SOUTHERN FOREST PRODUCTS ASSN., PO Box 52468, New Orleans, LA 70152

U.S. FOREST SERVICE, 1720 Peachtree Rd. NW, Atlanta, GA 30067

The word "kit" may give you pause. You might expect the most simplified pole structure—all utility and no fun. The California Pole Kit home shown in the accompanying three photos has an airy entrance halfway along the veranda. Like an old-fashioned apartment airway, the top glass, door, and fixed side light show a continuous etched-glass tree when closed—an appropriate design considering the view.

CORDWOOD, STONE, EARTH MASONRY

AND

Yes, the title is correct. And it tells you there is a lot more to masonry construction than bricks and mortar. Stone masonry, at least, should be no surprise. It is certainly nothing new. In fact, stone is the ultimate, durable building material, used through the ages in a continuum of stone-built Egyptian pyramids, Greek Temples, Roman villas, medieval castles, Renaissance cathedrals, and New England dry walls. But earth and wood masonry?

Prepare yourself for something a little different. This part of the book covers some of the least publicized but possibly most interesting options for home buyers, particularly owner-builders. Frankly, these alternatives are not as established and accepted for modern, residential building as log and pole and timber-frame construction. Their performance is not as well or as widely proven. And that is reason for caution. Also, there are not many experienced professional builders working full time in these alternative fields, compared to the growing number of handcrafters, small firms, and even kit manufacturers building log, timber-frame, pole, and even dome homes. Only a few of these professionals have much of a track record.

But you need not worry that these building alternatives are somehow crackbrained—inherently weaker or less reliable than better known systems. They're not. And in any case, better or worse is not the issue. But it is fair to say that earth and wood masonry

in particular are so distinctly different, so unexpected, that they could pose unique problems. I am not thinking only of initial stumbling and fumbling with unfamiliar materials and methods, although that will probably happen. It's one thing to build a home using building products and techniques that are unconventional, quite another to go all the way out to the fringes of alternative construction. On the other hand, as long as you are treading on new ground, learning about construction in detail and trying new designs, why not go all the way? At least try to keep your mind wide open and consider some of the unorthodox methods of building a house.

I hope you have not already heard a little voice saying that you would be crazy to go the alternative route. It's hard enough translating ideas into blueprints for a conventional, stick-built home. You can well imagine how the building inspector and the mortgage banker will look at each other, and look at your drawings, and look at each other again before one of them will let you know you are in for a long process of explanation and persuasion by saying something like, "Ah, I've seen a wood house, and I've seen a masonry house, but I don't believe I've ever seen a wood-masonry house." And then you can explain that it has walls made of seasoned cordwood ends with a masonry binder, and even insulated just like a regular wall, and that it is solid and strong and energy-efficient.

Building with cordwood is a lot like building with brick, except for wood's infinite variety, thermal efficiency, and low cost. Plus, there are not many sites where you come across large piles of bricks in the woods. Cordwood is indigenous almost everywhere.

It would be hard to find a wall more beautiful than this, simply a pile of wood with big logs and kindling mixed in, with a wall of quarter splits nestled together in the background. If you could seal up the gaps between the logs, tie the wall together, and use the natural insulating value of wood, why not make a house out of it?

"Ah, you mean you will building brick walls, but instead of brick you'll be using firewood?" Well, sort of, but not quite. And it may be harder yet, when you submit plans and an application for a permit, and tell the housing professionals about earth masonry.

"Ah, after you do all the sifting and mixing, and after you do all the tamping and smoothing, don't you suppose the whole thing might just wash away in the rain?" And then you can tell them about the magazine articles on 40-year-old rammed-earth homes, and about Rammed Earth Works, a firm that builds houses, such as a 400 square-foot home for $1,500 and a 3,200 square-foot, 4-bedroom, 3-bath home for $200,000.

Still, you may get a disappointing answer. "Ah, I'm not sure we have 30-year mortgage money available for your mud hut."

Stone-, earth-, and wood-masonry homes are unusual. Most of the designs are unique. And it feels different to live in them. But all three building systems (and some of the offshoots, such as adobe and clay-block construction) have three characteristics in common. First, you need only a few basic tools to build these homes. Second, you can build them using locally available materials. Third, you can build them inexpensively.

"Local cordwood" and "local soil" are not materials found on architectural spec sheets. But "local stone" is. It means that, within limits, the mason contractor

Local stone can be easy pickings, especially if your property is interlaced with old stone fencing originally unearthed the hard way—with horse and plow. Some would-be stone builders hunt for property with cannibalization of old stone walls in mind.

Below, rammed-earth walls shown in the background are composed of soil found on site and mixed with a binder. Here, ingredients for the walls are dirt cheap.

does not have to send out for the bluestone or a particular shade of slate. Depending on where you build, a definitive specification like that could mean sending across town or across the country.

"Local stone" allows the mason to use material found at or near the site, even stone turned up during the excavating. And you can assume that the spec writer (the architect or engineer) knows the variety of stone, that it is suitable for construction, and that it is available in sufficient quantity.

Using local building materials is economical. And, not surprisingly, "local stone" up off the ground in mortared walls and chimneys gives a home a very "local" appearance, in the best sense of the word. Combined with a well-conceived design, this approach is summed up in a phrase reserved for homes that complement their sites: It looks as if it grew there.

Now you get to discover just how open-minded you really are. I'll be honest with you and say that I had misgivings about cordwood masonry in particular. Building with stone or even with rammed earth makes sense to most people. You can imagine both systems being technically feasible even if you have never cut a stone or slip-formed a masonry wall.

But even before I got into the nitty-gritty details of bearing values, shrinkage, weatherization, and such, cordwood facades struck me as odd looking—not quite as pleasing as a stack of raw firewood, although a cordwood wall features log ends and is not quite as pleasing as a random-rubble stone wall strung together with careful mortaring, although the log ends are highlighted by the surrounding mortar. At first, cordwood construction did not look like shelter to me; it looked like fuel. For me, this initial reservation was a formidable barrier. At first, you might feel the same way.

Think of the English country farmer-inventor who first experimented with harnessing methane gas from pig manure to power his small car. (Yes, it's been done very successfully.) Can you imagine the guffaws as a group of farmers stands around the pen of squealing pigs, and the inventor-farmer points down at the muck and says, "You know what I'm going to do with that stuff? I'm going to run my car with it." (Oh, sure.)

You are unlikely to find a cordwood house with parquet floors, silk-brocade chairs, and crystal chandeliers. Cordwood is rustic, a euphemism usually employed to give backhanded compliments to structures that are roughly handmade and somewhat unfinished and inelegant. But cordwood is not a construction and design shortcut. It is composed of rough pieces, but the pieces are connected.

In other sections of this book, I have suggested that if you are going to be an owner-builder of, say, your own log home, it would be wise to run a full-fledged field test of tools, skills, time, and other components of the job. You can do this by building a small wood shed or an even smaller wellhead cover—some manageable piece of work where you will discover some of the realities (like what happens to your trick back when you lift the end of a 12-inch-diameter log).

This advice may not fit into your building plans. And, of course, you will be tempted to plunge into the project, to act on your enthusiasm, to see some results. But I urge you to try at least one of the many building courses, or even something as simple as a small retaining wall. You must get your hands on these specialized alternative systems. Even a small project is a fair test, because all these building alternatives translate to full-scale, even large-scale construction.

CORDWOOD MASONRY

DEVELOPMENT, DESIGN, AND COMPONENTS

The term "cordwood masonry" accurately describes the two parts of this building system, although you may also hear it called log-end, stackwood, cordwood, or stackwall construction. You can choose one of the several names and one of the many stories about the origins of cordwood masonry.

Historians, cordwood builders, and writers have thrown out numbers like 1,000 years old, along with ideas about stacks of firewood serving as the first raw shelters when cavemen came out into the open air, and the Vikings carrying the idea to North America. Jack Henstridge, one of today's major forces in cordwood construction, takes the commonsense approach. He assumes that since fire was precious to primitive man, a supply of firewood was just as important. Assume that the firewood is chopped into sections and stacked, and you can imagine that it would make a pretty good weather break. Take this image a few more steps, and you can assume that branches were laid to project from the stack to shelter the fire from rain, and that mud and clay were stuffed into the gaps between the cordwood to seal out the wind that could extinguish the precious fire.

Following this line of thinking, the pile of wood that

Jack Henstridge's "Ship with Wings" house was built in 1974. Jack reports, "It's still going strong. I've taken off the sod roof, didn't quite come up to my expectations—leaked like a sieve." But no problems with the cordwood.

protects the fire becomes a shelter. Henstridge and Rob Roy, another cordwood builder and teacher, have been prompted to a very specific hypothesis based on what appear to be rounded cordwood shelter ruins at L'Anse de Mere Meadows in northern Newfoundland, an area that may well have been visited by Vikings. The rounded structure is assumed to have been stacked, then packed with clay, which was fired (turned into a hard, weather-resistant material, the way roof tile and chimney liner is vitrified) by lighting a gigantic bonfire in the middle of the structure.

These and other suppositions are certainly plausible. Anyone who has tended a fireplace or wood stove has stacked wood without giving history, or the Vikings, a second thought. In any case, cordwood masonry has northern roots (for obvious reasons) and was used in northeastern Canada and, in the 19th century, in the Great Lakes area of the United States.

Cordwood is an indigenous building material, that is, regional, native, natural—almost innate. No, this does not mean we are all born with stackwall brain cells, although many of us do have a tendency to gather things into piles. We may not have a predisposition to build cordwood structures, but one creature certainly does—the beaver. (Actually, some bird species build nests that are fortified with mud paste, which is a lot like stackwall on a small scale.)

Scientifically, it may be impossible to link an animal building system with a human building system. But there is some relationship. Jack Henstridge thinks so. And in the newsletter he once published, he awarded the Cordwood Builder of the Year Award to Castor C. Beaver and his family. Jack reasons as follows.

The Beavers received the award this year in recognition of the fine work they've done in the construction of an addition to their dwelling place. The enlargement of the existing dam system has also provided more space for the Trout and Wood Ducks. Naturally, before the Award can be given certain

Below, a two-story cordwood home rests on a 30 × 40-foot floating slab and features a 408-square-foot solar collecting addition. The 14 cords of red cedar cost owner-builder Richard Flatau $350. Inset, right, is a handsome cordwood mansion styled after a medieval guild hall. Interior timber framing allows clear spans to the roof peak.

qualifications must be met. The Beaver Family, naturally, used local Poplar, Aspen and other such non-commercial woods—and used clay and water plants for mortar and chinking. You can't get too much more "Indigenous Material" than that! Again this year they are using the water insulation and transport system that they have used so successfully in the past. The canal system that was developed to transport the heavier timbers to the job site has other distinct advantages too [great fishing].

Cordwood houses are a bit more orderly looking than the beaver's handiwork, but the components are the same. Cordwood builders have not improved on the basic ingredient, wood, although cedar and some other varieties seem to work better than others. But there is no way to keep masonry technology out of the other half of the indigenous puzzle, the mud.

As it is with other types of construction, the elements of a cordwood home should be adjusted to site conditions, particularly temperature and humidity. And like

Cordwood walls within a heavy timber frame are used in Rob Roy's earth-sheltered house. With the wooden walls and soil protection, Roy uses three cords of medium-grade firewood per year for heating, figuring only a 50-percent combustion efficiency. And that's up near the New York/Canadian border.

The farmhouse above, in Hemmingford, Quebec, was built in 1952, and has been subject to severe weather. The lapped corner construction is "pure" stackwall, a completely wood log building, using dimensional lumber only for window and door frames.

other systems, stackwood is a system of building with components. Each one must meet certain standards and be combined with the others in a certain way.

Overall, cordwood-masonry walls have four pieces: one interior and one exterior mortar facing (roughly 4 inches thick); a structural connection between, and embedded into, the two mortar walls of cordwood (16-24 inches long or longer, depending on requirements for insulation and thermal mass); and an insulating core in and around the log ends between the two mortar walls.

Unlike stick-built structures, where there are many ways to enclose a space (with 2×4s, 2×6s, 2×12s, web trusses, etc.), the cordwood system is constant. Mortar is built up, logs are embedded, insulation is filled in, over and over again. It is a process that displays your progress more than most, particularly when you consider that you are building a complete wall, finished inside and out, weatherized, and insulated. Now, doesn't that seem like a nice idea for a change? No more specialty acts of foundation masons, framer carpenters, sheathers, roofers, trim carpenters, all of whom must pick up where the last one left off, as though the building process was handled by one person. Of course, that is virtually impossible, which is one of the troubles with new homes put together by an entourage of specialty subcontractors.

You have several choices of wood, although the quality of debarking, seasoning, and installation will play a greater role in the longevity of the structure

Wood can be cut from trees to make framing timbers, trim, furniture, and more. Full rounds, splits, square-hewn sections, and endgrains all can be combined to make homes that look like stone houses from a distance or like houses out of a fairy tale even when you get up close.

(within limits) than the species of wood you use. It figures that since raw wood cannot be doctored, there are only a few choices. The mortar can be doctored though. Even experienced cordwood-masonry builders alter their mixtures over time. Consequently, there are a great many choices, and no real consensus on the one or two best bets. That is one of the minor drawbacks to using an unorthodox, alternative method of construction. When you lose red tape and codes, you also lose thoroughly tested and monitored standards of construction.

There is one more choice that offers a wonderful series of endless possibilities: simply how you arrange the logs in the wall. It is the most exciting part of stackwall construction. More than in any other building system, the log-end options allow you to leave your imprint on the building. Log-end arrangement enlivens what might otherwise be just another repetitive building process.

Without a full cord of wood staring you in the face, the optional patterns may be difficult to visualize, but the variety and the possibilities are endless. Like clouds, which change their shapes, change their relationship to other clouds, and change their overall aspect against the blue-sky background, a stackwall facade can be enigmatic, a personal mural, a puzzle, a giant structural Rorschach inkblot that may look like butterflies in a bed of flowers, then faces of friends, then pastry and arrowheads, then something you never saw before—even after you have been looking at the wall every day for years. All this, and it holds up the roof, too.

And there is one more double-barreled benefit. Cordwood projects are manageable, physically and financially. Instead of 16- and 18-foot lengths of solid logs weighing 200 to 250 pounds apiece or even more (and it's quite a trick to get a log like that up on top of a cabin wall with help, much less by yourself), the largest log in a cordwood project rarely exceeds 15 pounds. Even a substantial cordwood design with 16-inch-thick walls will be made of pieces weighing 10 pounds or less. No cranes. No come-alongs. No gin poles. And nearly every piece of cordwood in the wall needs just one chainsaw cut. And these are rough plumb cuts, the straightforward right angle. No mortise and tenon. No saddle notches. No dapped poles and aligned bolts.

Can you imagine putting together a house good for 100 years, requiring no more maintenance than you would need on a conventional home, for something like $10 a square foot? That's enclosed living space, not an open carport. Depending on how elaborate you get with glazing and mechanical systems and interior finishing aside from the walls, that is a shockingly low (but achievable) cost.

Shame on you if you are thinking that if cordwood is so uncomplicated, so manageable, and so inexpensive, there has got to be something radically wrong with the system. Granted, with many consumer products carrying the promise of high quality and low cost, you would be wise to suspect one or the other or both. But don't let the "good, better, best" syndrome of marketing keep you away from the unusual case where one of the good systems is also one of the easiest and least expensive.

As you build a cordwood wall, you are conscious of the separate pieces: full-width logs, two rows of mortar, and one row of insulation. As you finish a wall section, the perception changes, and the wall becomes monolithic. Careful pointing, recessing the mortar face, highlights every log face.

One of the best ways to select cordwood patterns for a wall is to look at and photograph woodpiles. Let your eyes go just slightly out of focus, and all kinds of interesting patterns, symbols, and figures emerge.

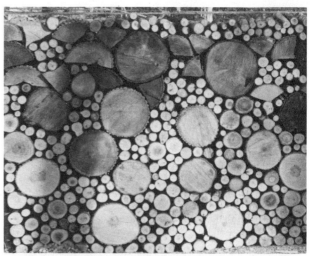

The cordwood pattern planner offers an organized way to lay out almost limitless wall patterns. Below, young Ryan Flatau, a future owner-builder, is surrounded by cordwood elements: planking on partition walls, timber frames, and log-end walls. End-grain checks are packed with oakum, which can be concealed with mortar or caulking. (Photo by Richard Flatau.)

CORDWOOD GRAPHIC PATTERN PLANNER

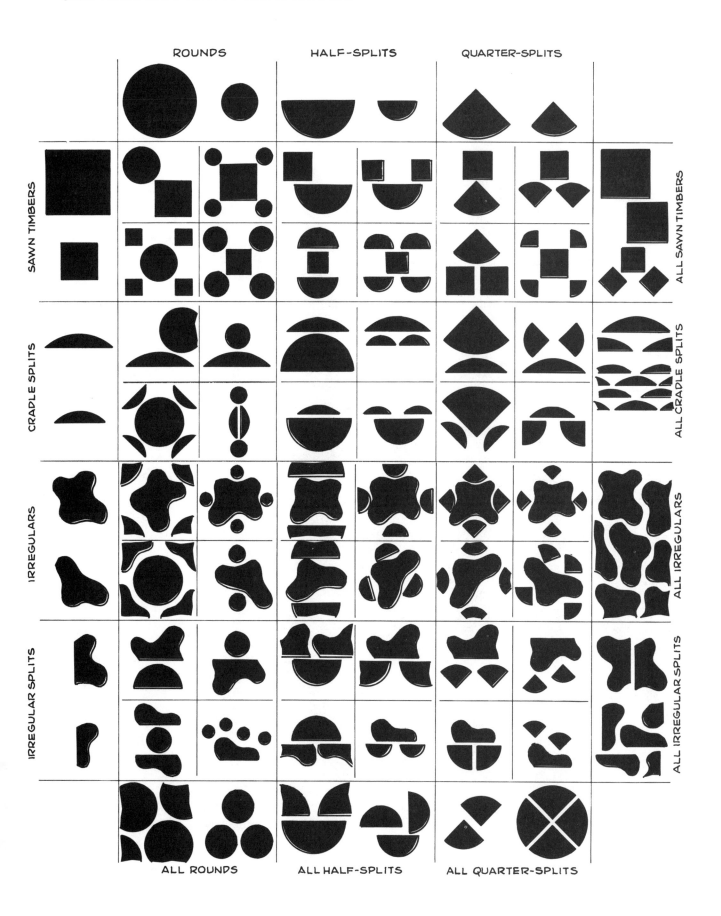

DESIGN
DECISIONS

There are stackwall Capes, gambrels, colonials, sheds, houses with log-end walls under modern, swept-wing roofs, and log-end facades protecting the exposed sides of earth-sheltered homes. And there seem to be even more stackwall homes with rounded walls, with dome roofs, with star skylights, with peculiar-looking turrets and arched openings that make the house look like a small-scale wooden model of a medieval castle.

These pictures show that with cordwood and other alternative building systems you are limited only by your imagination. This cordwood estate was built in 1929 as a guesthouse and theatre for an even larger estate. The master bedroom at right with adjoining bath shares the second story with two other bedrooms and baths, and a unique clerestory lightwell. Passersby mistake the house for stone. (This page photos, Richard Koser; next page, Neil Soderstrom.)

The alternative building–alternative lifestyle syndrome strikes again. It's wonderful to be inventive, to build the one-of-a-kind house, to include personal touches and idiosyncrasies, with colored glass bottles here and there for the stackwall equivalent of stained glass illumination. It's interesting and progressive to follow geometric formulas that demonstrate how much more floor space you can enclose with rounded instead of rectilinear walls. But all this (and there is much more) isolates stackwall from mainstream construction.

Good thing, too, right? Maybe not.

Glass bottles may look nice on the windowsill or holding candles on the dining-room table. But even though it is a nifty idea to use them as skylights, that is not a feature many home buyers are looking for. Neither is fortress-and-rampart detailing that makes you expect to see a longbow archer complete with chain mail pop up and ask you for the password. These details are fine for the owner-builder who exercises his imagination. But they do not help the idea of stackwall.

This is a heady issue. Should the emphasis be on producing more conventional designs, on lobbying for code approval, on pushing for standardized flame spread testing, on urging manufacturers to make compatible products? Or should the emphasis be (as it seems to be now) on encouraging the isolation and on preserving the low-cost system for free-spirited owner-builders working on no-code homesteads?

By comparison, the log-home industry decided on the first emphasis and pushed long and hard for mainstream acceptance. And, with their own trade associations and magazines and energy-efficiency reports, that industry now has quite a few companies spewing out "executive series" cabins with foam-packed, grooved, beaded, machined-to-death, laminated, faced "logs." You are bound to get products like that with rapid, industry-wide expansion; it's just one part of the spectrum. But you don't have to buy them. There are many log handcrafters and custom building firms turning out beautiful work who would not have had a market for their talents without that same expansion.

Commercialization isn't always all bad, after all.

In 1977, Jack Henstridge ended his book on cordwood construction with four pictures of a rather conventional looking Cape-and-ell home in Fredricton, New Brunswick. The house had 16-inch center rafters, a nice little porch under the eaves, double-hung windows, a stone chimney—practically right out of a storybook. There was an interesting caption: "The first cordwood home built within a Zoned area!"

I know there is a certain pleasure in being the only one on the block (or in the county) with the special house, in discovering the "new" old idea, the slightly experimental method. It puts you in a special place—out on the leading edge of appropriate-technology housing. That's a fascinating and worthwhile place to be, because in addition to building yourself a home, you confront issues of the environment, community, empowerment, bureaucracy, politics, money value, time value, and much more—unless you close your eyes and ears and focus exclusively on your own plan, your own lifestyle, your own home.

This 700-square foot cordwood addition to Pat's Gourmet, a restaurant in New Brunswick, cost only $16 per foot. Jack Henstridge used split blocks and large beams to create this large, rustic, clear-span space.

Malcolm Miller's house was the first cordwood home in a zoned area near Fredricton, New Brunswick. The 3000-square-foot house with interlocking corners is heated with a woodstove. (Photo by Jack Henstridge.)

It's a cordwood sauna—that's right, cemented full face on the inside, with the outside finished just like the walls of the house. When it's used this way, you can appreciate that cordwood, for all its unique appearance, acts like aggregate in concrete to make a solid wall.

Whether you're building a small bathhouse or a barn or a home, it is easier to fill in the spaces with cordwood than to create the boundaries. Use dry timbers (old ones if you can find them) to minimize joint openings due to wood shrinkage where mortar meets timber.

I hope more people will bargain with the building inspector, use stackwood in overbuilt and otherwise mainstream homes designed not to arouse the building department to cite obscure building codes, or the local planning board or even the local community to adopt exclusionary facade standards. It's a good fight to be in. And if you lose, you can always try across the county or state or province line, where there are fewer people, fewer houses, fewer banks, and fewer building inspectors. But why take the trouble to go out to the leading edge, to be part of an appropriate-technology spearhead, if you don't care about the spearhead direction?

Those who will reap the rewards of low-cost, manageable, indigenous-material stackwall construction would do well to make the system even more manageable and more acceptable to a growing number of owner-builders. It's not a crusade. It doesn't have to become your life's work. The idea is to do some prudent proselytizing, to have your individual project make a collective, if only a marginal, difference.

On a more practical level (although the difference between a round and right-angle house can produce a lot of practical side effects), you will have to make a choice between using pure stackwall and stackwall filler inside timber framing. Pure stackwall is the easier method by far. Using stackwall like a high-quality, appearance-grade, wattle-and-daub filler introduces another building system—timber framing—with some special quirks like extra width to accommodate the thick walls needed in cold climates.

On the other hand, the timber framing for a rough stackwall filler need not be the cabinet-quality joinery used in homes where the frame and not the filler is the heart of the structure. Still, it is a big piece of work on a good-size house, and to some extent it defeats the simplicity of the stackwall system. Framing does look great, of course, defining the architectural lines of a house more clearly than the piecemeal effect of cordwood. And if you are planning something special for the roof, such as earth sheltering, you might appreciate 6×6s or 8×8s under critical beams.

Since there is an entire section on timber framing earlier in this book, please refer back there for the nitty-gritty details—with one exception. With a stackwall filler, you will be laying wet mortar up against the timbers. Green frames will pull away from the drying mortar, and if headers are green as well, they

Old barn beams are ideal for cordwood frames. Rough or hand-hewn surfaces are fine, as any gaps or irregularities are bridged by mortar.

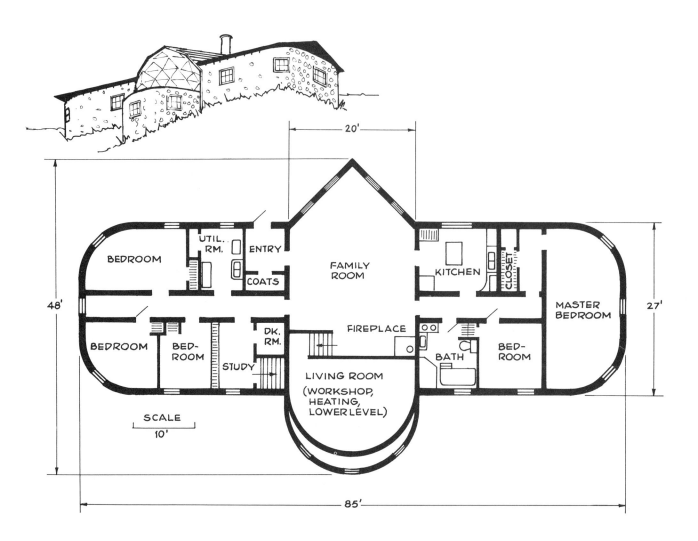

Jack Henstridge's house eliminates all right-angle corners from cordwood walls. Jack notes that the tight radius of the dome-covered living room/workshop area could have been improved by making the curved wall with a larger radius and minimizing the excess mortar on the outside wall face.

may apply twisting and racking forces that disrupt the cordwood-and-mortar filler. Diagonal bracing, a traditional feature of European timber frames, will help but will create more boundaries against which split sections of cordwood and mortar must be custom-fitted. It looks elegant, but it's time-consuming.

If you stick with pure stackwall, you stick with simplicity of construction until you get to corners. (That's one reason, aside from the increased interior floor space, that some builders like circular cordwood walls.) Full round or split logs nestle together quite well. But every pile of firewood needs some support at each end of the pile; otherwise, the pile spills out at the ends.

With pure stackwall, the solution to this problem is to interlace the corners. Instead of running cordwood lengths perpendicular to the wall facade, you turn lengths parallel to the facade alternatively, in the same way you would build a box-type campfire. This detail eliminates potential weak corners of a stackwall building. And the parallel lengths may be left rounded, or hewn, or sawn with a square face exposed to the weather to create a sleek, symmetrical corner detail.

The heavy-duty design decision is, first, to experiment with stackwall and, second, if you like it, to go ahead and build your entire house out of firewood and mortar.

Stackwall corners need support during construction until the walls meeting at right angles are fully interlaced. Quarter-sawn logs or square logs, long and short lengths, all can be worked into a pattern.

STACKWALL CORNERS

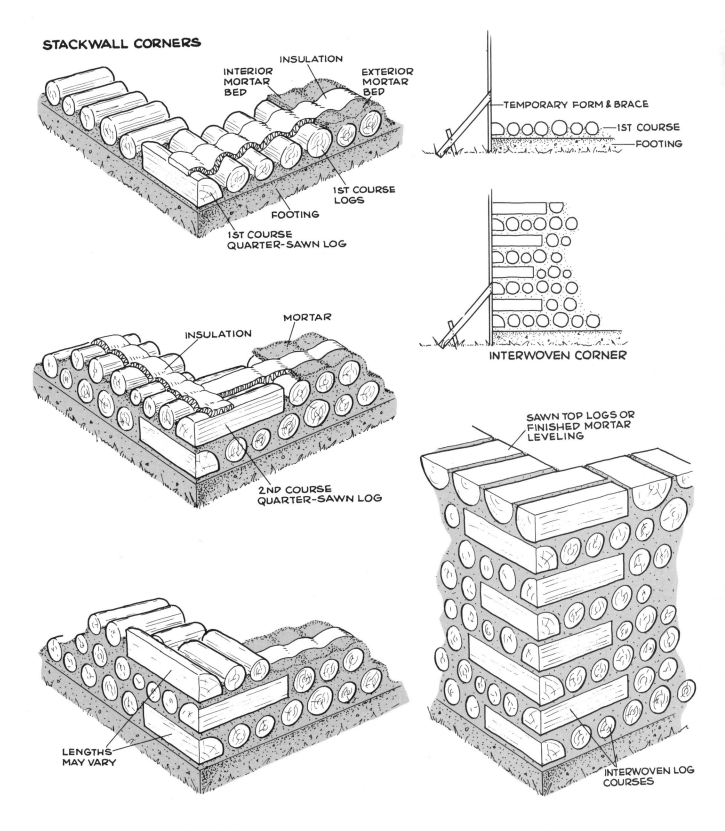

INTERIOR MORTAR BED

INSULATION

EXTERIOR MORTAR BED

1ST COURSE LOGS

FOOTING

1ST COURSE QUARTER-SAWN LOG

TEMPORARY FORM & BRACE

1ST COURSE

FOOTING

INSULATION

MORTAR

2ND COURSE QUARTER-SAWN LOG

INTERWOVEN CORNER

SAWN TOP LOGS OR FINISHED MORTAR LEVELING

LENGTHS MAY VARY

INTERWOVEN LOG COURSES

PREPARATION FOR CONSTRUCTION

If the width of your stackwall is in the ballpark of a normal foundation—say, anywhere from 8 to 12 inches wide—conventional concrete or concrete-block combination footings can be used successfully. The width of the stackwall should be a function of the insulating value you want. Mortar, which is a good conductor and a bad insulator, is not the problem, since stackwall mortar beds in and around the logs are set in two completely independent layers: one to the inside and one to the outside, separated by sawdust, fiberglass, or another insulator. But the logs are full width and can offer no more than their natural insulating capacity, which is not so bad.

With all the news about energy-tight houses and superinsulation, wood gets lost in the shuffle. Only when remodelers who buy aluminum- or vinyl-framed replacement windows discover that the glass stays dry but the frames sweat like crazy do they remember that this didn't happen on their old wooden-framed windows.

The exact R-value through a log is difficult to pin down. Generally, hardwoods with dense, tightly packed fibers transmit temperature more efficiently than less-dense softwoods. For most softwoods and poplar, it is reasonable to figure about R-1.00 to 1.25 per inch of length, even allowing for an increase in heat trans-

Cordwood is one more piece in the energy-efficiency puzzle. It provides reasonable R-values for above- and below-ground buildings. Rob Roy reports that with cordwood, earth sheltering, and a deep northeastern snow load, this house needs only three cords of wood for winter heat.

mission along the end grain. This is generally considered a bit less resistant to heat transmission than parallel grain—the way logs are placed in a typical cabin.

A little multiplication tells you that a modest, 8-inch stackwall would provide an R-10 through the logs; a 12-inch stackwall, roughly R-15; a 16-inch stackwall, about R-20. In a wide wall, the mortar layers may add no more than R-3.00 to the wall. But the inner core of fiberglass insulation (about R-3.5 per inch) could add R-21. Roughly R-20 through the wood and masonry components of the wall is pretty good. But don't make the mistake of designing and building only by the almighty R-value.

There are so many other variables. Any large pile of dry cordwood will have many pieces with end-grain checking, cracks that open up across the radial growth rings. Some of this checking may be skin deep, and some may greatly decrease the thermal resistance of the log; it may require a little handwork, stuffing loose-fill fiberglass into the recesses.

If the cordwood you use is not all that dry and shrinks as it ages, or if you do not do a thorough job of debarking the logs, so that insects underneath eat at the wood and the bark breaks from the solid wood, all kinds of cracks can open up, making the mortared stackwall more like a mortarless stack of firewood.

Finally, there is thermal mass. (See page 73 for details of mass gain in log walls.) A massive, 16-inch-thick cordwood masonry wall stores a lot of heat. Some of it is lost to the outside, but a significant amount is reradiated back into the living space. The masonry portion of the stackwall will take some time to heat up, but since it is backed by a continuous insulating layer, it will reradiate most of its stored heat back.

And there is yet another tradeoff. You might want to avoid hardwood like oak, partly because it takes a long time to dry but mainly because its high strength and density, otherwise attractive features, detract from a stackwall's thermal efficiency. Because the hardwood is more solid, denser, and heavier, it provides a mass-gain increase. Acting more like stone than firewood, it takes longer to heat up and holds and reradiates the heat more efficiently.

Playing this game by the R-values (heat-flow retardance) and U-factors (heat transmission) can get ludicrous. So why split hairs? A few more notches of energy efficiency is not the reason for using or avoiding stackwall. The quality of construction and the consistency of your mortar mix will, in the end, make more of an energy difference than whether you use oak or poplar. And if you think you have this figured out now, consider the excellent advice of Rob Roy, who puts cedar at the top of the list because of the balanced combination of thermal efficiency, mass gain, and the bonus of resistance to rot and insect damage.

Cordwood is as versatile as the diameter of the logs you use. With manufactured block, the form must conform to the material limits. With cordwood, you have a forest full of choices in diameter, shape, density, even color.

CORDWOOD

Different materials respond to stress and moisture and temperature and everything else at different rates. Also, each type of material has its own range of response, so you have some margin to play with. The idea is to maneuver two different materials (mortar and cordwood) to the ends of their respective response ranges, where their different qualities are most alike. It's sort of like arbitration, finding the common ground.

For example, by adding sawdust to your mortar mix, you can make it more stable. By using dry log ends (even if you have to fill checked openings), you can make the wood more stable. There is still a difference between the two materials, but if you are selective and careful, it is a manageable difference.

Some legwork may be in order: finding old, completely dried timbers from an unused barn (or from an antique-timber-supply firm—to use in your framed stackwall; buying cordwood that is well checked, down and cut for two years (maybe a year if the logs are split); or felling, cutting, and debarking the cordwood in the spring, when sap is running and you can get the bark to pop off in great chunks, then storing the wood to air-dry, protected from direct exposure, for a minimum of a year.

Depending on the appearance you like best, you can use bolts of wood roughly equal in diameter, or a wide variety of diameters nestled together, or all split sections, or a combination of splits and rounds. You can determine the amount of wood needed in face cords, a 4-foot-high by 8-foot-long section of wood cut to the depth of your wall. Simply divide the total square footage of all your stackwalls by 32 to arrive at the number of face cords needed. For instance, a 20×30-foot house with 8-foot-high walls has 100 linear feet of wall, or 800 square feet. Divide by 32 to get 25 face cords needed. If your walls will be 12 inches thick, you will need 25 4×8-foot stacks of wood a foot deep. Since you can get four of those stacks from a full cord of wood (4×4×8), the 20×30-foot house will require roughly (allowing for some waste) 6½ full cords of wood.

Finally, as a general rule, you can safely subtract 10-20 percent of this amount, allowing for the area in the walls occupied by mortar. With many large-diameter, full-round logs using a lot of masonry filler, figure close to 20 percent. With many split logs nestled closely together, figure 10 percent.

Don't dismiss the stackwall idea because you do not have access to the ideal cordwood. Hardwoods are okay, and although cedar offers special benefits, many low-density species such as spruce, tamarack, pine, and poplar have been used successfully.

MORTAR MIX

I am not going to send you after the one-and-only cordwood-masonry mix—there isn't one. I have found this to be true of chinking mixtures used by log builders as well. Slightly different mixes seem to be preferred not only with different wood species but also with different percentages of moisture content in the wood (this will affect how much moisture is drawn out of the mortar mix as it cures in place) and with different site environments. Most open-minded builders admit to discovering improvements in their ultimate formulas with each new home they build. That does not mean they were wrong the last time around, just that they are learning—a very important part of their craft.

Forthright admissions by Jack Henstridge are instructive. In his book, a special chapter "If We Had To Do It All Again" covers job improvements such as buying old telephone poles or dead elm to ensure dryness and eliminate time-consuming caulking and oakum packing around shrunken joints between cordwood and masonry. (Elm is tough to split, though, so you might consider using the logs whole.) The book mentions design improvements like going to 16-inch wall depth for stability, ease of construction, mass gain, all without increasing mortar; also, the money-saving idea of buying hydrated lime for the mortar mix in bulk.

Hmmm. Hydrated lime. Why not concrete? The lime acts to preserve the wood, for one thing. It also whitens up the mix, making it more visually compatible with wood tones. And it extends the time during which the mix remains tempered (easily workable without the addition of water).

Bulking the mortar mix with powdered sawdust makes the mix go further. Also, you can fiddle with the ratio of wood to mortar on the facade. The wall illustrated is just about in the middle. More mortar, and the wood gets lost. Irregular log shapes here could be nestled more efficiently to cut down on large patches of solid mortar in the facade. But skimp on mortar, and the cordwood will behave more like pieces of wood than a cohesive wall.

Sawdust is another component of the cordwood mix you won't find in concrete. Powdered sawdust increases the mix bulk (a kind of Hamburger Helper for mortar), stiffens the mix by absorbing excess moisture, extends the wet-cure time of the mix by slowly filtering moisture held in the powder back into the mix, and acts as a rudimentary air-entraining agent by shrinking as it gives up moisture to leave insulating entrapped-air spaces.

For continuity, you should wet-mix ingredients before adding sawdust. And whether you use a sand base or gravel base, screen the ingredients so you won't have to be bothered picking oversize stones out of nooks and crannies between logs as you build. Here are two mixes, both tried and tried again by two of the most experienced cordwood builders.

Henstridge: 2 parts Portland cement; 1 part hydrated lime; 10 parts pit-run gravel (through a 2-inch sieve); water to make a very wet mix; 10 parts sawdust.

Roy: 7 parts sand; 3 parts masonry cement; 1 part hydrated lime; all dry mixed (through a ½-inch sieve) with 5 parts sawdust; then mixed gradually with water.

In all cases, the final consistency should be stiff enough to ball up, to shape roughly in your hands. The typical soupy consistency of concrete mix will not support the logs and will make a mess of your nice, clean cordwood.

It makes sense to tinker with the proportions on a small scale. The amount of moisture already in the aggregate base of the mix (wet or dry sand), and the temperature, humidity, and exposure to direct sunlight as you mix and work on the wall can gang up to make the textbook numbers produce gloppy or crumbly mortar. You have to get your hands in there, make mud pies, and add small amounts of water or sawdust as needed.

This cordwood workshop addition (inset) has 18-inch walls, and requires only minimal electric heating. The 2000-square-foot cordwood building below is a Farm Market in Mangerville, New Brunswick. It's also used for neighborhood barn dances. The cordwood is so energy efficient that exhaust from the market's refrigeration units provides the heat, with a fan boost on winter nights. (Photos by Jack Henstridge.)

CONSTRUCTION

You can imagine that the footing for a stackwood wall must be one large, monolithic hunk of reinforced concrete—at least 24 inches wide and 12 inches deep. Following the builder's rule that footings should be two times wall thickness and as deep as the wall is wide, a 16-inch stackwall requires a monstrous pour 32 inches across and 16 inches deep—about 3½ cubic feet of concrete per linear foot of wall.

A very modest 20 × 30-foot house (100-foot perimeter) with 16-inch stackwalls would use 355 cubic feet of footing concrete, about 13 yards. Add a 6-inch slab (20 feet × 30 feet × 6 inches) for another 300 cubic feet, about 11 yards. That's quite a ready-mix order, definitely not a mix-your-own job. Even if you have a power mixer, you'd never be able to keep the concrete uniform. And even if you have help, you would probably run out of aggregate or cement or sand or gas or strength or daylight before you got the pour leveled and puddled. (A rosy picture, isn't it?)

According to the Indigenous Materials Housing Institute (IMHI), stackwood is built most economically, although by no means exclusively, directly on a well-graded slab. What they have in mind is more elaborate than a batch of concrete dumped on the ground, however. Here it is from the bottom up:

1. Excavating down to level, compacted subsoil or ledge.
2. Leveling compacted gravel to form a 6 to 12-inch base, depending on site drainage.
3. Installing plumbing drain lines below slab level with slope into the gravel.
4. Installing perimeter-foundation drains, depending on site characteristics.
5. Embedding 4-inch-thick, 4-foot-wide, rigid-styrofoam insulating panels in the gravel around the perimeter.
6. Laying 4-mil plastic for a vapor barrier. (I'd suggest that 4 mil may tear as you lay it, as you set up rebar supports and as you pour, unless everything is handled tenderly.)
7. Laying a network of welded wire (reinforcing mesh) or ⅜- to ½-inch steel rod 8 inches on center on stone supports. (I'd consider commercial rod

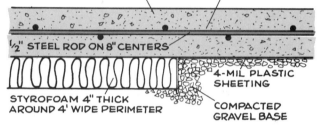

REINFORCED SLAB

2400 PSI CONCRETE 6" THICK

STEEL HELD AT MID SLAB BY PLACING ON ROCKS OR ROD CHAIRS PRIOR TO POURING CEMENT

½" STEEL ROD ON 8" CENTERS

4-MIL PLASTIC SHEETING

STYROFOAM 4" THICK AROUND 4' WIDE PERIMETER

COMPACTED GRAVEL BASE

This is the slab system recommended by Jack Henstridge: a base of compacted gravel over compacted or undisturbed soil with adequate bearing pressure, perimeter rigid insulation, vapor barrier, and a reinforced slab made of 2,400-psi concrete, 6 inches thick.

hangers, which are somewhat more stable than stone and may help you to maintain critical reinforcement level in, instead of at the edge of, the pour. Also, stones tend to kick out under the force of a pour from a ready-mix truck.)

8. Pouring a 6-inch-thick slab of 2,400-psi concrete.

While you allow the concrete to cure (and that's wet cure with one of the proprietary films or by the less-reliable method of covering with burlap that is periodically soaked), take this time to organize your tools and materials.

You will be able to use the time profitably, because all but the most experienced builders generally underestimate the enormous bulk of a houseworth of cordwood, the amount of effort but, more important, the time needed to fell it, haul it, cut it, debark it, stack it, season it, clean up the cut ends, and get it into position by the stackwall.

Remember the 20 × 30-foot house requiring almost six full cords for 8-foot-high walls? That's six stacks of wood measuring 4 × 4 × 8 feet, 128 cubic feet per cord, 768 cubic feet total—a 10 × 10-foot room of wood nearly 8 feet high. No forest is that dense. It's a lot of trees.

If, like most owners and builders, you do not try to build the house literally from scratch (from standing trees to finished walls), there are still some potential

random pattern is a pattern) as you go, as the logs and your intuition dictate. As you take the opportunity during curing time to transport logs to the building perimeter, look them over. If you spot half a starfish or a quarter moon, put it aside and wait for the other half to show up. It's probably in there somewhere.

The best analogy I can make for building a cordwood wall is shooting pool. Every time you lay a piece of wood in the wall, you've got to be thinking about the next piece; when you pick up one piece of wood, you have to look for the combination shot—the second piece it will work with to make a cradle for the third piece. When you get done with a wall panel, you can stuff in the last little bits of mortar, or cheat a little and lay the top course of mortar, then add the frame, shown below.

roadblocks. Buying 6 to 10 full cords of wood is not like buying a pack of gum. It's not neat and tidy. To get a decent price, you may have to travel, arrange hauling, and deal with people who are not accustomed to dealing with owner-builders. You must buy in the right season for debarking and, even with seasoned wood, leave enough time for hauling and building before the weather shuts down your job.

Once you start building the walls, it seems like a distraction to go back to the pile for more logs. You're busy. You're making progress. You should be done with the carrying. It's bad enough that you have to stop to mix more mortar. That's part of the job, of course. But I'm not talking about a rational feeling. It's just that you are likely to tire of the mixing after a while because, ideally, you should do it exactly the same way every time, and that's boring. You can't mix too much at once, or it will set up before you can use it, and that breaks the boring job of mixing into several boring steps.

The cordwood, on the other hand, is all there, all ready to go. Each piece is a little different. Each log may be turned inside for outside and placed in one of several possible locations. Each piece is just enough of an adventure. It's really fun! You can treat the job like a free climb where you look for handholds as you go. You can make up the log-end pattern (and even a

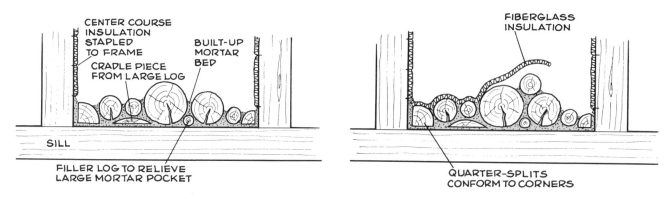

CENTER COURSE
INSULATION
STAPLED
TO FRAME

CRADLE PIECE
FROM LARGE LOG

BUILT-UP
MORTAR
BED

SILL

FILLER LOG TO RELIEVE
LARGE MORTAR POCKET

FIBERGLASS
INSULATION

QUARTER-SPLITS
CONFORM TO CORNERS

Some cordwood builders take the trouble to stack a preliminary selection of logs, just to get an idea of how they work together and how the wall will look. As you build, tack insulation to the frame (a few staples will do) and turn major checking openings down away from vertical, where they would otherwise trap water.

MORTARING

There is no way to specify just how much mortar you should mix. But try starting small, particularly if you're building under the summer sun, which dries out the mix. Even after you get underway, you should not lay up more than about 3 feet in elevation at one time (like staged lifts with poured concrete), because further loading will compress even stiff mortar.

Jack Henstridge recommends 2 to 2½-foot lifts with four or five hours' set time before further loading. I wouldn't get stuck on the 2½-foot lift limit. If a few large logs take the wall up to 3 feet or so, no problem. But I would not reload the wall until the next day. Don't worry, if you do hit the lift limit before dark, there will still be plenty to do.

Since the high compressive strength and the great weight of the finished wall will keep the structure in place, it is not necessary to add an elaborate anchoring system. Bolts, spikes, even protruding cut nails will probably just get in your way. But if you have the space between the logs, a few such anchors wouldn't hurt.

For 12-inch or wider stackwalls, you start with two 4-inch-wide furrows of mortar separated by a 4-inch (or greater, depending on log length) insulated airspace. Contrary to the pillowing technique used with block construction, where you run a groove down the center of the mortar bed, in stackwall the initial mortar beds should be full-bodied and uniform. Ideally, you should leave at least an inch of mortar between logs and more like 2 inches, at least, between the first logs and the slab or foundation.

Before we get too far along, though, be warned that working hour after hour with cordwood and mortar will tear up your hands, even if you are used to construction. Once the ends of your fingers get raw from the water and abrasive aggregate (and it can be brutal), the fun will fade, believe me. So even though you may associate rubber gloves with doing the dishes, get a pair and use them right from the start.

True to the promise that cordwood is low-cost and manageable, the few other tools you will need are a chainsaw or cutoff saw to make even log ends, a big mortar box, a mason's wheelbarrow for extra mixing and carting, trowels, shovels, and other earth-moving and basic masonry tools that you probably have already.

Remember that the mortar is not glue to make the pieces of wood stick together. The mix is more of a filler than a binder, a bed for each log to lie in. Even though you work the wood in 2 to 3-foot lifts around the perimeter, you will find that large-diameter logs make more of an impression, and sink into the mix more deeply, than small-diameter logs. That's no surprise. But if the mix gets squeezed out of a joint between logs, it won't be as strong, weathertight, or durable.

You can minimize this settling with a process called cradling. The idea is to split a large-diameter log near its edge, making a piece of cordwood 4 to 6 inches or more wide across the flat and only an inch or so thick, then lay it flat-side-down for maximum bearing against the mortar bed. When mortar is laid on the cradle piece, it helps to support a large log with increased resistance to compression.

While heavy logs can sink in too far, light logs usually need a little help—a shove or a shot with the heel of your hand to embed them. Even though it takes more patience and more touch to work with odd sizes and with both rounds and splits, the variety has more potential than do logs of roughly equal diameter. And think of the waste of time and material in eliminating large logs from the tree trunk and small logs from the straight branches.

BASIC APPROACHES

There are three basic plans you can use to guide your work on the stackwalls, three approaches that affect construction time and appearance.

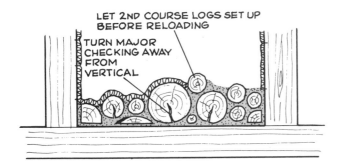

LET 2ND COURSE LOGS SET UP
BEFORE RELOADING

TURN MAJOR
CHECKING AWAY
FROM
VERTICAL

The plan for Rob Roy's cordwood sauna building provides some guidelines for the relationship between timber frames and cordwood filler walls; also, note the interesting use of a central piece, a large log that focuses attention on the wall.

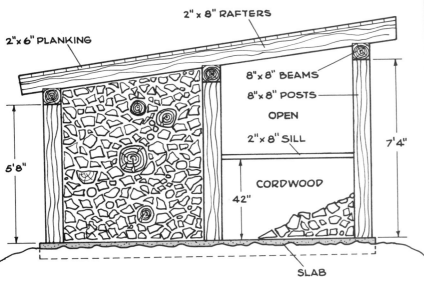

2" x 6" PLANKING

2" x 8" RAFTERS

8" x 8" BEAMS

8" x 8" POSTS

OPEN

2" x 8" SILL

CORDWOOD

42"

5'8"

7'4"

SLAB

Building for structure. With this approach, you concentrate on the support system, on cradling large logs and maintaining consistent mortar margins. When you reach back for another log, you look for one that best fits the mortar bed. You should wind up with a random pattern, possibly a bit lopsided in sections, but a unified appearance.

Building for pattern. With this approach, you emphasize the pattern of logs, the special combinations that please you, centering large logs, or running rows of logs with increasing diameters to get a wave effect, for example. Expect to spend more time with this idea because the preconceived plan, particularly if it is complex, will at times conflict with the most sensible structural solution. You'll be tempted to crunch some

of the logs together and spread others apart with widened mortar pockets to maintain the pattern. A little stretching is okay as long as you do not stack mortar joints. Cap each masonry pocket with a log.

Building for random mix. By putting cradle splits near large rounds near small rounds near half splits—really mixing them up—you have a lot of freedom to pick the best log for the structure, and the result counteracts the piecemeal effect of cordwood. If your eye does not get pulled to a specific pattern, for example, or to a lopsided group of logs at the bottom of the wall, you tend to see the entire wall as a cohesive unit. "Random" is not really the right word, because a plan to avoid a specific pattern produces its own overall pattern.

Ideally, you can use all three approaches. Given a choice, and remembering that you don't have the rest of your life to lay out all the possibilities before adding the mortar, I'd suggest you keep the preconceived designs to a minimum and let the logs fall where they want to—at least on your first cordwood building.

INSULATION

The Northern Housing Committee (NHC) of the University of Manitoba (one of the groups paying attention to stackwall) allows for 3-inch-wide mortar beds, leaving 10 inches of insulation in a 16-inch stackwall cavity. Rob Roy, Jack Henstridge, and others use 4-inch mortar beds for added strength. In his book *Cordwood Masonry Houses*, Roy suggests that the sawdust included in the mortar mix compensates for the loss of 2 extra inches of insulation, and is a positive trade-off.

Building stackwall within timber frames, Roy tacks up layers of fiberglass to the frames and wraps fiberglass in and around the log ends. Vermiculite and other loose fill insulators also work well. But the most interesting and economical insulator is sawdust.

The NHC suggests a sawdust mix of one shovelful of hydrated lime with each wheelbarrow of dust. This buffering increases resistance to insect damage and fungus. The resulting sawdust-lime mix is rated from R-3.0 to R-4.0 per inch, so that 8 inches inside the two, 4-inch mortar runs of a 16-inch stackwall rates between R-24 and R-32. (That last number, like the highway mpg numbers on new cars, is a bit optimistic.)

Jack Henstridge has experimented with sawdust-lime insulation—soaking it with water and packing it down into the insulating cavity. That may seem like the last thing you would want to do to dry insulation, but Jack's method does give a long, wet cure to the mortar as the sawdust dries. I might not try that, though, because I'd worry about the effect of the wet environment on the logs, and also about the final configuration of the bulked and soaked sawdust in the cavity after it dries. But what an interesting idea.

WINDOW AND DOOR OPENINGS

Stackwall works best when it is continuous. Of course, that makes the house kind of dark and hard to enter or leave. But each time you open the wall for a window or door, you create the same condition that exists at each end of a long stack of firewood—the tendency to spill out.

The NHC wants you to include lintels and sills. Over a big window in a 16-inch stackwall, that means some-

Openings can float in the cordwood wall or tie into a timber frame. This cordwood-and-frame house combines a traditional square frame, heavy-timber diagonal bracing, fixed glass angled to complement the diagonals, and right-angle vent windows above.

thing like three 6×6s glued and bolted together running at least a foot into the cordwood-masonry walls on each side of the opening. The tendency to spill logs off the stackwall is accentuated at door openings, because the interruption, while not that wide, nearly severs the cordwood construction (say, 7 feet of door height in an 8-foot-high wall).

Simple 2-inch nominal, dimensional frames will not do the job; they are likely to bow into the opening. You can attach wire mesh to the mortar side of the frames and set half-driven spikes to grab the mortar mix, but if the frame bows, it will pull chunks of the mortar loose along with it. You might get by (many have) with the protruding nailheads plus spikes driven from the inside faces of the frame into the cordwood surrounding the opening. But there is no substitute for a 4-inch or greater frame.

Now the expedient choice to use pure stackwall instead of stackwall within a timber frame doesn't look so expedient after all. If you plan for a lot of doors and windows, you'll use a lot of timbers and spend a lot of time on frames, so much so that it might seem more sensible to go ahead and frame the darn house, then fill in with cordwood masonry.

As usual, the choice is not that clear cut, because cordwood plus masonry is a lot stronger than a raw stack of firewood, and because there are many ways to add support to the openings.

First, even if you build a stackwall 16 to 24 inches

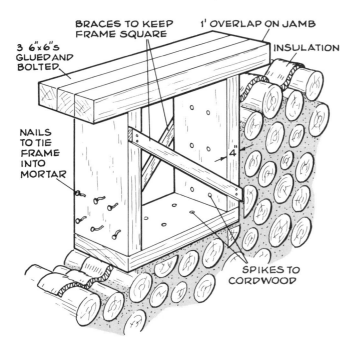

BRACES TO KEEP FRAME SQUARE

1' OVERLAP ON JAMB

3 6"x6"s GLUED AND BOLTED

INSULATION

NAILS TO TIE FRAME INTO MORTAR

4'

SPIKES TO CORDWOOD

The National Housing Committee of the University of Manitoba recommends heavy-duty framing. The frame must be squared and well braced to resist bowing as logs and mortar bear against it. Nailheads outside the frame can grab the mortar, while spikes through the frame (hidden by stop) tie into the cordwood.

Dimensional timber (limited to 2×12, or $11\frac{1}{2}$ inches actual) limits applications where you want a full-depth frame on thick walls. Recessed frames can be made in heavy timbers if they are centered under the load. By strengthening the mix around the window or door opening with rebar, or by adding a lintel on larger openings, minimal finishing frames can be built to custom sizes.

STACKWALL OPENINGS (SECTION VIEWS)

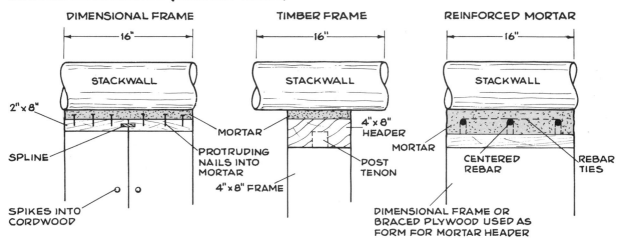

DIMENSIONAL FRAME

16"

STACKWALL

2"x8"

SPLINE

SPIKES INTO CORDWOOD

MORTAR

PROTRUDING NAILS INTO MORTAR

4"x8" FRAME

TIMBER FRAME

16"

STACKWALL

4"x8" HEADER

POST TENON

REINFORCED MORTAR

16"

STACKWALL

MORTAR

CENTERED REBAR

REBAR TIES

DIMENSIONAL FRAME OR BRACED PLYWOOD USED AS FORM FOR MORTAR HEADER

deep, you do not have to build headers and frames and sills to the full thickness. Structural loads are not sneaky. Say you have a loaded rafter over a 16-inch stackwall with a doorway cut underneath. The load will not wriggle past the header to the last outside edge of the cordwood. If you center a 4×8 header on 4×8 posts on a 4×8 sill in the middle of the stackwall, the loads from above will find the header and use the posts to get down to the ground. This means your framed openings will be recessed, which has the advantage of keeping all the joints a bit out of the weather. And there are other options:

SETTING-IN rigid, mortise-and-tenon timber-frame openings that float in the stackwall without connections to an overall structural timber frame.

CUTTING keyways into the mortar side of dimensional frames matched to keyways formed into the stackwall as you build, then tying the two elements together with a spline (like 1×4 stock).

FRAMING with splined, 2-inch dimensional timber (two 2×8s or two 2×12s), wrapping three rows of rebar (bent around the frame corners) a fat inch away from the frame in a wider-than-normal band of mortar.

BUILDING arched openings (Palladian windows for stackwall, no less!) with tightly fitted cordwood in minimal mortar beds, where the "keystone" is a large-diameter, quarter-section split. The drawback to this nifty idea is the custom windows needed for the custom openings.

CORNER LACING along the vertical walls of openings (you still need a header), to create the same effect as laying the occasional deadman railroad tie in a retaining wall.

As in most alternative building systems, there is room for finessing, experimenting, and splitting the difference, when you are up against a problem, to make a positive compromise.

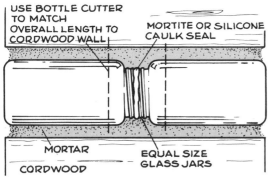

USE BOTTLE CUTTER TO MATCH OVERALL LENGTH TO CORDWOOD WALL

MORTITE OR SILICONE CAULK SEAL

MORTAR

CORDWOOD

EQUAL SIZE GLASS JARS

Where bottles are used for intermittent light access through the wall, try Roy's method rather than simply laying soda or beer bottles in mortar. He cuts bottles down to 4 or 5 inches in length, then butts two of equal diameter together with Mortite. It's kind of a thermopane glass bottle miniskylight.

Colored bottle ends are used instead of small-diameter filler logs on Rob Roy's Long-End Cave—do-it-yourself glass block. This facade also features a continuous header that ties all vertical frames together.

BUILT-INS

Piece-by-piece construction coupled with great compressive force in a stackwall gives you the opportunity to add some specialized pieces that cantilever past the wall facade on one or both sides.

But first, consider that all the logs can project beyond the mortar—in other words, be shown in relief by recessed mortar. When you point around log ends, and that's an essential step in construction, you push hard enough to recess the mortar anywhere from ¼ inch (just enough to notice) to nearly 1 inch on wide walls made of large log ends. This relief pointing accentuates the logs and makes the stackwall more like a three-dimensional sculpture.

Pointing is not just a cosmetic operation, although it improves appearance. Without pushing so hard that you send mortar into the insulation cavity, a little compression around the log ends gets the mix a bit

denser, improves the bond between mix and log, and decreases the amount of patching and caulking you'll do in following seasons.

And pointing is a nice way to wind down from a day of rough work as the light fades. I could write you 10 pages on how staggeringly important little touches like saving some pointing for the end of the day can be: a time to catch your breath and catch up on what all the effort amounts to, what it says about you and how you feel; what's up for tomorrow; there's a spot that could use a touch more mortar mixed in; what an amazing pattern in those radial rings; shoulders a little sunburned today; a really purple sunset tonight; time to wash up the tools; smells like hamburgers and onions for dinner.

You will put a lot of yourself into the wall. Don't miss the opportunity to pull some very special things back out.

The most obvious, extremely useful cordwood cantilevers are small-diameter branches. Like Shaker pegs set into wood strips all around the room for hanging coats, tools, and even chairs, they make the ultimate indigenous-material hat rack. And don't be too quick to toss away that peculiarly gnarled and twisted branch stem. Why not curved pegs, too?

If you have the patience to plan ahead for split-log shelf supports, even half-log stair treads, you can incorporate a lot of add-ons into the structure. Projecting a run of stair treads (difficult enough to lay out in the shop), or even three or four split cantilevers 3 feet apart and level enough to hold a shelf or a mantle, can destroy any preconceived cordwood pattern. This is the time to let form follow function, to build a strong wall that holds dishes and coats and books, and holds up the roof.

CURVED WALLS

In other parts of this book, I have not hidden the fact that I am not a big fan of curved walls. I am dead against them when they are curvilinear architectural fantasies handed to the builder, who must then execute them with rectilinear building materials. (One experience springs to mind—a nightmare that included back-kerfing lattice strips to make them lie down around a curved soffit built with custom-cut 2 × 8s and built-up layers of ⅛-inch laminate board.)

But cordwood in the round does make some sense— even to me. You do away with corner lacing and cold corners in winter. As long as you keep the curve gradual, you will not even notice the discrepancy between tight mortar joints inside and wide mortar joints outside.

Remember, if you build a 24-inch-thick wall section with only a 6-foot radius to the inside wall, for example,

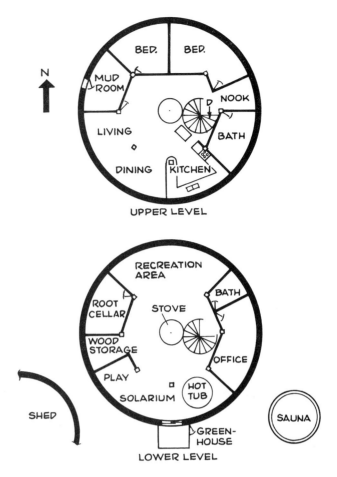

Earthwood is the name given to Roy's most recent and ambitious project. It serves as house and school for cordwood and earth-shelter students. Curved walls make for unconventional floor plans. Here, you can build partitions only as you need them. Note that large areas on both floors are left open.

the log ends better be close together on the inside face, because on the outside face the radius is 8 feet, and the surface area is greater, even though the log diameter stays the same. The increase in taper on 24-inch logs can help if you set all the slightly smaller ends to the inside, but the help is negligible.

Minimize this effect by using soft curves, 10 to 12-foot (and greater) inside radii even on bubble extensions—the stackwall version of a big bay window, for example—and use smaller-diameter logs in general, so the discrepancy between inside and outside mortar joints in any given surface area is divided up more times among more log ends.

If you are thinking about 32-inch stackwalls for extreme climates, where you want the 24 inches of sawdust or fiberglass (R-84 or thereabouts) in the wall, carefully lay out 4-, 6-, 8-, and 10-inch-diameter logs on paper, and on the ground at full scale to gauge the correct proportions between the inside-outside joints, the radius of the curve, and the diameter of the logs.

You can prove it to yourself on paper, but you should lay out logs on the ground to get a look at inside and outside facade mortar. Large-diameter logs in tight curves can easily change a nice, 1-inch mortar joint inside the wall to more than a 2-inch joint outside, dramatically altering the amount of mortar used and the appearance of the wall.

Anyway, hats off to the innovative owner-builders like Sam Felt and his 40-foot-diameter, arched opening fortress in Georgia, and to Jack Henstridge and his octahedron half bubble over the living room of the "Ship with Wings" house in New Brunswick.

ROUND WALL MORTAR PROPORTIONS

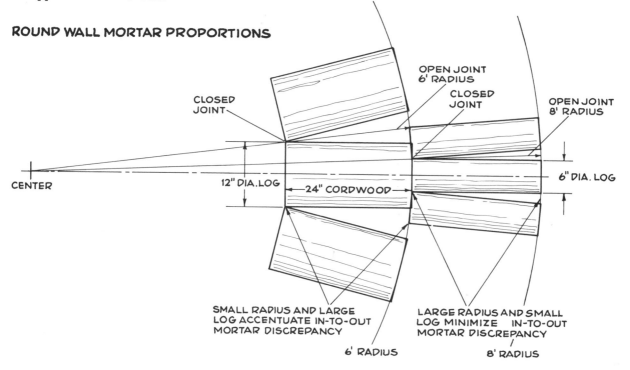

Work in progress at Earthwood is a building course in itself: stacked cordwood ready to go, near the wall; curved walls eliminating corner lacing and heavy timber frames; a gentle radius to minimize inside-to-outside-face mortar discrepancy; stepped-earth sheltering along the back wall, insulated and ready for waterproofing; free-floating window frames on the second story braced against cordwood and mortar loads; protection for the working course until construction continues, and more.

INFORMATION SOURCES

You have become familiar with two names in this section on cordwood masonry. There are several reasons why I have singled them out. First, though it is not exactly a compliment, there are not many to choose from. More importantly, both Rob Roy and Jack Henstridge are builders and teachers. Both have put their work, including the successes and the mistakes, into print. That is unusual and admirable.

In most cases, you would be right to question a book ending on an author's note that some of the information you just read has been superseded by new thoughts and discoveries. But I think you have to forgive it here. Changing and updating is symptomatic of truly alternative construction.

If you build with cordwood now, might you be too far ahead of your time and miss some of the second thoughts and new ideas that turn out to be important? Maybe. No promises here. But I don't think you will miss anything crucial. Cordwood works. It's not on Main Street, but it is in Quebec, and Nova Scotia, and Manitoba, and Georgia, and New York, Nebraska, Wisconsin, and growing.

BOOKS

Building the Cordwood Home, Jack Henstridge (RR #1, Oromocto, New Brunswick, Canada E2V 2G2). One of the two books you must read before building stackwall. Jack is a friendly, informal writer who covers the complete building process using a case-history approach. The artwork and a lot of handwritten notes can get hard to decipher. But the text is loaded with hands-on details and a forthright account of the work and the ideas behind it. Order from Jack. Enclose $1 with inquiries.

Build It Underground, David Carter (Sterling Publishing Co.). Cordwood is only a part of this book, but there is interesting coverage of a cordwood house in Niobrara, Nebraska, that uses cordwood below grade. Lots of ideas, designs, and modest looks at different structural styles.

Cordwood Construction: *A Log End View*, Richard Flatau (typewritten). This includes a construction summary of Flatau's own cordwood-in-timber-frame house as well as general tips and reprinted drawings and photos from books by Jack Henstridge and Rob Roy. Worth a look at $6 postpaid from the author at W4837 Schulz Spur Dr., Merrill, WI 54452.

Cordwood Masonry Houses: A Practical Guide for the Owner-Builder, Rob Roy (Sterling Publishing Co.). Also available through Rob Roy's Earthwood Building School (address below), this is the other must-read book on cordwood. Some of the information has been updated in *Underground Houses* and in *Money-Saving Strategies for the Owner-Builder*, also by Rob Roy. This book covers several designs and projects, concentrating on the step-by-step work and development of the Log-End Cottage and Log-End Cave, an earth-sheltered classic enclosing 910 square feet of usable space that was built for $7,760—about $8.50/sq. ft.

Money-Saving Strategies for the Owner-Builder, Rob Roy (Sterling Publishing Co.). Consider this book the updated companion piece to Roy's first book on cordwood. You've got to love a how-to cordwood book with a bibliography that includes *Siddhartha*, *Walden*, and *Drive It Till It Drops*.

Stackwall: How To Build It, Northern Housing Committee, Dept. of Engineering, University of Manitoba, Winnipeg, Manitoba, Canada. A thorough treatment of cordwood with specific recommendations. More academic, naturally, and less hands-on than the books from the builders, but a solid third source.

ASSOCIATIONS AND INSTRUCTION

EARTHWOOD BUILDING SCHOOL, RR1, Box 105, West Chazy, NY 12992. The Roys' center for day-long and week-long workshops and hands-on training primarily in cordwood and underground construction. Write for the current brochure, as course offerings and schedules may change. Solid instruction: You'll know what you're doing by the time you leave. INDIGENOUS MATERIALS HOUSING INSTITUTE (IMHI), RR1, Oromocto, NB, Canada E2V 2G2. Jack Henstridge offers vacation camping privileges on Institute property in exchange for two hours of labor a day. He calls it a learning vacation with week-long, hands-on cordwood seminars (and other subjects). You get Jack's expertise and unique, wide-ranging philosophy. Write for seminar details.

ARTICLES

Over the last several years, cordwood has been covered in *Harrowsmith, Mother Earth News, Outdoor Life, Home Energy Digest,* and *New Shelter* (and I have undoubtedly missed a few good articles). For a quick look at cordwood, find them through the periodical file at your library or by writing directly to the magazines.

If, after all this, you still don't get the feel and the sense of mortar and wood, think about a trip to the

After seasoning the 12½-inch-long logs for a full year, owner-builder Richard Flatau built cordwood walls in 8-foot wide sections bounded by post and beam framing. Each 8-foot section required about 12 hours to fill with logs (mostly full round), two independent mortar beds, and an insulating core. (Photos in this series by Richard Flatau.)

Below, Flatau's solar addition measures 21 feet along the eaves wall with triangular cordwood extensions at each end. Enhancing the natural temperature control provided by the log and mortar walls, a raised planter in front of the glazing shades the addition in summer with 6-foot-high hollyhocks and 4-foot foxglove. In colder seasons, the dormant plantings allow passive solar gain.

Ottawa River Valley in Quebec Province, where cordwood homes proliferated in the 1700s.

This is a building system that may have been used as long ago as 1200 B.C., a system that worked for homesteaders in eastern Canada and that was brought into America during the American Revolution by troops who occupied Montreal. It was used in New England, in northern New York, and in regions around the Great Lakes. Cordwood moved west with settlers to Manitoba, Wisconsin, and Iowa.

Then, like so many regional and native building skills, it was forgotten in the flurry of expansion and migration, left like so much other handwork, replaced by sawmills, Tabitha Babbitt's circular saw (yes, it was invented by a Shaker woman), manufactured nails, and balloon framing.

Cordwood masonry is not new, although a lot of people never heard of it. It's not high-tech, although some versions qualify as superinsulated. It's not expensive, or too technical, or too time-consuming. Cordwood masonry may have been forgotten but, as you've seen here, it's not lost anymore.

Borrowing a bit of energy-efficient design from colonial timber-frame houses, the Flatau's Hearthstone stove vents into a massive central chimney column. This is one more heat storage and reradiation system that works with the cordwood on exterior walls to provide a naturally regulated, temperate environment inside the house.

Inside the solar addition, a potbelly stove supplements passive solar collection. The stove backup system helps on cloudy days. Further heat storage is provided by a cobblestone floor laid over 1-inch T&G extruded styrene set directly on a layer of compacted sand. The addition perimeter rests on sono tubes.

STONE MASONRY

INTRODUCTION

Man built log houses, pole houses, timber-frame houses, mud- and earth-formed houses, houses on stilts, and underground shelters. But stone shelters were on earth even before man.

Picture yourself walking along, a crude wooden spear in your hand, an animal skin draped over your shoulder

Stone cannot be molded, planed, trimmed, sanded, and generally adjusted as easily as wood. But you can make stone walls conform to your whims as you lay the stone. For this there are two principal options: First, you can create slip forms, as below, that move up the wall in lifts as the work progresses. Second, you can lay stone freehand, as shown on the next page.

to keep out some of the cold, your eyes and ears alert to the ever-present dangers that confronted primitive man. As the light fades and the temperature drops, you look around for shelter: a tree, a gully, someplace just a little out of the weather where you can be warmed by a fire. And then you spot a hole in the rocks.

It's really not much of a contest. Do you sleep and live outside with the critters and the weather, or do you go inside the stone cave, where you can close down and protect the entrance, get out of the weather, more easily maintain a fire, and (although it sounds like a very 20th-century thing to do) get some peace and quiet and a good night's sleep?

Think of all the things we have discovered about stone shelter and how all of it must have dawned on the primitive fellow who wandered into the cave to get out of the cold. Stone is exceptionally strong. (When a tree falls on your stone house, usually the tree breaks, not the house.) Stone is exceptionally durable. (When you go hunting for two days, it's still there when you

get back.) It has great compressive strength. (When an animal charges into it, the animal gets hurt instead of the house.) It is fireproof. (When the night fire blows out of control, you get hot but you don't go up in flames with the house.) It provides energy-efficient mass gain. (When the fire dies and a raging snowstorm would have frozen you solid outside, you survive.) Stone is an indigenous material that can be used with appropriate technology by owner-builders. (Rocks are all over the place, and you can gather them, stack them, and live inside the shelter they create.)

What man knows now, even though we use the information to produce varied and sophisticated structures, man knew then. You could consider the first rearrangement of stones and boulders into a cohesive shelter to be the beginning of architecture—about 10,000 years ago, when nomads settled down and began the cycle of growth from single dwellings to villages to towns to cities. Over 4,000 years ago, Minoans constructed four-story masonry houses on Crete.

Large scale stone structures and single homes were built roughly the same way for thousands of years, with only a few substantial changes, such as the arch. You can still see stone walls and stone foundations and stone houses, but almost all the new versions are made of concrete. And there's a good reason for that, a sensible explanation that you should consider before starting a heavy-duty house of stone.

"Heavy" is the key word, even though several books emphasize the point that, taken one at a time, the rocks may be no more than 15 or 20 pounds each—that's from the ground to the truck to the site to the wall, maybe a thousand times. The double dome of St. Peter's in Rome is 137 feet across and weighs some 450 pounds per square foot. By contrast, the concrete double dome of the Centre National des Industries et des Techniques, a huge exhibition hall in Paris that is five times the size of St. Peter's, weighs 90 pounds per square foot. And concrete is a lot easier to gather, transport, and place.

Still, there is something magnificent about unearthing rocks or wedging them out of a ledge, then arranging them in a completely unique pattern to make a home. It is awe-inspiring once you're done, even if the result is rougher and slightly less grand than the 2,300,000 limestone blocks (about 6½ million tons of stone) of the Great Pyramid at Gizeh.

You can use stone in arches and curved fireplace facades (below) and other forms as long as you abide by the premier rules of working with stone: Always provide a firm (flat, if possible) base, and always try to distribute loads and tie the wall together (above) by bridging seams. (Fireplace photo; below right, by Timber Log Homes.)

DEVELOPMENT, DESIGN, AND COMPONENTS

Let me relate to you one of the most vivid impressions I have of stone building. Oddly, it comes off the television, and because of the way it was filmed, it undoubtedly made more of an impression than if I'd seen it in person.

This was a documentary about a stone mason in Pennsylvania who built dry walls. The camera and the interviewer were in the front seat of his pickup, talking and filming as the mason drove along. He talked about a pasture he walled in one county over, a wall he built last year, another he built 10 years ago. As he continued to talk about other walls and other jobs, the pickup rumbled along. The backdrop outside the window was a stone wall bordering the road, blurred a bit because of the pickup's speed. The mason sort of nodded out the window, adding, "And this one here." The camera

held the shot in silence a moment more, just in case it had not already hit you—the speed of the truck, the time spent talking while the wall blurred by, the number of rocks, the skill, the effort, the time.

To counteract this awe-inspiring image, here are six reasons to put in the time and effort to work with stone, the six tenets of Lewis Watson and family, who built a 45 × 28-foot, three-bedroom stone house in one summer for $2,090, and wrote about it in *House of Stone*.

A stone house has historically been considered a home for the rich and powerful, the upper crust building for posterity. With local stone and slip-form construction (and quite a bit of sweat equity), you can get a 1,260-square-foot house like this for under $5 a square foot. Lewis Watson, shown atop his house, built his for $2,090, in the late 1970s.

"1. THE CASH SAVED IS SIMPLY STAGGERING! By spending a few months building a personal home of average size and quality, you can save over $100,000 by TOTALLY DODGING (1) today's wildly inflated base prices, (2) impossibly high mortgage interest, (3) assorted closing costs, (4) mandatory insurance, (5) 20 to 30 YEARS of state and federal income taxes on this earned amount!

2. EARLY PERSONAL FREEDOM AND SECURITY. Within one year from right now you can have a TOTALLY PAID-FOR RURAL HOME.

3. AN IMMEDIATE HUGE INVESTMENT VALUE.

4. A DWELLING OF YOUR OWN PERSONAL DESIGN.

5. SELF-DISCOVERY THROUGH THE BUILDING PROCESS.

6. YOU CAN START TODAY! No more saving for a huge down payment. JUST PICK UP YOUR TOOLS AND BEGIN!"

That kind of enthusiasm is all through the Watsons' book, a wonderful case history of their project showing how to realize an impossible dream of a family building their own house in a summer in the late 1970s for 60 cents a square foot. (Yes, 45 × 28 feet is 1,260 square feet, divided by $2,090 is .6028 . . . 60 cents.)

STONE

One of the most interesting new-tech construction systems is called surface bonding. The process capitalizes on the compression of concrete blocks stacked in a wall, holding them together with a fibrous coating troweled on the surface. There is no mortaring, no buttering edges, no pointing joints. All the blocks are set close against each other, and set dry.

That's the way stone masonry began. In school you probably studied some of the most impressive examples: the pyramids, to be sure, but also decorative, elegant Greek buildings and monumental pieces of Roman engineering. Roman architecture may appear heavy compared to the Parthenon. But the Romans were more engineers than architects, building roads, aqueducts, bridges, and building at an amazing rate and scale—a lot of it with dry stone.

Mortar was not used commonly until 200 B.C., and early mixes were weak. The Romans eventually improved mortar quality, which diminished the art of stonecutting. Before the cushion of mortar was available to take up discrepancies between blocks, the blocks had to fit precisely.

Most of the ruins (unfortunately, that is the correct description in almost all cases) I remember in and around Rome showed such erosion—even on the stones

You're wrong if you were thinking that the only way Lewis Watson (shown on previous page) could build a house for about $2,000 in one summer was to work with a gang of big, beefy masons donating their time. This is his construction crew.

left intact and in place—that the once-clean and precise edges were pitted and rounded.

Originally, they were cut as finely as hardwood trim. Around Rome, masons were fortunate to deal with a soft stone called *cappellacio*, formed through the ages out of volcanic dust. But they soon discovered the Catch-22 of stone: Hard stone, which is difficult to quarry and haul and cut and place, lasts a long time, while soft stone, which is easy to quarry and haul and cut and place, lasts a short time (by comparison, at least). *Cappellacio* crumbled under constant exposure to the weather.

Roman masons introduced and developed other skills. Using harder volcanic stone from the Alban Hills and the Apennine Mountains, they discovered that the stone was most durable when laid up in a structure the way it lay naturally in the quarry. And they used a joint-tightening process in stone that I use regularly in trim work with wood—backcutting, so the joint that shows is the tightest edge-to-edge connection. Roman stone masons called the process *anathyrosis*, cutting a shallow concave face into the sides of a block so that only the finely honed edges would meet adjacent blocks

ROMAN BUILDING BLOCK

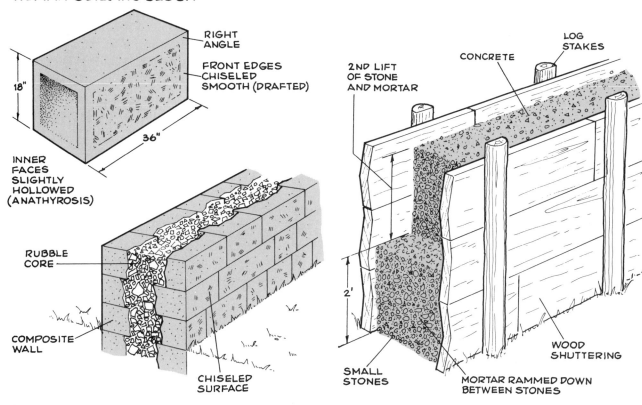

Time and money being so far out of focus in home-building today, who has the patience, much less the skill, to hand-carve stone block? Since ancient times, this is the way it was done, and done by the hundreds of thousands—cutting butt edges slightly concave for a tight fit, and drafting the face perimeter to permit alignment.

in the wall. (It's the same idea as hollow-grinding blades on ice skates, so that only the razor-sharp outside and inside edges hit the ice.)

It was common for the masons to leave the inside face rough where it abutted a rubble core of a composite, two-faced wall. The exposed face was drafted—a process of fine chiseling a strip around the perimeter of the face so that stones could be accurately aligned and plumbed. The major portion of the face may have been dressed down to a rough convex shape, but squared and finished only on important buildings, as opposed to walls and other structures outside the city. You see this treatment of drafted edges on many brownstones.

Most of the ancient techniques, which work as well today as they did then (stone is still stone, just a little older), made structures more elegant and more finished. But their central purpose was structural—to build a tighter, stronger wall and to keep out water.

What an apprenticeship you could have served, starting as a hod carrier instead of a spear carrier. Over time, Roman masons experimented with a cohesive mortar called *pozzolana*, made of two parts vol-

The Romans also used small stones in a compacted mix to make concrete walls. Set in lifts much the way concrete is poured today, a 1- to 2-foot layer of stone was set in forms. Then a layer of mortar was poured in and rammed down to flow between the rocks. This staged construction was repeated until the desired height was reached.

canic ash (prehistoric ash from particular sites was found to be greatly superior to fresher ash) and one part lime. They used rammed mortar and stone, with built-in lifts of concrete rubble behind wooden shuttering (forms). At the Pozzuoli amphitheatre, massive concrete arches exposed to the weather after the brick facing fell away have remained intact for 2,000 years.

There are many kinds of building stone, different types found in different regions. Since stone is heavy and, at commercial rates, about the most expensive and labor-intensive type of construction, owner-builders cannot afford to pick and choose types of rock like different brands on a supermarket shelf. The pharaohs had slaves to bring 15-ton blocks up the Nile and drag them across the desert. And when there is enough money and prestige on the line, big-name architects can convince corporate clients to pay for a particular shade of facing stone from a particular quarry in South Carolina, even if the skyscraper is in Seattle.

To build a stone house without breaking the bank, think of stone as one of the indigenous material possibilities. Use the type that you can gather around the place where you live.

If you do have the option (and the money) to pick and choose, I suggest you work through *Architectural Graphic Standards* and *Stone Masonry*, which has a lot of good information for owner-builders too, and, most importantly, make contact with a firm quarrying and selling rock.

CHARACTERISTICS OF COMMERCIAL BUILDING STONE

Stone	Compressive strength (psi range)	Toughness scale range-average		Wear-resistance scale range-average	
Granite	7,700–60,000	8–27	13	44–88	61
Limestone	2,600–28,000	5–20	7	1–24	8
Marble	8,000–50,000	2–23	6	7–42	19
Quartzite	16,000–45,000	5–30	15	....	
Sandstone	5,000–20,000	2–35	10	2–29	13
Serpentine	11,000–28,000	....		13–111	47
Slate	....	10–56	....	6–12	8

MINIMUM DESIGN LOADS FOR COMMERCIAL BUILDING STONE

Stone	Road (lb per cu ft)
Granite	96
Limestone	95
Marble	95
Sandstone	82
Slate	92

PERMEABILITY OF COMMERCIAL BUILDING STONE
(rate for cubic inch per square foot per hour at ½ inch thickness)

Stone	Pressure (psi) 1.2		50	100
Granite	0.60–	0.80	0.11	0.28
Limestone	0.36–	2.24	4.20–44.80	0.90–109.00
Marble	0.06–	0.35	1.30–16.80	0.90– 28.00
Sandstone	4.20–174.00		51.20	221.00
Slate	0.006–	0.008	0.08– 0.11	0.11

PREPARATIONS FOR CONSTRUCTION

If you stick to the stone you are likely to use, local stone, it simplifies the choices somewhat. It's certainly the most efficient way to get sweat equity in your stone home. But it depends a lot on where you buy land.

For all the technical information about stone, it is hard to go wrong if you use your good judgment. When you find rocks that you can hit with a hammer, drop, soak, then hit again, and feel their strength ring through the hammer into your arm, use them.

You should make sure there will be enough of them. However, Lewis Watson notes that staggering estimates of 50 or more pickup loads for their 45 × 28-foot rectangular home were wildly pessimistic. It took only 18 loads, with Lewis working alone, doing three full loads a day—short work, even considering that he was collecting free rock of modest diameter, weight, and density.

What might you find? Rubble, which is a little more rounded and irregular than quarried rubble. Maybe enough in 8 to 16-inch diameters that would work in a slip form. Maybe fieldstone or riverstone, which is sleek, smooth, and slippery looking. Where you find the stone is also important. Riverstone may be easy to collect and clean, and fieldstone may be all or partially in the ground, covered with lichen, moss, dirt, and discolored from exposure to impurities.

Quarried rock may wind up looking this way, but when it is blasted or wedged out of the ledge, it is fresh—clean, clear, and literally untouched. So, depending on what rock you find and where you find it, you may have to do more or less hauling and a fair amount of cleaning and wirebrushing. These kinds of considerations (multiplied by the number of stones you collect) wind up being the determining factors, not that you used a particular grade from a particular quarry.

A word about quarrying: Don't. Not unless you have professional experience in the field. The same goes for blasting, even if local law allows it. I know, what about the mammoth stump, the outcropping ledge on the homestead in the middle of nowhere? Folks that far out and that self-sufficient usually find a safe way to do things. Their solitude makes them prudent. You would be prudent not to experiment with operations where there is a lot of kinetic energy that can be difficult to control when it is released all at once.

So as you look at property and look at rocks, you should keep a kind of checklist in your head.

WORKABILITY. If this is crucial, you want limestone or sandstone. They are lighter and easier to cut and shape.

DURABILITY. The ultimate choice here is granite. It is heavy and has great color variety; irregular seams make it difficult to cut; it is not porous; great slow-motion mass gain.

UNIFORMITY. Not so important for piece-by-piece custom work; take a ruler or template to check maximum diameter (in any direction) against the predetermined width of a slip-form wall (illustrated on upcoming pages); roughly 10-12 inches is all you need for 8-foot-high walls.

SPECIAL ROCKS. Uniformity is wonderful, but keep an eye out for good cornerstones and, if you're lucky, for lintels and arch keystones.

SURFACE INTEGRITY. Aside from dropping rocks on other rocks and hitting them with hammers, look for soft, porous, pulpy, crumbly surfaces, and when you find them, leave them.

TEXTURE. Although silky-smooth riverstone is stunning and unmistakable, rougher surfaces (as you find on quarried rubble) grab the mortar better.

CONFIGURATION. Big, beautiful, round stones are a big problem, because as the surface recedes into the wall, a cavity is left that, when filled with mortar, makes the wall appear to be more mortar than rock. Look for two relatively flat sides (even if they are curved), so that stone makes the facade.

BEST ROCKS. Short of finding a field full of pavers, look for at least two full faces, flat or close to flat, with a bigger face than back where you can use extra mortar and wedge rocks out of sight; with a top edge that is flat or sloped back down into the wall. If all this starts to sound like your basic rectangle, you're right. Bricks are easier to stack than rubble.

CONSTRUCTION

Here is a parable of caution: My mother-in-law was teaching school, American history, the Civil War, and, as they say, to make the history lesson come alive, she brought in a genuine Civil War cannonball, held it up for the class to see, and, yes, dropped it on her toes. She was out of school for a week, having brought the class to life with a prolonged, piercing scream.

You may build in the heat of summer, but you should still be wearing boots (for ankle support, and with steel toe caps if you can stand them), leather gloves, dungarees or some other sturdy fabric, and, if you are working with large rocks that you wind up "hugging" as you move them, something heavier than a tee shirt on top.

You have to take care of yourself and, if you're doing your own hauling, your truck. If you want to save the box, lay a sheet of plywood down, and don't try to get so much per load that you bottom out the springs. When I was driving my 4wd, ¾-ton Chevy, I got stuck short on a job and went for a full boxload of gravel. The truck groaned but seemed okay—until I started off and found that the front wheels were barely touching the ground. When I turned the wheel, the truck would act like it was thinking about turning for a while before it decided to go ahead and give it a try. It was weird; just a little backheavy.

SLIP FORMS

This is the way to go for owner-builders. With slip forms, you build only a limited series of forms, moving them up the wall as you set rock and mortar in lifts. This wonderfully simple and economical system eliminates the main drawback of working with concrete, mortar, and stone—spending so much time and money on wooden formwork that you start to think you could have framed the house instead.

To give appropriate credit, slip forms should be called Flagg-Nearing forms. Ernest Flagg, an architect who worked on homes in New York, developed the system for use. Helen and Scott Nearing, authors of *Living the Good Life*, further popularized it and offered how-to help and other alternatives.

There are many varieties of slip forms, although they all work on the same principle. On long, straight walls, forms might be 16 to 24 inches high and run the full length of the wall. As the height of openings is reached, box forms can be set into the space between the forms. On walls with several openings, particularly with large openings, it can be more economical to build slip forms to enclose only the portions of continuous, unbroken wall, and to enclose the ends of the forms.

You can make slip forms out of plywood, 1-inch dimensional timber, or rough planking. Purists may not want to use plywood, but it works pretty well. Full 4×8-foot sheets are ripped into three equal sections and backed by 2×4-inch framing top and bottom, with additional 2×4-inch verticals 2 feet on center.

To resist the outward thrust of rock and mortar inside the forms, use a conventional system of spacer blocks and wire ties. With ¾-inch plywood and good nailing to the 2×4 bracing, wire ties run through the forms can be replaced by a series of 2×4 uprights running from the footing to the wall top behind the forms.

Have you ever seen pump jacks? They are an ingenious scaffold system capable of "climbing" walls. To use them for applying siding or for painting, doubled 2×4s are set about 3 feet away from, and parallel to, the wall. They are pinned to the wall face with a series of angled brackets. That is the entire scaffold superstructure. The operating part of the system consists of a pair of cam-activated (pumped by foot) climbers from which a horizontal bracket extends that carries the scaffold planking. The system is easy and inexpensive to set up and use. Slip forms work essentially the same way.

Set up a series of single 2×4s, pin them in place at the foundation, and brace them apart (or use looped wire ties) where they extend above the height of the finished wall. In the Nearing system, two sets of narrow forms were built, so that the slip part of the operation was really a piggyback sequence. Although it required twice the form material, one set served as the anchor for the next, and leaving the first set in place overnight provided extended curing time; the mortar was protected and out of the air for a longer period.

WATSON SLIP-FORM WALL (TOP VIEW)

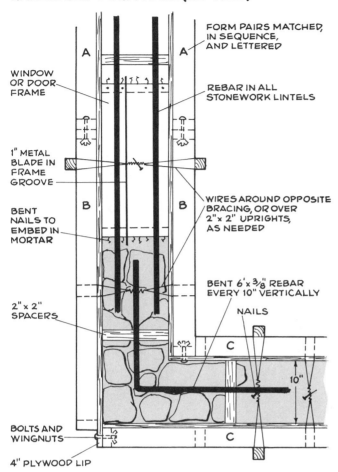

WINDOW OR DOOR FRAME

1" METAL BLADE IN FRAME GROOVE

BENT NAILS TO EMBED IN MORTAR

2" x 2" SPACERS

BOLTS AND WINGNUTS

4" PLYWOOD LIP

FORM PAIRS MATCHED, IN SEQUENCE, AND LETTERED

REBAR IN ALL STONEWORK LINTELS

WIRES AROUND OPPOSITE BRACING, OR OVER 2" x 2" UPRIGHTS, AS NEEDED

BENT 6'x ³⁄₈" REBAR EVERY 10" VERTICALLY

NAILS

10"

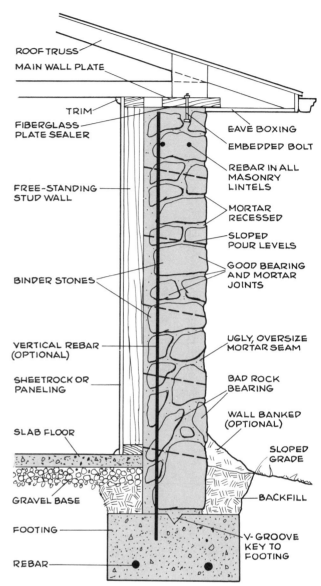

ROOF TRUSS

MAIN WALL PLATE

TRIM

FIBERGLASS PLATE SEALER

FREE-STANDING STUD WALL

BINDER STONES

VERTICAL REBAR (OPTIONAL)

SHEETROCK OR PANELING

SLAB FLOOR

GRAVEL BASE

FOOTING

REBAR

EAVE BOXING

EMBEDDED BOLT

REBAR IN ALL MASONRY LINTELS

MORTAR RECESSED

SLOPED POUR LEVELS

GOOD BEARING AND MORTAR JOINTS

UGLY, OVERSIZE MORTAR SEAM

BAD ROCK BEARING

WALL BANKED (OPTIONAL)

SLOPED GRADE

BACKFILL

V-GROOVE KEY TO FOOTING

FLAT TRENCH BOTTOM, BELOW FROSTLINE

Lewis Watson used the basic, inexpensive slip form above to produce the 10-inch-thick walls, above right. The trick is to invest in enough forms so that you can keep working, but not so many that you have empty forms awaiting stone at the end of the day.

The basic and low-cost system below is described by Christian Bruyère and Robert Inwood in *Country Comforts*. Here, steel strapping and top spacers hold the forms against temporary spacers pulled out as the stone is laid.

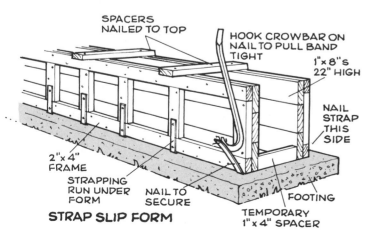

SPACERS NAILED TO TOP

HOOK CROWBAR ON NAIL TO PULL BAND TIGHT

1"x 8"s 22" HIGH

NAIL STRAP THIS SIDE

2"x 4" FRAME

STRAPPING RUN UNDER FORM

NAIL TO SECURE

FOOTING

TEMPORARY 1"x 4" SPACER

STRAP SLIP FORM

There are many little twists to this system, but one nice piece of work you can see is in the wonderful book for homesteaders, *Country Comforts*, by Christian Bruyère and Robert Inwood. They illustrate a slip-form system using 1×8s backed by 2×4s, secured across the top with wooden spacers and across the bottom with removable steel banding. When one lift is poured and set up, the banding is pulled out (or cut, if it gets locked in with the mortar), and the forms are moved up and aligned for the next lift. It's interesting, but no match for the system with 2×4 verticals.

You invest time in plumbing these 2×4s and bracing them only once. From that point on, they are like rail-road tracks for the forms. You don't have to worry about plumb. Just raise the forms, bolt or tack them to the verticals, and away you go.

I like the idea of two sets of slip forms, but I don't think two are necessary. And there are a few drawbacks. For one, leaving the first set in place may help the cure, but it will give the mortar that inevitably spills over onto the rock faces you selected with such care a chance to dig in. Instead of scraping mortar from the rock faces the same day, you set the stone (waiting half a day or so, to prevent slump) and then attend to the cleaning with light wirebrushing; later, you may have to come back with extensive brushing and muriatic acid. Besides, it is reasonable to devote some of the day to heavy-duty lifting, mortar mixing, and setting, some to gathering materials for the next day, some to moving forms on one wall, some to cleaning, and some to pointing. It is highly satisfying to bring all the pieces of the job along at once. And one set of forms is a lot cheaper than two.

By now you should be prepared for more possibilities and other interesting twists on form systems. Here is a quick look at just a few of them.

Lewis Watson's simple plywood-and-2×4 slip forms (above right) can be staked at ground level using the footing as a level base. Below, serving like commercial pump jacks (climbing scaffolds), the same forms are released and reset on 2×4 uprights as each lift of stone is laid. Uprights must be plumb.

Flagg form. Although Ernest Flagg was a prominent commercial architect (in 1908 he designed the Singer Building in New York City, the tallest skyscraper in the United States at the time), his specialty was small, affordable homes. In 1921, he published a book about his designs and the slip-form system called *Small Homes, Their Economic Design and Construction.* Here are the steps to the original Flagg system:

1. Dig the foundation trench below the frost line, and fill tightly with mortarless rubble to within 2 foot of grade. (From there on, use mortar.)

2. Lay sleepers (4 × 4s) on grade, projecting past the wall line on each side about 8 inches, tapered and greased to permit removal later on.

3. Set 4 × 4 uprights a few inches off each wall, face-pinned to the sleepers, drilled for form board holding pins every 6 inches.

4. Set an alignment truss at the top of the verticals, so that when wire ties pull the verticals toward each other, the truss will stop them when they reach plumb.

5. Hold three 2 × 10s on the outside with wooden release bars tucked between the forms and the verticals, while spreaders inside keep the forms open.

6. Set flat-faced stones tightly along one face of the forms, and fill in the 15-inch wall with concrete and rubble stone.

7. Pour 24-inch lifts per day. Release the form pins the following day, point and clean the stone, and transfer the form boards up to a new set of pins.

Nearing form. Helen and Scott Nearing came up with the piggyback forms, each one an independent framework of 1 × 6 form boards backed by 2 × 3s. Their ideas about hand-built, low-cost houses, slip forms, and a lot more about building developed on their homesteads since the 1930s can be found in their book *Living the Good Life.*

Magdiel form. Two brothers, Dan and John Magdiel, patented their first Wall Building Machine in the 1930s, when the Depression had a lot of people thinking about low-cost housing, and particularly owner-built housing. The machine was metal-sided, 13 inches high and 4 feet long. What a great idea! Set the form in place, pour the wall rock in, tamp down a stiff mortar mix, then, working with two machines, jump the first over the second, or, with one machine, release the closed end and slide the metal form along the wall to work on the next section. In *Stone Masonry*, Ken Kern, Steve Magers, and Lou Penfield offer a modified version of the Magdiel form made of ½-inch threaded pipe for the stiffening frame and 14-gauge galvanized

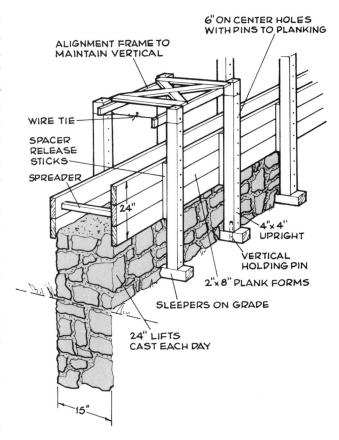

The Flagg form system, above, was rudimentary by construction standards 60 years ago—simple, efficient, even minimal. By today's standards it is an elaborate version of a simple idea.

This modified version of the complex Magdiel Wall Building Machine was designed for manageable fabrication by owner-builders. It has been used with success on projects presented in *Stone Masonry*, by Ken Kern, Steve Magers, and Lou Penfield. This particular version uses ½-inch pipe bent to produce a curved wall. As stone is set, a long form can be pulled along the wall or, with two or more forms, piggy-backed and coupled to the front of the line.

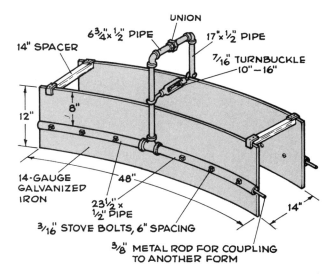

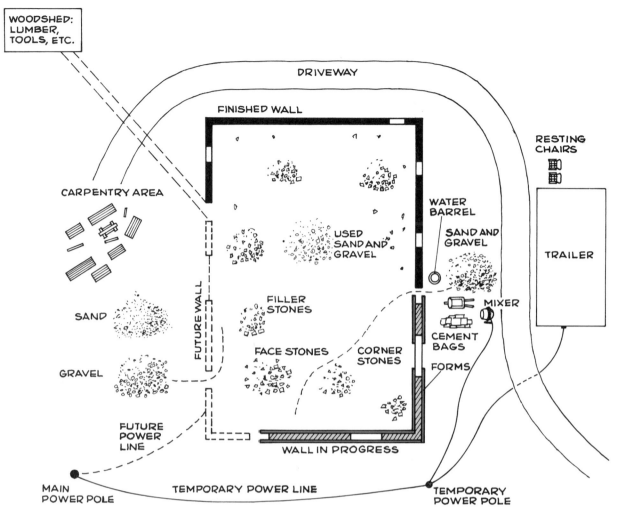

WOODSHED: LUMBER, TOOLS, ETC.

DRIVEWAY

FINISHED WALL

RESTING CHAIRS

CARPENTRY AREA

WATER BARREL

SAND AND GRAVEL

TRAILER

USED SAND AND GRAVEL

FUTURE WALL

SAND

FILLER STONES

MIXER

GRAVEL

FACE STONES

CORNER STONES

CEMENT BAGS

FORMS

FUTURE POWER LINE

WALL IN PROGRESS

MAIN POWER POLE

TEMPORARY POWER LINE

TEMPORARY POWER POLE

Your site may not wind up being this organized, but Lewis Watson's site plan includes a few crucial points not to miss: segregated piles of stone *inside* the building lines to save work; mixer set between the building and access road, where trucks can get in with sand and gravel; and don't forget the "resting chairs."

iron for the form. It's as appropriate as appropriate technology can get, reasonably and inexpensively built on site.

FILLING THE FORMS

If you like issues to be black and white, you'll enjoy the mechanical side of homebuilding—the plumbing, heating, and electrical systems. They are more like mathematics. You're either right or you're wrong. The copper pipe has only one very specific job—to deliver water. Either it does or it leaks. Straightforward work like this has little to do with wall building, where you consider structure, appearance, time, money, mechanical access, energy efficiency, and more, and usually at the same time.

Here are a few of the choices. You could build the stone wall without slip forms—that is, stone by stone, set by hand. The results might be a little neater that way, too. Slip forms make the job a bit easier, because there are guidelines (the forms), and quite a bit faster,

because you can build 2-foot lifts at one time. (Flagg estimated that slip forms saved 50 percent on labor.)

Even stiff mortar will compress and slump under a 20-pound rock. Without forms, mortar would be squeezed out of the joints between rocks. And as you add more rocks and more weight, the rocks would find each other, creating not only a weak wall with only patches of mortar, but also joints where there was little or no mortar.

But just because you have forms in place, you cannot dump in a bunch of rocks and mortar and expect good results. Actually, you could get a strong wall that way, but it would look pretty bad. If you fill the forms with mortar and rock haphazardly, you'll get a concrete-faced wall with an occasional rock poking through.

You have to set the rocks with care, orienting the faces and the joints. And you must decide if one good face (the way most slip-form walls are built) is good enough. If you want two good faces, you have to find very large and very regular rocks (like find them in a quarry for a price), or build a thick composite wall with two runs of face stone and a rubble-and-mortar core.

WALL SEAMS

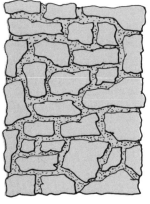
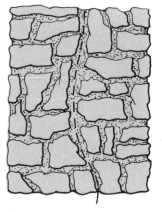

COHESIVE WALL WITH BRIDGED SEAMS

WEAK-LINK WALL WITH UNINTERRUPTED SEAM

There is a world of difference between the above two wall patterns. On the left, most seams are bridged, and the thrust of each stone is near plumb and directed onto another stone. On the right, a continuous, un-bridged seam has created two walls connected only by a mortar joint.

Below, bearing is even more critical in an arch, where loads are split at the keystone and travel laterally to the ground.

Above, an electric mixer speeds work and ensures thorough mixing. Below, resist the temptation to "pour-in" rocks. The forms support the rocks as the mortar sets, but they don't assure the good looks of the wall facade. The forms are only a guide. Selection and placement are in your hands—and, as the bottom photo shows, this can be quite a handful.

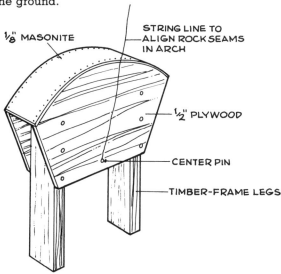

⅛" MASONITE

STRING LINE TO ALIGN ROCK SEAMS IN ARCH

½" PLYWOOD

CENTER PIN

TIMBER-FRAME LEGS

ARCH FORM

KEYSTONE

STRING LINE LAYOUTS

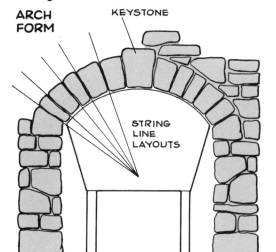

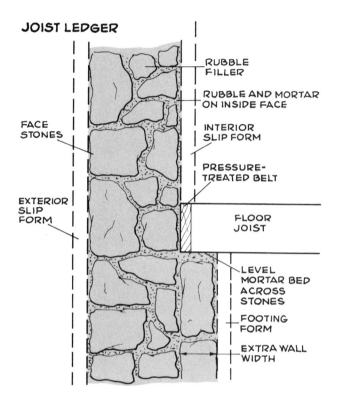

JOIST LEDGER

RUBBLE FILLER

RUBBLE AND MORTAR ON INSIDE FACE

INTERIOR SLIP FORM

PRESSURE-TREATED BELT

FLOOR JOIST

FACE STONES

EXTERIOR SLIP FORM

LEVEL MORTAR BED ACROSS STONES

FOOTING FORM

EXTRA WALL WIDTH

At a 10-inch wall thickness, you may have trouble finding enough stones that present two satisfactory facades. A second facing used at the foundation, or up to the second story, can be leveled with a bed of mortar to form a ledger for joists.

For roughly twice the effort expended on a thick wall you can build a cavity wall, which adds insulation value with dead airspace, or a narrow cavity wall with a film vapor barrier, rigid insulation, or both.

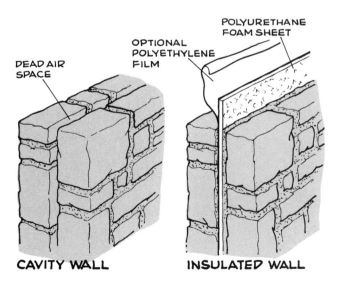

DEAD AIR SPACE

OPTIONAL POLYETHYLENE FILM

POLYURETHANE FOAM SHEET

CAVITY WALL　　　　**INSULATED WALL**

Slip forms can save you time and money. It is not a magical system that will arrange the rocks for you and produce an elegantly jointed, interlocked pattern. Like a computer that can't tell you something that has not been fed into the program, slip forms will help you build economically and efficiently, buy they can't supply the strength, the patience, and the expertise.

Here is the updated slip-form sequence from the bottom up:

1.　Dampen the concrete foundation or footing floor between forms, and pour in about 2 inches of stiff concrete mix, roughly 3 or 4 linear feet at one shot, if your rocks are laid out and ready to set.
2.　Embed only enough stones to fill the few feet with faces against the outside form.
3.　Set filler stones and the mix in and completely around the face stones, adding wire reinforcement around corners.
4.　Add stones to the lift over more mix, keeping joints close but not allowing the stones to nestle directly against each other.
5.　Keep the mix stiff. Very wet mixes provide a high-strength cure but cannot resist loading.
6.　Allow 24 hours setup time in the forms, with the lift covered by wet burlap.
7.　As Lewis Watson advises, "Don't *ever* move a stone once it's mortared in place and has begun to set even for a few minutes."

You will read different advice about different sequences. The key to this general outline is that you combine the expediency of slip form with the care of piece-by-piece building, placing stones to bridge other stone joints in the facade—really, trying to assemble a wall within the forms that could stand on its own without mortar and without the forms.

There is more general agreement on the mortar mix: one part cement to five parts sand and aggregate. However, the Nearings had success with a sparse 1:9 ratio. And some builders have done well with just one hour of set time before moving the forms. You can push economy to the limits of structural security, or you can make life a little easier by running a 3:2 sand-to-gravel ratio in the mix to make it more workable. The emphasis has a lot to do with your temperament.

Finally, there is also agreement on pointing: When the forms come off, clean the rocks and use a water-resistant, rich mix (like 1:3 cement to sand) to fill up joints to a uniform recess, if only a quarter inch, across the facade.

No system answers all the concerns of efficient, low-cost construction without a few trade-offs. Local-stone slip form comes pretty close, but you can appreciate some of the questions.

1. If you use one small, portable form, you get stuck waiting around for the mortar to set (or take a chance on mix slump by pulling the form off after only a few hours).
2. If you build one long, not-so-portable form, you get to spend more time on the rocks and less time moving and realigning forms, at least until it's time to move the 16- or 24-foot form up the vertical supports for another lift of rocks.
3. If you leave forms in place overnight—or 48 hours, as recommended by some builders to get a strong, water-resistant wall—it costs you extra time and effort in cleaning to get an attractive wall.

Tough questions. Some of the concerns will be resolved by events—where you buy land, how much and what kind of rock you find. Some will be resolved by the house plan—if it has long, straight, only minimally interrupted walls (clerestory windows, for example) or only 6- or 8-foot sections of full-height walls between large windows and doors.

Slip forms, below, allow you to lay roughly 2 feet of wall elevation at one time. That's about what the wall below can take, and what you can take, which leaves time for pointing.

Photo above: If you are pleased with the character of the stone you collected, you can feature it by deep pointing, cutting back mortar pockets wrapping around the edges of curved rocks.

HOW TO LAY STONE FREEHAND

Big stones, little stones, even irregular carved brownstones recycled from old buildings can be fitted together for foundations, retaining walls, and house walls. The most helpful rule of thumb is to build mortared walls as though they were dry walls, stacking stones across vertical seams with optimum surface-to-surface bearing across mortar beds. (Photo sequence by the author.)

If one or two hundred rocks don't provide enough possibilities, you can double your options by swinging a 2-pound hammer. Stonecutting is a craft learned only by years of experience. That should not stop you from scoring stones with a cold chisel to control splitting, or knocking off corners to improve the fit. *Caution:* Wear eye protection.

STONE CUTTING

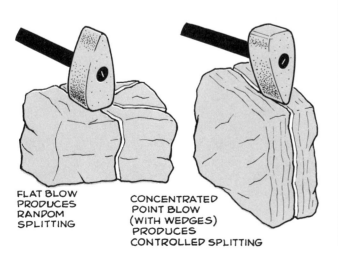

FLAT BLOW
PRODUCES
RANDOM
SPLITTING

CONCENTRATED
POINT BLOW
(WITH WEDGES)
PRODUCES
CONTROLLED SPLITTING

Endless stone shapes and sizes provide limitless combinations for wall construction. You can't search through every rock on the property, but you'll build a better, more beautiful wall if you spread out a good selection nearby. The trick (and the fun) is to match an empty space in the wall with a rock from the collection—like a jig-saw puzzle with many solutions.

Think twice before excluding large, highly irregular stones from the wall, just for the sake of uniformity. A unique block, like this carved lintel support from an old, inner city brownstone, focuses a random pattern. And how satisfying that the craft of an unknown stonecutter, working a century ago, survives the wrecking ball.

A test fit is possible if you are careful not to disrupt surrounding stones already embedded in stiffening mortar. As a stone is set in fresh mortar it can be "worked" (gently wrestled) into the right position. But too much pushing and shoving will squeeze out the mortar, make a mess of the wall facade, and produce a weak, leak-prone joint.

When mortar does not fill irregular contours, remove the stone and build up the bed instead of cramming in mortar from the side. Inexperienced builders can get by laying excess mortar and working the stone to squeeze out the excess. Ideally, you form the bed by hand and trowel, and set the stone carefully with minimal adjustments to layout lines.

Even overbuilt retaining walls cannot withstand the relentless force of hydrostatic pressure—groundwater pushing against the masonry. Relieve the pressure with weep holes, allowing groundwater to escape through the wall without damaging it. Provide for drains during construction, creating a water-collecting, gravel-filled pocket on the inlet side.

Although some soil will get into the drain, clogging is minimized by screening the drain inlet, and covering the gravel area with burlap before backfilling. The exposed drain outlet can be placed unobtrusively between stones just above the footing (even if you have to widen a mortar joint), or covered with a small grate recessed into the mortar.

INFORMATION SOURCES

Luckily, at least a few of the innovators in stone-built, slip-form construction not only lived their ideas but also wrote about them. Even so, slip form and its many derivations are little known. If the system really excites you, make the effort to read some of the books that trace its development, working through search services in libraries, some of the publishers listed below, and Stonehouse Publications (Sweet, ID 83670), which I am happy to plug because you can get a lot of valuable information from the Watsons, who run it, and a good selection of books, too.

If you want to get into stone building from the geological perspective, I'm afraid I have not been able to find a good book for you. Engineering texts that decipher igneous, sedimentary, and metamorphic stone, that delve into the tensile and shear strengths of diorite, obsidian, and chert, tend to be utterly confusing (to me, anyway) or utterly soporific.

Finally, I'm not sure why I feel the tendency to zero in on two sources, two books that could really make the significant difference in your project, but I do. Maybe I feel it is unreasonable to steer you to three, four, and five indispensable sources on three, four, and five of the alternative systems that seem to be real possibilities. That's 20 to 25 books.

So, of all the sources that follow, I especially commend two to your attention: *The Owner Builder's Guide to Stone Masonry*, because it is so thorough and balanced, covering piecework, slip form, linear laid stone looking so much like the ground-hugging lines of Wright's prairie architecture, and more; and *How To Build a Low-Cost House of Stone*, the Watsons' wonderful case-history book that covers only one specific method but does so with so much common sense and enthusiasm.

Building Stone Walls, John Vivian (Garden Way Publishing). A typically thorough book from the author of one of the definitive books on wood heating; excellent material on homesteading.

Build Your Own Stone House, Karl and Sue Schwenke. A detailed building manual with a lot of how-to slip-form details.

How to Build a Low-Cost House of Stone, Lewis and Sharon Watson (Stonehouse Publications, Timber Butte Rd., Sweet, ID 83670). A detailed, practical slip-form construction project; a personal account with detailed, commonsense help for owner-builders.

Stone Shelters, Edward Allen (MIT Press). An elegant look at the variety of indigenous stone structures in a specific region of southern Italy.

The Forgotten Art of Building a Stone Wall, Curtis P. Fields (Yankee Books). Concentrating on flat-stone dry walls, this book has about the best information on splitting and shaping granite, a difficult and often frustrating job.

The Owner Builder's Guide to Stone Masonry, Ken Kern, Steve Magers, Lou Penfield (Charles Scribner's Sons). A thorough treatment of many kinds of stonework with a wide variety of construction and architectural styles.

TECHNICAL BACKGROUND

Modern Masonry, Clois E. Kicklighter (Goodheart-Wilcox, 123 W. Taft Dr., South Holland, IL 60473). This textbook (originally intended for vocational schools) covers brick and concrete block in addition to stone; methodical and filled with formulas.

Concrete Masonry Handbook, Randall and Panarese (Portland Cement Assn., 5420 Old Orchard Rd., Skokie, IL 60076). This engineering handbook has nothing on local natural stone, some information on ashlar and cut fieldstone, and many pages on the nitty-gritty details of mortar mixes, admixtures, accelerators, and other commercial factors.

EARTH MASONRY

INTRODUCTION

If you have ever turned over earth to plant a rose bush or sink a fence post or dig a foundation pier, you know that some earth is soft and loose, some is hard and tight. You might be lucky enough to have rich topsoil that is thick, full of nutrients, and as easy to dig as sand. As you go deeper into compacted soils and subsoils, you may need a pick and shovel. On hardpan and claypan, you may need more than a pick.

Rammed-earth masonry is made of soil, but it is specially prepared, mixed with a little binder, damp-ened, compressed into forms, and cured, all before you get extremely low-cost, fireproof, termite-proof, energy-efficient, and hard walls, all probably in some shade of brown.

Rammed earth is similar in many ways to poured-concrete construction, complete with forms, foundations, and rebars. And it is just an updated tag (with some special twists) for adobe; in fact, rammed earth is sometimes referred to as California adobe. That's provincial but not inaccurate.

Right now there are very few owners, or owner-builders, or professional builders using rammed earth.

From a distance, rammed earth can look nondescript, although literally earth-toned. The closer you get, the more interesting the wall, which becomes a bit lighter with age, maintaining varying tone and texture.

Oh, you can find the odd house in New York, Florida, or Illinois, but the boom (if you can call it that) in rammed-earth construction was in the '40s, when just about every smidgen of raw material went for the war effort, not for private housing. But every site had some dirt.

That is the beauty of rammed earth. Excavate for footings, a crawl space, or a cellar; then save and screen the dirt, mix in some cement, and build your walls with it. It's the ultimate cut-and-fill solution—a problem for road-building engineers who must calculate a route figuring how much dirt will have to be cut from the side of a mountain to fill the ravine next to it. With a rammed-earth house it pays to excavate.

Here are two free-standing rammed-earth monoliths. Earth and binder are mixed dry and then compacted dry within wooden forms so that forms can be removed immediately. Just two forms are needed to complete the monoliths for an entire house.

There are builders setting rammed-earth walls from the footings up to the rafters, one firm working full time in rammed earth that uses concrete footings, rammed-earth wall segments, and a concrete-bond beam across the wall heads to tie everything together. There is a lot of technically oriented how-to information available from places such as VITA (Volunteers for International Technical Assistance) and the United Nations on using rammed earth in block form. (Most of the technical and third-world, appropriate-technology literature calls it soil-cement.) There are a lot of possibilities, not only in the kind of soil-cement mix you use and how you form it in place, but also in how much of the house should be made with the mix.

The two biggest issues, however, are cost and determining whether rammed earth would be a reasonable substitute for concrete. There is little doubt that it will be less costly to build a house almost entirely out of materials already at the site. Big surprise, right? But when the house becomes a little less indigenous, with concrete footings and elaborate formwork, saving money becomes less and less of a good reason for using rammed-earth construction.

Obviously, if your building site is so remote that you can't bring in a ready-mix truck or a cement mixer, rammed earth looks pretty good, particularly if there are no trees for lumber around. Rammed-earth proponents rate rammed earth above concrete in category after category. An excellent article on rammed earth that appeared in the October/November 1982 issue of *Fine Homebuilding* was subtitled, "An ancient building method made easier by new technology." Made easier, yes, with pneumatic tampers and bucket loaders and rototillers to mix and set great quantities of soil-cement at one time. Made more economical, no.

There is an interesting little piece of physics called the Heisenberg principle, named after Werner Heisenberg, the German physicist who formulated the idea. In language I can understand, the principle says that when you go to measure finitely something indescribably small like the exact position of an electron, any measurement you get is uncertain because you changed what's happening when you interfered by measuring it.

Heisenberg talked about the uncertainty principle in subatomic physics. But there is an application to all the indigenous, alternative, appropriate building systems: Watch out when you try to enhance and improve and update a small, simple idea, because in "improving" it you may change it so that you can no longer be certain of all the benefits that attracted you in the first place will still apply.

DEVELOPMENT, DESIGN, AND COMPONENTS

Given all the aluminum frames, glass-curtain walls, steel decking, poured concrete, vinyl siding, asbestos shakes, and other materials found in structures today, it is still easy to pick out the three basic materials that have the longest track record and the greatest potential for owner-builders: stone, logs, and soil.

Stones and logs are preformed in manageable units. Even in their natural state, practically by accident, they can be found in arrangements that provide reasonably efficient shelter. You can't say that about plastic. Soil is the next best thing to stones and logs and is available where there are not enough rocks and

With forms removed, rammed-earth walls reveal every pattern and texture of the material that held them. Forms lined with hardboard (photo right) produce consistently smooth rammed earth. Voids, below, can be caused by rocks or clumps of soil that sneak through the screening process.

not enough trees. It's malleable, too, so that it can be shaped like stone into blocks, or shaped into continuous, interlocked walls the way trees go together to make a log cabin.

Hand-formed earth has been used for centuries all over the world. In areas with wet climates, of course, even tightly compacted soil-cement structures have eroded. But in dry climates, there are structures standing that were built with formed earth over 2,000 years ago.

In the modern era, technical research in soil-cement combinations was generated by engineers designing and building highways and airports, projects normally requiring huge amounts of concrete. It has to do with something you've probably heard before: Necessity is the mother of invention.

Studies undertaken in the 1920s found that "stabilized" soil was an excellent building material. Earth blocks were rammed with hand-driven machines. Researchers tried adding silt, loam, cement, lime, asphalt,

The Cinva-Ram block press revolutionized low-cost rammed-earth construction because the machine itself is inexpensive (about $400). It uses no fuel, has few moving parts, is easy to operate, produces uniform blocks of reliably uniform structural value, and is portable. The ram is available from Schrader Bellows, listed later.

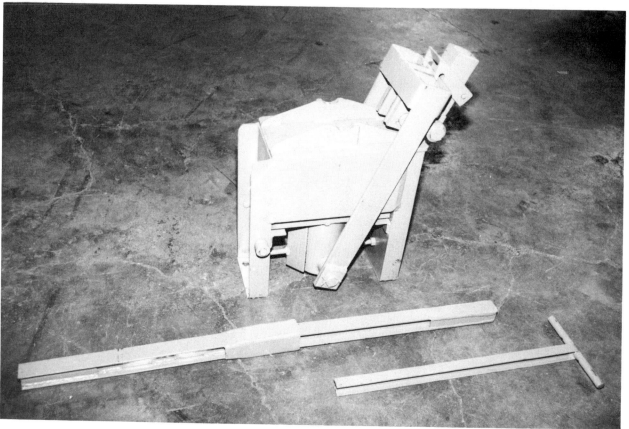

and other mixes. The best results came when a small amount of cement (under 10 percent) was damp-mixed with the soil, formed into building blocks under great compression, and cured for at least eight days before loading. The resulting strength came from the migration of cementitious fibers in cement and from the compression. (In a snowball fight, the loosely packed missiles that splatter on your back are easy to take. It's those hard-packed snowballs that really do the damage.) Airfields made of these cement-mix blocks and subjected to frequent and concentrated stresses of countless aircraft (and the weather) are still in use after 60 years.

In addition to the interest generated by the shortage of building materials during the war, rammed earth got a boost from the Cinva-Ram block press. This efficient, hand-operated building machine was developed by Raoul Ramirez, a Chilean engineer-inventor working with the National University of Chile and the University of Illinois Housing Mission to Colombia.

Patents granted in 1957 included a series of forms to be fitted inside the press that could turn out solid, hollow, and specially keyed block patterns. Samples tested by the U.S. National Bureau of Standards with soil stabilizers in an extremely sparse ratio of 1:20 compared favorably to brick (NBS Building Materials Report BMS-78). A Cinva-Ram block of soil and sand with 8 percent silt, clay, and cement mixed in, rated compressive strength to 800 pounds per square inch, transverse loading (wind loading) to 112 pounds per square foot, and thermal resistance roughly equal to poured concrete.

By now, the Cinva-Ram is in use in close to 50 countries. And although it costs close to $400, the building blocks cost about 1¢ each to produce. Including soil sifting and mixing, a two-man team can produce as many as 500 blocks per day ($4 \times 6 \times 12$ inches each).

Consistent with the best application of this technology—hand-powered for the owner-builder on his own home—there are also simplified field tests to use in conjunction with the machine, tests that determine soil type and provide mix-ratio guidelines.

The emphasis with the block press is pure appropriate technology, directed largely to developing countries. That doesn't mean it won't work for you, too. Most of the current research is concentrated on simplifying the mix, adapting the process to different climates and soil types, searching for inexpensive chemical catalysts, and in all cases sticking with sweat equity. No fuel. No forms. No waste.

Even though cement is a key ingredient in rammed earth, rammed earth is not concrete. It is an inevitable comparison, because for most builders there is a choice. First, rammed earth (with cement) does have advantages over adobe, which differs in that it is formed without compressive packing and is sun-dried. Yet, there is a process started by a company in Arizona (the right place for it) that uses cement plus adobe ingredients stuffed into burlap sacks, which are stacked up to make walls, connected with long lengths of rebar driven down through the sacks, then thoroughly soaked in place to activate the cement and harden the structure. That's a lot like reinforced concrete.

Concrete incorporates aggregate. When the right amount of water is added, the binding fibers in the cement migrate through the mix, wrapping around every little hunk of stone. Even though the binding paste may be only 22 to 34 percent of the mix, a 5- or 6-gallon-per-sack mix produces exceptionally strong concrete. The familiar words are Portland cement, named by the inventor Joseph Aspdin after quarries turning out stone of a similar color in the Isle of Portland.

Where compressed rammed-earth blocks can reach up to the 800-pounds-per-square-inch compressive strength, 5-gallon-per-sack concrete (with adequate wet curing) reaches 1,000 pounds per square inch in one day, almost 3,000 psi in three days, over 4,000 psi in seven days, and 6,000 psi (close to its ultimate strength) in 28 days. The ingredients are widely available for portable power mixing on site and, for large pours, via the ready-mix truck.

Concrete walls offer most of the bonuses supplied by rammed earth: fireproof, termite-proof, rot-resistant, and more. Then there's the stinger—the forms. That's one reason a lot of people opt for the piecemeal approach of block and mortar—to avoid the time and money needed to set up the space for concrete.

Rammed Earth Works, a California firm building (and offering complete building plans for) rammed-earth homes, uses an elaborate set of forms—really well designed and incredibly strong—with 1⅛-inch plywood form panels with reinforced corners, 2×10 horizontal whalers on each side of the form, all pinned with ¾-inch pipe clamps. They are as elaborate as any concrete forms.

Sometimes, owner-builders design forms for poured concrete based on the idea of holding the mix in position until it sets up and begins to cure. They forget that in addition to the obvious loads, there is a more concentrated stress as the concrete pours out of the ready-mix chute. It's the difference between leaning up against a stuck door and standing back to deliver a karate kick on the lock with the heel of your foot.

With rammed earth, forms must withstand the stresses of soil stacked up into a wall and the concentrated forces as the soil is tamped. "Tamping" sounds a little delicate, but it's not. "Rammed" is a much better description of the process and of the forces exerted on the forms.

FORMS FOR RAMMED EARTH

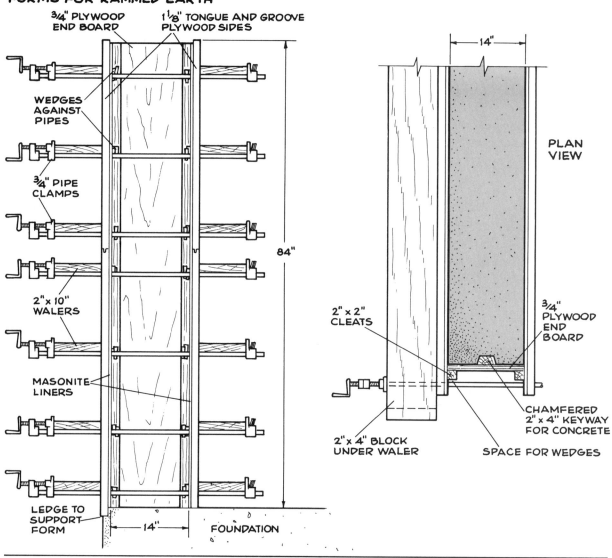

3/4" PLYWOOD END BOARD

1 1/8" TONGUE AND GROOVE PLYWOOD SIDES

WEDGES AGAINST PIPES

3/4" PIPE CLAMPS

2" x 10" WALERS

MASONITE LINERS

LEDGE TO SUPPORT FORM

84"

14"

FOUNDATION

PLAN VIEW

14"

2" x 2" CLEATS

3/4" PLYWOOD END BOARD

2" x 4" BLOCK UNDER WALER

CHAMFERED 2" x 4" KEYWAY FOR CONCRETE

SPACE FOR WEDGES

CORNER BRACING

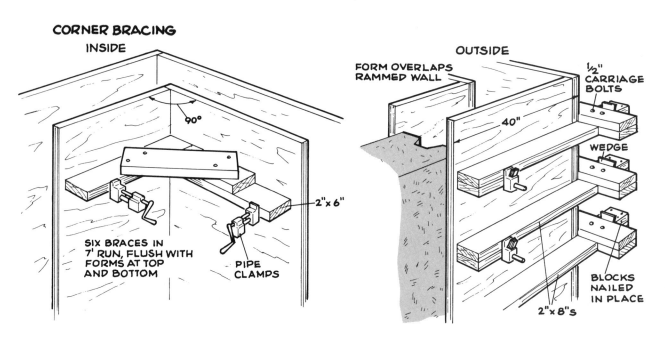

INSIDE

90°

2" x 6"

SIX BRACES IN 7' RUN, FLUSH WITH FORMS AT TOP AND BOTTOM

PIPE CLAMPS

OUTSIDE

FORM OVERLAPS RAMMED WALL

1/2" CARRIAGE BOLTS

40"

WEDGE

BLOCKS NAILED IN PLACE

2" x 8"s

CORDWOOD, STONE, AND EARTH MASONRY **298**

Rammed Earth Works, an experienced and specialized rammed-earth design and construction firm, uses forms, like this corner form, that can be built and assembled by owner-builders. The great strength provided by the forms, pipe clamps, and horizontal walers is not required to hold the earth itself in place, though it is needed to resist the force of tamping.

The two types of forms (above and below) together cost less than $500. At the time these photographs were taken, the forms had been in use for over three years on many homes and were still in good condition.

DESIGN DECISIONS

Like poured concrete, rammed-earth walls reflect the quality of the forms, although the finished product is usually smooth and brown, with almost a patina you'd expect from richly finished wood. Even so, early rammed-earth homes were stuccoed, sometimes inside and out. A good mix is really too good looking to cover up. The material is terrific once you have it in place. The question is, how do you go about it?

Continuous or block construction.

Continuous or monolithic rammed-earth walls require forms—elaborate assemblies, such as those used by Rammed Earth Works, or the more primitive, home-grown variety, such as 2 × 4s pointed and driven into the ground, plumbed with wire ties, and lined with planking. A block press is not that big an investment, considering that it eliminates all formwork expenses. On a homestead or a farm, it can be used again and again for the house, barn, well, retaining walls, interior and exterior pavers (inserts also allow you to produce 2 × 6 × 12-inch tile), and more.

All-rammed-earth construction.

Some builders have used formed rammed earth throughout the wall, including the footing. But in every such case history I've found, the builder also notes that the wall is protected from the weather with a large overhang, and that the wall is not in a position where it must repel groundwater. Even though it is tempting to use the low-cost system start to finish, most rammed-earth houses and builders (and building codes) call for a more conventional concrete footing.

Rammed earth and concrete.

If you use concrete in the footing, for key load-bearing columns, and even as a 7-inch-thick bond beam tying the wall sections together and providing headers for all openings at the same time, rammed earth may be more filler than structure. That's okay. Rammed earth can be used inside a timber frame, too, although you can imagine that it gets a little difficult to tamp when you run out of room at the top of the frame. You can think of it as an elaborate, elegant wattle and daub. But if

Rough forms (below left) for continuous rammed earth can be made with planking supported by 2 × 4 stakes driven into the ground and braced for extra strength. Below right: Continuous walls can be presented even though sections are formed and tamped separately (it saves on forms), by covering major seams with trim or by interrupting wall sections with full-height framed openings.

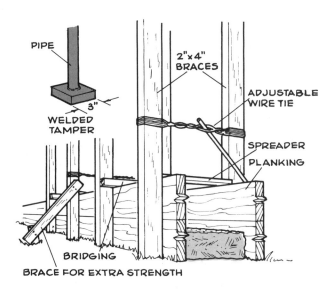

you do, let go of the notion that you will be building for pennies per square foot with the indigenous material winner of all time.

Adobe. This has become a fashionable word. (Back to basics, finding roots, and all that.) Some builders who have been swept away by the concept have had their hogan or their "Santa Fe" facade home complete with authentic vigas (rafters) swept away, too, because they weren't built in an area with Santa Fe's climate. Anyone going back to the native Americans for their architectural wisdom should remember that the Indians were wise enough to limit the use of adobe to areas of dry climate like the Southwest; they used longhouses (an arched structure of bent and interwoven trees) in the Northeast.

Adobe differs from rammed earth because it is simply an accumulation of clay and mud and one of several binders (straw or dried grasses) that is not extraordinarily compressed, just formed and left to dry in the sun. Blocks formed to $4 \times 8 \times 16$ inches, $4 \times 12 \times 18$ inches, and other sizes can develop compressive strength of 300 to 500 psi with two to three months of storage time.

Monolithic adobe walls and adobe block should be considered only for relatively dry climates. If you don't push the design limits, the right mix in the right place can be solid and durable. One of the oldest documented adobe buildings is a small house in Sialk, an oasis in Iran, that dates to 4000 B.C.

Adobe granaries built by the Pharaoh Ramesseum are 3,400 years old. (The Pharaoh was not much of a

Resurging interest in adobe includes traditional forms with exposed vigas (rafters) in contractor Albert Parra's home in Albuquerque (above), as well as passive solar designs (below) incorporating full glazing walls and a sun room. (Photos by Adobe News.)

Spaces between wall sections on this Rammed Earth Works building are built by a kind of slip-forming—moving expensive forms from area to area instead of forming the entire job at once. The stage platform is fed by a small bucket loader, shoveled in place, and set with a pneumatic tamper.

Poured concrete bond beams tie adobe walls together. After curing, forms are stripped, and concrete is covered with plaster or adobe veneer. UL-rated Romex lines run through the beam and through adobe courses to recessed electrical fixtures. (Photos by Adobe News.)

Adobes laid in unformed adobe mud produce uniform expansion-contraction rates in a one-mass wall. A 4-inch overlap is required by state code in New Mexico. Asphalt-emulsion stabilizer produces waterproof exteriors. Interiors, as shown, can be rough finished or plastered. (Photos by Adobe News.)

do-it-yourselfer, so other folks no doubt took care of the construction details.) In the United States, the Acoma Pueblo is about 1,000 years old. Besides being old, adobe is useful, low-cost, and indigenous. I'm thinking of the number of adobe buildings put up in East Germany when there was no postwar construction or lumber industry, and the country was literally in ruins.

So why is it, with many examples of incredible longevity and utility; that adobe is not even mentioned in the Uniform Building Code? You'd think it would be there, since Ronald Reagan's California ranch home is adobe. And so is Supreme Court Justice Sandra Day O'Connor's home in Tucson. Between the two of them you'd think the Code could have been changed.

I think the big issue for code writers is the fear of newspaper headlines after an earthquake (and their real concern for safety)—something like, "New Adobe Condo Collapses," with a subhead, "Minor Damage Elsewhere in City." That's overstated, but code people do have a legitimate point. There are solutions, though—rebar, wire ties, propylene mesh, and bond beams.

Fashionable friends of adobe may change the image, which may in turn spur changes in the Code. As it stands now, although the National Bureau of Standards

reports that adobe is 30 percent more efficient than conventional housing for cooling, the U.S. Department of Housing and Urban Development won't build with adobe on Indian reservations.

Sulfur block.

Because sulfur is one of the unwanted by-products of petroleum production, nearly 30 percent of the world's sulfur is produced (in a backhanded way) by the petroleum industry. In addition, it is locally available in many parts of the world, where it is used largely for fertilizer. Sulfur also has some promising characteristics for construction: good compressive strength when mixed with aggregate, impermeability to water, and good insulating value. Sulfur concrete provides a different twist because it becomes liquid when heated, and it can be poured into molds that shape various patterns of interlocking block.

Cost is estimated at a bit less than one-third that of concrete block construction, including tools, materials, and molds, which you can use again and again, or sell. A fascinating and thorough report on sulfur-concrete construction, including information on test structures built in Canada, is available from the Minimum Cost Housing Group of the School of Architecture at McGill University (see "Information Sources").

The time-honored three-block mold system uses a wooden form with a flush top for screeding the adobe mud. Modern, multi-form machines with hopper feeds can produce 32 blocks in one lift. This sun-cured adobe stack awaits delivery in Humberto Camacho's Adobe Facility in Mountainair, NM. (Photo by Adobe News.)

Sulfur block molds must be precise to turn out high-quality, uniform blocks. This form produces interlocking dovetail block. Sulfur molds can shape inexpensive roofing tile, floor tile, and virtually any structural shape required for construction.

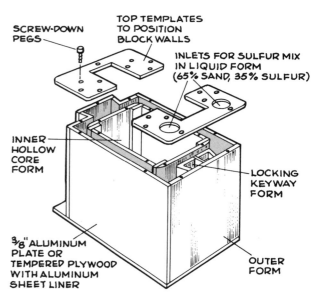

SCREW-DOWN PEGS

TOP TEMPLATES TO POSITION BLOCK WALLS

INLETS FOR SULFUR MIX IN LIQUID FORM (65% SAND, 35% SULFUR)

INNER HOLLOW CORE FORM

LOCKING KEYWAY FORM

⅜" ALUMINUM PLATE OR TEMPERED PLYWOOD WITH ALUMINUM SHEET LINER

OUTER FORM

PREPARATION
FOR CONSTRUCTION

Rammed earth in monolithic wall or block form is heavy. For stability and thermal efficiency, walls are typically at least 12 inches thick, commonly 15 or 16 inches, and at heights over 8 feet (for two-story construction), up to 24 inches thick. A lot of earth. Too much to import.

The central attraction of rammed earth is that the site is the prime building material. You should read that phrase over again, because it goes to the very heart of efficient and appropriate owner-builder construction, and it's true of very few structures.

Since there is enough earth for construction on all but the most unusual sites (hanging on a rock ledge or perched on the edge of a beach, where you wouldn't want to use rammed earth in any case), the only question is about soil type: what you've got, and what you have to do to it and add to it to make sound, rammed-earth walls.

You can take samples of soil to the laboratory for analysis, if you want to get technical. Or you can use simple field tests to approximate mix proportions. In any case, you should find the time before construction to test your numbers and ratios in a block or small molded form to simulate wall construction, curing, and loading. I would not transplant someone else's numbers, no matter how successful, to a different site.

Metalibec Ltda, manufacturer of the Cinva-Ram block press, offers the following field test, developed by the Inter-American Housing and Planning Center in Bogota, Colombia. As a general rule, they say most soil will make good rammed-earth blocks if it is reasonably free of vegetable matter. In most cases, the earth below surface vegetation and roots excavated for foundations or basements will be suitable following guidelines of the accompanying Soil Suspension Field Test.

SOIL SUSPENSION FIELD TEST

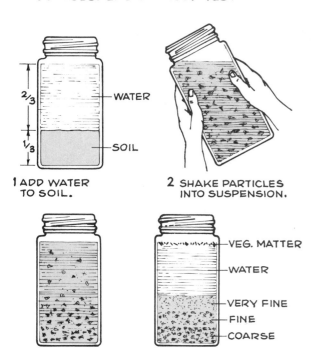

1 ADD WATER TO SOIL.

2 SHAKE PARTICLES INTO SUSPENSION.

3 ALLOW PARTICLES TO STRATIFY AND SETTLE.

4 EXAMINE SOIL GRADATION.

1. Fill a straight-sided glass jar one-third full of the soil sample.
2. Add water to fill the jar two-thirds of the way.
3. Cover the jar and shake vigorously until all the earth is in suspension (swimming around in the water instead of caked at the bottom of the jar).
4. Allow the earth to settle (about 30 minutes) with the jar at rest, so you can observe the segregation and stratification of particles in the soil.
5. Particles can range from coarse at the bottom, to fine in the next layer up, to very fine particles in the layer above that, with slightly cloudy but clearing water above, and any vegetable matter floating at the top.
6. The only unsuitable soil will show one uniform layer in the jar—all particles of the same size. The most suitable soil has three layers in bands of coarse, fine, and very fine, with no organic matter. If you do come up with an unsuitable soil, you may be able to rescue it by adding coarse sand. You can check by retesting the new mix.
7. For the best block, coarse particles should be not less than one-third or more than two-thirds of the total soil component in the jar.

One of the most efficient ways to batch-mix soil for rammed-earth walls is to make adjacent piles of screened soil and binder, then combine them in a large mixing pit with a power tiller. This mixture can be transported to, up, and into forms with a bucket loader.

To take a reliable sample, dig down to excavation depth, where you will be getting the bulk of the soil. Then screen the sample (as you will the entire load) through ¼-inch screening to remove larger aggregate. Metalibec and the Housing Center engineers specify only a range for cement-to-soil proportion: 5 to 10 percent. As you move from hotter, drier climates to colder, wetter climates, you might adjust that ratio upward to 7 to 12 percent.

Realistically, when you work with large piles of soil, bags of cement, and large-scale mixes, 1 or 2 percent may blow away in the wind. Well, almost. The point is that these are guidelines, not formulas that blow up if one of the ingredients is off a hair. And if it is more practical to use lime than cement, you can substitute 2 parts lime for 1 part cement and double curing time.

As you start to test different mixes, use water sparingly after dry-mixing the soil and the stabilizer. The final mix should ball up when squeezed, so that you can break a compressed handful in two without crumbling the mix apart. To get a true test of compressive strength, wet-cure the sample (the way you would treat concrete) for seven days. Like concrete with a 28-day full-cure period, the compressed-earth mix reaches

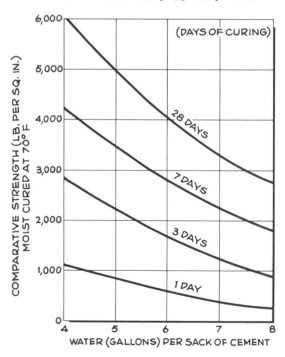

CONCRETE CURING STRENGTH

Graph: y-axis "COMPARATIVE STRENGTH (LB. PER SQ. IN.) MOIST CURED AT 70° F" from 0 to 6,000; x-axis "WATER (GALLONS) PER SACK OF CEMENT" from 4 to 8. Curves labeled (DAYS OF CURING): 28 DAYS, 7 DAYS, 3 DAYS, 1 DAY.

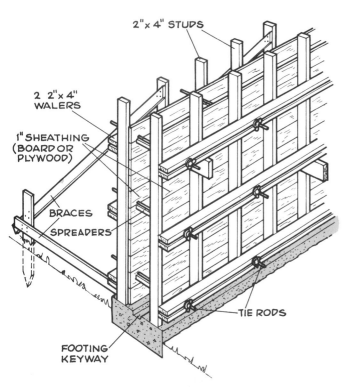

2"x 4" STUDS

2 2"x 4" WALERS

1" SHEATHING (BOARD OR PLYWOOD)

BRACES

SPREADERS

TIE RODS

FOOTING KEYWAY

The argument for poured concrete is simple: Whether you use proprietary or site-built forms, once they are in place, why not pour all the walls into a monolithic, reinforced structure at the same time?

within a few percentage points of its ultimate strength in 30 days.

If you ask around about footings for rammed-earth walls, you will get different answers. Being reluctant to compromise the simplicity of rammed-earth by adding concrete to make a composite wall, I have struggled with the prospect of forming for poured concrete below grade, then switching over to rammed earth above grade. I think I hear the guy on the ready-mix truck asking why I want to go and do that since he's here

with concrete, and my answers aren't terribly convincing—not with the forms all built and the truck full of concrete a few feet away. The switch is a compromise of sorts. On the other hand, I would look at site drainage and exposure and the water table oh so carefully before compacting soil for a footing—even though soil-cement footings can work well when they are not up against a flow of water.

I suggest that you study both sides of the question, as it is a decision that could determine if you go with rammed earth or stick with more conventional masonry from footing to sill. Check Chapter 9 in the Bruyère-Inwood book, *In Harmony with Nature*, for a first-hand, case-history report on rammed earth in retaining walls of a foundation (60:40 sand to clay, with 7 percent Portland). Even in this case, the builder writes, "The retaining wall around a basement which is set into a slope like ours can be made of rammed earth (soil cement) since there is no danger of runoff to wash it away."

How much material will you need? A 1,200-square-foot house with 9-foot-high walls built 12 inches thick will use roughly 58 cubic yards of soil-cement and about 200 cubic feet of cement at 8 percent by volume. If there is some question as to the amount of soil you have on hand (it should be the last thing you run out of) or need to prepare, you can accurately figure soil-cement volume in monolithic, rammed-earth forms by allowing for a 50-percent volume reduction due to compaction.

For example, the average rammed-earth wall section (each one is bordered by reinforced-concrete columns) built by Rammed Earth Works is 6 feet long, 7 feet high (it's capped with a concrete-bond beam), and 14 inches wide—a volume of 50 cubic feet. Each section takes roughly 100 cubic feet of soil mix with 3½ cubic feet of cement, figured at 7 percent by volume of the wall, not the raw soil mix.

CONSTRUCTION

Between the lines of mix ratios and psi's of compression, there is another message: study, experiment, search. Don't accept the first option at face value because you will find others. Don't waste time waiting for the solution with no problems and no drawbacks.

One of the things I get to do aside from writing books and building is host two consumer, how-to radio programs every week in New York. I talk to a lot of people and get to do a few interviews, one of which is appropriate here. In the spring of 1984, I talked with Stuart Brand from the *Whole Earth Catalog* and *CoEvolution Quarterly*. I asked Stuart this: If he were beginning a building project, starting from scratch with a new method, where he would concentrate his time and attention to have the most positive and far-reaching influence on the job. His answer: books, with a distant second place to the telephone for easy and efficient access to varied information. So between the lines, Stuart's message is track down sources, call them up, talk to them, and read, read, read.

SOIL ANALYSIS

The jar-and-particle-suspension test illustrated earlier is a good indicator of soil composition. But you can be more precise by taking some fine measurements and working through the tables on soil characteristics, shown on the next page.

MAKING BLOCKS

The block press is a versatile machine. Estimates of 300 to 500 blocks per day from a two-man crew are borne out by experience internationally. Paying for about 100 pounds of cement per 150 blocks and getting the soil by your own labor makes the total cost minuscule. But unlike sulfur-block systems, which use molds to produce keyed, interlocking blocks, block-press, rammed-earth building blocks have to be mortared. The other option, which I have not yet tried, is surface bonding.

Since blocks are uniform coming out of the press, mortaring is straightforward. In fact, the Cinva Center in Bogota developed a hand-drawn mortar spreader to use with the blocks that lays a uniform, ⅝-inch-thick bed not quite as wide as the block. When blocks are set, the load closes the mortar seam about ⅛ inch, fills out the joints, and still keeps the block clean.

Technical specs for the block-press system further suggest a sand-cement mortar on the first two courses of block to increase waterproofing, and then a mortar for the rest of the building of 1 part cement, 2 parts lime, and 9 parts of the same soil used to make the blocks. This mix should be stiffer than the cement-sand mortar. After a week (depending on weather) for drying, the mortar is surface-coated with a cement wash brushed thoroughly into any cracks.

To protect the blocks from erosion, one of several surface applications is recommended: a clear silicone coating (particularly for rainy areas); a lime wash that is renewed every year; or three coats of exterior cement wash.

This Cinva mortar spreader is based on the same kind of technology found in the block press—appropriate technology. The spreader works like a controllable hopper. It's easy and cheap to build. Used with the extended side to serve as a running guide and with a steady hand, it produces reliable and uniform mortar beds between block courses.

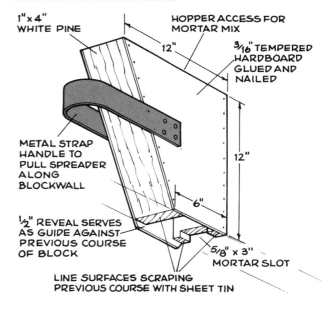

1" x 4" WHITE PINE

HOPPER ACCESS FOR MORTAR MIX

12"

3/16" TEMPERED HARDBOARD GLUED AND NAILED

METAL STRAP HANDLE TO PULL SPREADER ALONG BLOCKWALL

12"

6"

½" REVEAL SERVES AS GUIDE AGAINST PREVIOUS COURSE OF BLOCK

5/8" x 3" MORTAR SLOT

LINE SURFACES SCRAPING PREVIOUS COURSE WITH SHEET TIN

The Cinva press works like a mason jar with a handle. (1) With the handle swung back and the cover mold hinged open, the block press is hand-loaded. (2) The "lid" is set over the block form. (3) The handle is drawn down to the opposite side with a hard pull and leverage that is increased by the Cinva cams.

SOIL DESCRIPTION

Soil	USDA classification	Suitability for construction	Clay mineral analysis
1	sandy loam	excellent	vermiculite and kaolinite
2	sandy loam	good	mica and kaolinite, both well crystallized
3	sandy clay loam	good	kaolinite, mica, and trace of vermiculite
4	sandy clay loam	good	kaolinite, vermiculite, and trace of mica

SOIL PARTICLE SIZE (mm)

Soil	Sand >1.0 %	1.0–0.5 %	0.5–0.25 %	0.25–0.1 %	0.10–0.05 %	Silt .05–.002 %	Clay <.002 %
1	6	39	6	16	7	16	10
2	2	13	2	24	23	21	15
3	6	13	7	27	12	11	24
4	12	38	4	3	1	18	24

SOIL MECHANICAL PROPERTIES

Soil	Moisture equivalent %	Liquid limit %	Plastic limit %	Shrinkage limit %	Plastic index %
1	7.8	15.0	14.6	15.1	0.4
2	14.7	20.3	16.7	17.9	3.6
3	15.6	26.6	17.4	22.9	9.2
4	13.1	21.6	16.4	18.3	5.2

Key to soil types
1. Top soil: 70% sand, 23% silt, 7% clay
2. Heavy soil: 30% sand, 60% fines
3. Moderate soil: 3% gravel, 53% sand, 44% fines
4. Sandy-clay soil: 70% sand, some large gravel, 25% fines

(4) Two-man operation simplifies loading and unloading. (5) With the form mold hinged back, full release of the handle raises the finished block. (6) Block should be removed with end pressure as shown to minimize damage. Block is set on edge to cure.

Drawings below right: Buying or making your own form inserts for the block press provides versatility. Your own small but diversified block factory can produce standard rectangular building block, floor tile, lintel block, and more.

Now you have apples and oranges to compare. Rammed blocks are easy to produce, and there are no elaborate forms to build. Construction costs can run as little as 5 percent of conventional masonry. But you have to mortar the blocks in one at a time. Monolithic rammed-earth walls are also inexpensive, but they do require forms—extremely strong forms at that. Once the forms are built, soil-cement goes in like concrete, as a one-piece wall with no mortaring, pointing, and cleaning.

Some of the best information available on soil-cement blocks is in an old report of extensive tests done on blocks by USDA engineers. These and other tests reveal some interesting characteristics and side effects.

1. The type of soil used has just as much effect on final compressive strength as the percentage of cement in the mix.
2. Compressive strength does not automatically increase in proportion to increased cement in the mix.
3. Tests on cured block showed that a 1:10 soil-to-cement ratio produced a compressive strength of 326 psi (averaged); a decrease in cement proportion to 1:12 produced an increase in compressive strength to 449 psi; and a further decrease to 1:14 produced the expected loss of compressive strength to 266 psi.
4. Stability on 7-foot and higher walls, was greatly increased by adding a bond beam (poured concrete or two 2 × 6s lapped at corners and tied with bolts).
5. Measurements with an impact-rebound hammer (it tells how hard materials are, not how hard you hit them) inside and outside every block in a rammed-earth block test house one year after construction revealed that the inside block surfaces had softened and the exterior surfaces had hardened, except for the two courses closest to the ground, which showed spalling.

SOLID BLOCK   **FLOOR TILE**

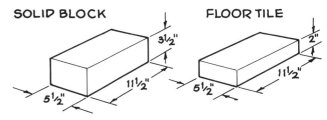

WILL ACCEPT
STANDARD WATERPROOFING CHEMICALS

COMPRESSIVE STRENGTH 500-1000 PSI
(DEPENDING ON SOIL AND PERCENTAGE OF CEMENT

EROSION RESISTANT

LINTEL BLOCK

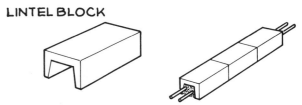

TIED WITH REINFORCING ROD AND CONCRETE FILL
SPANS UP TO 7'

USED IN GRILL
PATTERN AS
STRUCTURAL
VENTING WALL

COMBINED WITH FLOOR BLOCK TO
MAKE DRAIN TILE

USED AS BOND BEAM ON TOP OF
COMPLETE BUILDING PERIMETER

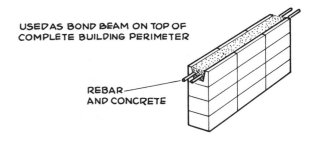

REBAR
AND CONCRETE

MONOLITHIC WALLS

David Easton, president of Rammed Earth Works, explains the use of pneumatic tampers and bucket loaders and reinforced concrete and other bits of technology used on rammed earth this way. A study of rammed earth by the USDA in 1938 found that a three-man crew working with hand tools (shovels, buckets, and tampers) could complete 70 square feet of 14-inch-thick wall in one day. Easton's crews (five people instead of three), using specialty forms that go up and come down efficiently, rototillers for mixing in bulk, bucket loaders for delivering mix into the forms, and pneumatic tampers, produce 23 wall-unit sections in a day. That's all the walls for a 1,000-square-foot house in 16 hours.

Under ideal conditions, it should be possible to slip-form a rammed-earth wall. But it is risky to dismantle the forms soon enough to make reusing them on the same job a practical matter. At Rammed Earth Works, forms stay in place only until the last lift of soil is tamped into place. But their forms are held with a system of walers and pipe clamps, so that the interior plywood panels can be peeled away from the wall without any banging and prying.

There are many types of proprietary forms made for concrete (many can be rented). That's the problem. To get the most materials for the money, form systems are commonly made with only moderately strong walls helped by frequent supports and ties. Recessed tie-rod holders and snap-off connectors are cleared once the forms are down, and all the regularly spaced holes in the wall are patched. You can't do that with rammed earth, where untamped voids are a structural problem, and patched finished walls are an appearance problem.

To preserve the unmistakably rich tone and texture of soil-cement walls, form boards are supported, but not tied through the earth. Using substantial forms with a smooth face inside, ties can be eliminated by supporting the walls with a system of horizontal stiffeners made of dimensional timbers called whalers (scaffold planks, since you will probably be climbing around on them).

Rammed Earth Works combines forms for footings (below left) and walls associated with concrete work with rammed-earth wall sections in a structural concrete framework. Below right, narrow vertical keyways between wall sections are used for mechanical access and are reinforced with rebar (note tied overlaps from rebar in the keyway to rebar protruding from the footing). The dimensional timber forms become a permanent part of the wall facade after the concrete post is poured.

Slabs are poured, puddled, and screeded (as shown), then cured, after the rammed-earth walls are compacted. Some weather extremes (rain, heat, freezing temperatures) warrant use of plastic curing chambers, as shown.

You could fill forms by hand, but a vertical wheelbarrow (the bucket) makes it oh-so-much easier. The same goes for pneumatic tamping. Your effort is expended more for guiding the tool and controlling compaction instead of pounding, pounding, pounding.

Another nice touch used by Rammed Earth Works is pattern printing on the slab. Professionals (above left) use a metal stamper, but low-cost plastic stamping tools are available from concrete supply houses. Above right, the finished, imprinted floor will stay free of stress cracking, and there's no grout to fall apart.

Rammed-earth forms are not that different from concrete forms (except for all the holes). But Rammed Earth Works has added some refinements that tailor the forms to earth masonry.

Adding a narrow strip of wood at the top edges of the footing forms produces a rabbet in the finished edge.

Richard Day Forms: Over a cured concrete footing 2″ × 12″ formboards are laid up with conventional ties and spreaders. Threaded rods (⅜ or ½ in.) are tightened through 4″ × 4″ uprights with wingnuts over washers. Formboards are held apart by the footing at the bottom of the wall, and by 2″ × 2″ spacers at the top. (Photo series by Richard Day.)

This also builds in a water table of sorts. But the purpose is to guide the bottom edges of the forms, to unload them at the same time, and to eliminate the need for wooden spreaders inside the base of the form.

By using stock Pony pipe-clamp fittings (on ¾-inch black iron) to pin 2 × 10 whalers, the forms stay rigid

Rammed-earth walls can be built with sophisticated, timesaving (and expensive) tools such as mini backhoes and pneumatic tampers. It takes longer by hand, but the finished walls are just about as solid. You have to be patient, working in lifts (or layers) of soil, tamping the loose earth as you fill the form.

despite the pressure from pneumatic tamping. And the 2×10s are wide enough to make comfortable scaffolding.

To get extra strength and produce a finished, rammed-earth surface, Rammed Earth Works uses 1⅛-inch T&G plywood formboards lined with Masonite.

Rammed-earth walls are compacted in lifts, 6- to 8-inch layers at a time, which compresses to roughly 3 or 4 inches. Working in large lifts, the top of the wall would be dense but the bottom would be loose. Since it is impossible to measure the compaction down in the form, you have to go by feel—not how the earth feels, but how it feels to you when you hit it with a tamper.

It's not quite as easy as working with a concrete vibrator, but if you've used one, it will be familiar. Concrete vibrators whine and strain at the mix. You know it is time to move on when the noise level subsides and relaxes—like a car straining the engine uphill then tooling along when it gets over the crest.

With hand tampers (the heavier the better, but light enough so you can keep at it for more than a few minutes at a time), you feel the force of the blow sink into the soil. The energy you provide gets used up in compaction. When the lift is solid, the energy will ring back through your hands and arms. Move on when you feel the unfamiliar and somewhat bone-jangling kinetic rebound.

With a pneumatic tamper, it is more difficult to feel the compaction but, like a concrete vibrator, easier to hear it. You'll get used to it, just like shifting gears

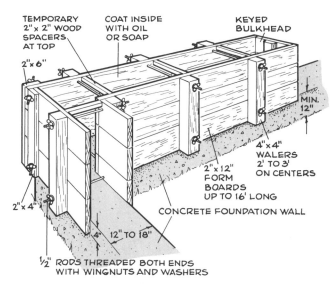

This is the rammed-earth corner form devised by Richard Day, who set up the photos below.

Hand tamping is tough work, jolting, like driving a car with bad shock absorbers. The trick is to work vertically, getting maximum body weight and some back and shoulder muscles behind your arm motion. You can see the soil firm up, and also hear and feel the progress, with the earth going from loose to solid.

Fully compacted rammed-earth walls will reproduce a mirror image of the form boards, including hairline ridges from the seams between boards—even knot and grain patterns. Holes left by tie rods can be packed with soil, and the entire wall can be "sanded" to a more uniform appearance with a brick.

This corner form consists of a rigid box inside the main form. The box displaces earth, creating the L-shaped corner, and is secured in place by the earth being compacted.

according to engine sound instead of tachometer reading. You can't hurt the soil by hitting it again when it's already compacted as much as your energy and tools allow. But you can hurt the structure by quitting too early.

When the screening and sifting and mixing and com-

In regions where zoning permits, solid timbers can frame openings where no earth walls are built. But in zoned areas, building officials may insist on posts and beams of reinforced concrete.

pacting are complete, remove the forms carefully, and wet-cure the earth walls for two weeks, leaving them unloaded and protected from the weather. You can use burlap sprinkled down three times a day or so, which is okay if you plan to be at the job all day anyway. If not, well-sealed plastic sheeting will let you attend to other business.

Cured, rammed-earth walls can be plastered, stuccoed, painted, wallpapered, paneled, drywalled. But why on earth (sorry about that one) would anyone use soil-cement, then stucco on the outside and paint or paper on the inside? Earth mixtures produce wonderful colors with subtle gradations not available in paint or paper—the difference between flat-brown hair color you get out of a bottle of dye and natural brown-auburn-umber-amber hair that changes shades with the angle of the sun. It's not something to bury, although a clear sealer on inside surfaces (something like Thompson's Water Seal) prevents surface dusting and acts as a partial vapor barrier.

Cured rammed earth will hold galvanized nails (hot-dipped if you can get them), which allows virtually any framed window or door treatment. Weather joints between a wood frame and a soil-cement wall can be closed by forming a keyway in the end of the wall. Attaching a nailer to the end panel of the form creates

Here are the prominent components of a rammed-earth project: screed board at ground level of patterned slab, main slab formed for access of thermal siphon loop, single-lift wall sections under openings, rebar set for verticals, and a bond beam.

REBAR

ACCESS FOR THERMAL SIPHON LOOP

MAIN SLAB

SCREED BOARD

the pattern. Even a small slot to accommodate most of a 1×2 spline, with a portion rabbeted into the frame, prevents air infiltration.

The final step with monolithic and block-formed soil-cement walls is to tie the walls together with a bond beam. A reinforced-concrete pour, structural wooden box, or even a double 2×6 plate can be used. Unsupported rammed earth won't work.

It makes sense to add to the walls only one element that does the jobs of tying the walls together and heading off openings. Rammed Earth Works pours concrete over four ½-inch rebars inside 2×8 forms on top of the walls. These hook up with keywayed, 6-inch-wide, poured-concrete posts between rammed-earth wall sections. It is an exceptionally strong detail that enables the rammed-earth walls to be formed in limited sections. It also adds more formwork to the job (the 2×8s are finished and left in place, covering the seams between concrete and soil) and adds a second concrete pour in addition to the footing.

Full-height rammed-earth walls are formed to accept a concrete bond beam poured horizontally, with concrete columns keyed between wall sections. Recessed wire-rod ties hold finished form boards together. After pouring, the countersunk holes in the boards are plugged, and the wood is left in place covering the seams between the concrete and the rammed earth.

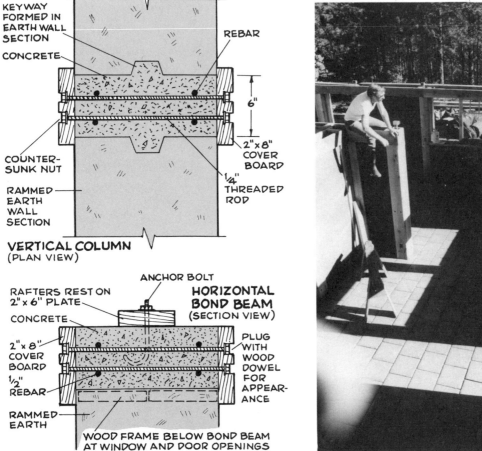

VERTICAL COLUMN (PLAN VIEW)

KEYWAY FORMED IN EARTH WALL SECTION
CONCRETE
REBAR
6"
2"x8" COVER BOARD
¼" THREADED ROD
COUNTER-SUNK NUT
RAMMED EARTH WALL SECTION

HORIZONTAL BOND BEAM (SECTION VIEW)

RAFTERS REST ON 2"x 6" PLATE
ANCHOR BOLT
CONCRETE
2"x 8" COVER BOARD
½" REBAR
RAMMED EARTH
PLUG WITH WOOD DOWEL FOR APPEARANCE
WOOD FRAME BELOW BOND BEAM AT WINDOW AND DOOR OPENINGS

To finish a house with rammed-earth walls, you can go your own way. But the less you work the walls once they are up and cured, the better off you'll be. For instance, instead of furring strips, incorporate a nailer in the foundation or the bond beam. Instead of toe-nailing rafters down into rammed earth, install a top plate with anchor bolts, so you can toenail wood to wood. Leave a liberal overhang, like 2 or 3 feet, so the earth walls will be protected from rain.

One of the biggest benefits of rammed earth is the mass gain, particularly the tempering effect as the walls operate their own energy cycle in counterpoint to the environment. Walls 12 inches and more thick work on roughly a 12-hour time delay. On cold days, it takes the full wall roughly half the day to fully soak up heat produced inside the house, which is then re-radiated (some inside and some outside) during the night. On hot days, the converse again works to your benefit, as cool nighttime temperatures stored in the walls keep the house from overheating during the day.

Bond beams of concrete can be left uncovered. But it is generally handy to have wood at the point where rafters cross the wall. When the bond beam is covered, paneling is added; with light fixtures, sinks, and such, a rammed-earth house could be any house.

INFORMATION SOURCES

Alternative building systems that have become or are in the process of becoming mainstream residential systems go through a process. There are experiments, rediscoveries of old buildings and manuscripts, then some projects and investigations of the results, then case-history reports, magazine articles, and on and on until the subject becomes fully popularized. There may not be all that many people building the houses at that point, but the information about the system is widely available, digested, and synthesized.

You won't find a *Reader's Digest* manual on rammed earth. You will find some off-beat sources, a lot of reports, technical papers, and such, because rammed-earth, soil-cement, California-adobe, earth-block construction (it doesn't even have one recognizable name yet) won't be mainstream for a while.

Adobe-Build It Yourself, Paul Graham McHenry Jr. (University of Arizona Press, Tucson, AZ 85722). A guide to adobe construction; thoroughly covers block production, installation, bond beams, viga design, plastering, and more.

A Sketchbook of Projects for Rammed Earth, David Easton (Blue Mountain Press, Wilseyville, CA 95257). The title tells it; a lot of ideas for rammed-earth owner-builders.

Concrete and Formwork, T.W. Love (Craftsman Book Co., 542 Stevens Ave., Solana Beach, CA 92075). If you are inexperienced with formwork (building negative space can be complicated), this traditional how-to look at plank forms will help.

In Harmony with Nature, Christian Bruyère and Robert Inwood (Sterling Publishing). One chapter is devoted to rammed-earth footings and foundation walls. The case-history approach covers one method, not construction and design options. It's only one chapter, but I think you'll enjoy the others.

Making Building Blocks with the Cinva-Ram—A Supervisor's Manual, Volunteers for International Technical Assistance (VITA) (College Campus, Schenectady, NY 12308). A typed and duplicated field manual that tells all you would need to know before deciding to use the block-press, mortared-block system. I got my copy from Schrader Bellows

(see below), the Cinva-Ram U.S. distributor. The report mentions two other foreign block-press manufacturers, listed without U.S. representation: Landcrete (a great name), Trans-Atlas Ltd., 15 Duke St., Dublin 2, Ireland; and Winget, Winget Ltd., Rochester, England. VITA seems to have relocated.

Making the Adobe Brick, Eugene H. Boudreau (Random House). Like the Watsons' personal account of their slip-form stone house, this case history covers the Boudreaus' adobe home, from excavation through hand-formed blockmaking and construction, using a system with rebars running between footing and bond beam inside the walls.

Research Results on Soil-Cement Building Block, Kent, Liu, and Teter (USDA, Agricultural Engineering Research Division, Beltsville, MD; also through Schrader, see below). An engineering report on all kinds of stress and environmental tests on rammed-earth blocks and block structures, plus a reference list of 15 other technical papers and reports on soil-cement.

Soil-Cement, United Nations Publications, New York, NY 10017. Third-world orientation here emphasizes low-cost, on-site, soil-cement block production and installation.

The ECOL Operation, Ortega, Rybcznski, et al. (Minimum Cost Housing Group, School of Architecture, McGill University, Montreal, Canada H3C 3G1). A fascinating and thorough look at sulfur block construction: mix design, molds, interlocking mortarless assembly, with followthroughs on compatible, ecologically sound (and esoteric) life-support systems.

The Rammed Earth Experience, David Easton (Blue Mountain Press, Wilseyville, CA 95257). Despite the very California title (it can't just be rammed earth or building with rammed earth, it's got to be an experience), this is an instructive book on the details of monolithic rammed-earth wall construction by one of the few professionals in the field right now with a lot of experience, David Easton of Rammed Earth Works and also a consulting firm called Earth Resource Technology.

PERIODICAL

Earthbuilder, published twice a year by Adobe News, Inc. This large-format, thoroughly illustrated magazine offers complete coverage of traditional adobe construction as well as innovations. In addition to the most complete listings of adobe suppliers and builders, *Earthbuilder* covers owner-builder projects. To subscribe or to obtain details on periodic, hands-on courses on adobe and rammed-earth building, write *Earthbuilder*, Box 7460, Old Town Station, Albuquerque, NM 87194.

FIRMS

RAMMED EARTH WORKS, Blue Mountain Rd., Wilseyville, CA 95257. David Easton's firm builds rammed-earth houses regionally. Write for their brochure, which includes information on construction details and rammed-earth building plans.

METALIBEC LTDA., Apartado Aereo 11798, Carrera 68-B, No. 18-30, Bogota, D.E. Colombia. Manufacturer of the Cinva-Ram block press.

SCHRADER BELLOWS, ATTN: K. Easterling, PO Box 631, Akron, OH 44309. The U.S. distributor for Metalibec's Cinva-Ram. They sell about 150 units per year, keep about 50 on hand, do not advertise the product at all, but do provide excellent and voluminous literature to serious, potential buyers who take the trouble to seek them out. The Cinva-Ram sells for about $400. The 140-pound press (155 pounds shipped) has a gross cubic measurement of 3.5 cubic feet.

EARTH-SHELTERED

HOUSES

Plato wrote, "Behold! Human beings living in an underground den, they see only their own shadows, or the shadows of one another, which the fire throws on the opposite side of the wall." It's a limited, claustrophobic picture. Underground could not be a good place to live. It's where miners get black lung. It's dark and damp, a place to grow mushrooms. It's where people are buried.

If you had to choose, would you rather live down in the basement, with all the dampness and mildew and only a few vent windows, or upstairs, where it's drier, lighter, cleaner, and healthier? Some choice.

It took a long time to discover the benefits of living up in the light, to build atriums and colonnaded overhangs full of fresh air, even if it did get a bit chilly. In recent decades, the people who went into the cellar to the *rec* room finally came back upstairs into the *family* room. (Names do make a difference.) Most people now let the cars live down there with the furnace and the water heater.

So why should we go back underground? Even with curvilinear forms, and reinforced concrete, and skylights that don't automatically leak (even though many will leak), and waterproofing emulsions, and rigid insulating board, and clever indirect lighting to compensate for the lack of 360° exposure, it still amounts to living under a pile of dirt. Isn't that how moles live?

People who favor earth sheltering don't like to hear this. And a lot of professionals in the field don't want to be dragged back into an argument they consider old and tired. To them, earth sheltering is another building system with definite advantages and few drawbacks.

On the positive side, earth sheltering seems to have moved past the public misconception associated with so many alternative and nonmainstream building systems: the misconception that you really don't belong in a log cabin or a stone house or a cordwood home unless you quit your very conventional job and raise goats by the hydroponic garden. Earth sheltering is just another building system. And when professionals in the field work on this uncomplicated premise, you're more likely to get help than hype.

On the negative side, many professionals in earth sheltering (and in the broad field of solar-efficient design, which overlaps with it more than with many other alternatives) are a little too touchy. They can turn you off by acting as though there must be something wrong with you if you don't see earth sheltering as the obvious answer. The touchiness can take the form of a "with us or against us" attitude. It's needlessly aggressive, particularly when you realize that it is understandable for many potential buyers and builders to be worried about giving up windows and a view. People hate to lose fresh air, sunshine, and the ability to watch the clouds float by. Even if you use one of the modifications of earth sheltering, maybe partial earth berming or an atrium design, it is difficult to get direct outside exposure (much less windows on two walls of the room) in every room in the house.

Instinctively, you may feel the loss of above-ground amenities. There is nothing wrong with you if you do feel it. But if you go through this section of the book and then read some of the books and other materials in the "Information" section, and you still feel that you'll be giving up a lot to get the energy efficiency and environmental benefits of earth sheltering, don't do it. Or at least think hard about it. Your home should not be a compromise crusade. Your home shouldn't have to prove a point.

But try this: Put aside all the preconceptions about dark, damp cellars, about windowless rooms, and give earth sheltering a clean slate. And let some of the professionals sell you.

MALCOLM WELLS, architect, one of the most innovative designers. Wells doesn't have just one or two bonuses for you. He cites protection from natural disasters such as fires and hurricanes; using the earth to moderate temperature; using the earth roof to begin the gradual percolation of rainwater back into the ground without erosion and undermining; building an environment free from noise and vibration.

ROB ROY, designer, including the Log-End Cave, and

ISSUES INFLUENCING RESIDENTIAL EARTH-SHELTERED BUILDING

Factor	Owners (mean)[a]	People considering earth shelter (mean)[a]
Reduced heating load	6.65	6.67(1)[b]
Reduced cooling load	6.35	6.40(2)
Maintenance reduction	6.25	6.16(4)
Enhanced alternative energy potential	6.06	6.26(3)
Storm protection	5.93	5.97(5)
Improved lifestyle	5.46	4.92(11)
Personal privacy	5.27	5.41(7)
Environmental noise reduction	5.13	5.40(8)
Insurance reduction/ elimination	4.71	5.21(10)
Demonstration/ experimentation	4.62	4.55(12)
Security from vandalism/crime	4.57	5.48(6)
Land preservation	4.37	5.22(9)

[a] Rating scale used: 1 = not important; 7 = very important.
[b] Rank ordering of factors for those considering building earth shelters is shown in parenthesis. Rank ordering for owners is indicated by factor listing in descending order of importance.

author. Roy concentrates on two powerful, practical advantages—high-efficiency heating and cooling and low-cost construction for owner-builders.

JOHN BARNARD, designer, including Ecology House on Cape Cod. Barnard brings out advantages that will thrill homeowners who are sick of recaulking, repainting, reroofing, and more—a nearly complete absence of traditional, time-consuming, money-eating, weekend-killing home maintenance. And underground, below the frost line, nothing will freeze, including plumbing pipes. But you will have to put up with the endless jokes about mowing your roof.

DAVID CARTER, consultant and writer. Carter, like Roy, thinks earth sheltering is great for owner-builders. He points out that it is flexible and can be expressed in different forms for different sites—berming on flat sites with high water tables, enveloping to make the least impact on a unique site.

TRI/ARCH ASSOCIATES, design firm. Collectively,

in their *Earth-Shelter Handbook*, these designers stress the complete and harmonious integration of building and design concerns, how earth sheltering can offer a unified expression of materials, structure, and aesthetics, with cost efficiency, energy efficiency, and minimal environmental impact.

That's how some of the professionals feel. How about owners and potential owners? In his report, "Earth Sheltered Structures," Lester L. Boyers provides an interesting table (shown at left) of factors influencing the decision to build underground.

One of the reasons it is important to deal forthrightly with concerns about living underground—and not to sell earth sheltering exclusively on its energy efficiency—is revealed in this study. Note that "improved lifestyle" ranks sixth on the list for owners, the people who are living there and who are dealing with everyday problems and benefits. But this issue ranks eleventh for people interested in, but not yet living in, an earth-sheltered home. Combine this with the fact that "land preservation" ranks last for owners but ninth for potential owners, and you start to get a picture of second-generation owners who are committed to environmental concerns (I include energy efficiency) and financial concerns (include energy efficiency here again), even if their expectations for the quality of life in an earth-sheltered home are not as high as they might be.

There are so many good reasons for earth-shelter construction. There has been an explosion of information, in design, waterproofing (where things change

Winter and summer energy efficiency is the number-one reason most people build sheltered housing. The low slope of Rob Roy's cordwood-facade house faces a courtyard with sunken entrance. Low-slope rafters give enough lift over grade for secondary light and ventilation.

quickly), and energy efficiency. For example, people at the Underground Space Center are just starting to get an accurate picture of thermal performance when you put reasonably conventional building materials under 1 or 2 feet of sod on the roof and 6, 12, 100, or more feet between the back wall of the house and the other side of the mountain. There is so much new data, in fact, that one of the most difficult jobs for consumers is locating practical, up-to-date information that won't take months to digest.

After you read the books and the reports, you'll find there are only a few basic types of earth-shelter sources. First, there is the back-to-basics approach: Gee, we all know earth sheltering is the best thing going; let's get some grass growing up on the roof. It's all so simple—until the walls crack and the roof leaks.

Second, there are the technocrats, who communicate in charts and tables, therms and BTUs, R-values and U-factors, mean daily temperatures, and on and on. The technocrats are almost as useless a source as the first group, even though their work is sincere and high-minded. Their publications start to sound like text-books. And unless you love to play with numbers, the dry, technical approach has no heart.

You can rely on the back-to-basics approach for enthusiasm and the technical approach for data and then rely on your common sense to take the good from each. And you should look for sources that recognize who you are, sources that do more than preach to a captive audience, sources that have practical experience and technical know-how, sources that are sensible enough to temper one with the other.

DEVELOPMENT, DESIGN, AND COMPONENTS

Some forms of earth sheltering can be found in all ages of architecture all over the world. If you have the time, you can track down sketches and pictures of centuries-old atrium houses cut into the soft stone of Matmata, Tunisia, homes with several rooms and 15-foot-high ceilings opening onto a sunken courtyard. They have been there for centuries. But of all the historical examples that should convince you that this system has something to offer, including 2,000-year-old courtyard houses in China and eight-story underground complexes in Cappadocia, Turkey, my favorites are more recent.

The earth-sheltered (really rock-sheltered) complex in the Loire Valley in France truly represents an ingenious redefinition of living space. In addition, these houses involved quite a marketing job, not unlike the masterful maneuvering of U.S. entrepreneurs who bought obsolete underground missile silos from the government, then turned around and charged the government and private firms for the privilege of dumping toxic wastes into the concrete holes.

The Loire Valley is the site of quarries that supplied massive amounts of stone for the region's great country estates. The quarried, negative space cut into the hillside left cavernous shelters with only one exposed facade. Through the marvels of real estate development, the caverns have been transformed into fashionable and luxurious weekend retreats complete with plush carpeting, rich paneling, and all the advantages of "country" living.

Although the earth-sheltered "courtyard" homes in China may look primitive today, this 2000-year-old design is highly energy efficient. Like a modern condominium, these homes shared a community courtyard that provided more than one home's worth of light and air.

Much of the substance of modern American earth sheltering is based on the pioneer's history of using the building materials at hand. Yes, this is another pioneer building system—from the first earth-sheltered homes, very temporary shelters called dugouts on the East Coast, through all combinations of stone- and wood-faced earth-sheltered homes in Pennsylvania, Ohio, Nebraska. Your state historical society will likely have pictures or drawings of 18th- or 19th-century earth-sheltered homes built in your area. Thinking back to a family outing to Valley Forge, I recall the log shelters for Washington's troops—more log than earth shelters really, but dug down 2, sometimes 3 or 4 feet. There were thermal benefits, no doubt. But without waterproofing and drainage, the earth floors became stagnant, disease-breeding pools. The winters glorified in history books hit you differently when you stand in the tiny little shelters, your feet in the mud, and wonder how the soldiers got from their slat bunks to the door or to the small fireplace without slopping through a foot of ice water and waste.

More recently, earth sheltering has produced some very interesting permutations. Last summer I built a big deck for the owners of a fairly standard development house. It had ramps, diagonal 2 × 4s, and other nifty features that required a solid frame and solid connections to the existing house. When I got inside their cellar to see what I was going to hook into, I was surprised to find a lot of windows and an elaborately finished ground floor with halls, bedrooms, closets, carpeting, full trim details, and more.

I guessed at the date—about 1948? I was close. This was one of a wave of postwar "basement" houses, built in stages as availability of materials and time and money allowed. The idea was to excavate, build, finish, and live in the basement, complete with mechanical systems, bathrooms, a roof (obviously), and a poured-concrete entry well with steps down to a full-height entry door protected by a stick-on dormer. Given enough time, many of these earth-shelter owner-builders learned how to do wiring, piping, plastering, etc., finished the basement, and saved up for the rest of the house above.

Probably the most bizarre twist on earth sheltering was the rash of bomb shelters built in the late 1950s and early '60s—that's backyard bomb shelters. I remember the civil-defense drills in school. Crouching down under your desk was supposed to protect you from glass sent flying from the A-bomb shock wave. (Don't look toward the windows, remember?) But what to do with the family caught at home?

At that time in my small hometown on the Con-

Early earth-sheltered homes in the U.S. were commonly protected by the earth, and made of it, too. Although shielded from cold winds, these homes were damp and dark. (Photo by Nebraska State Historical Society.)

necticut River, I had not started learning about construction. I remember vividly the summer our next-door neighbor (a piano teacher) spent hand-digging a bomb shelter in his backyard, one wee shovelful at a time.

It probably was not as big as I remember it; childhood memories are generally out of scale. But it seemed like an immense hole in the ground, taking up most of his yard. He had to cut down a lot of his carefully pruned bushes. God knows where he put all the dirt. For months there was digging in the morning and after dinner, a frenzy of dirt and shovels and wheelbarrows, then forms, then concrete from the portable mixer, then concrete block. Then the shelter seemed to disappear. Amazing.

I never saw the inside. But the idea was to scurry on down to your very own shelter when the sirens warned of incoming missiles, button up the hatch, then subsist on your stores of water and canned goods (what happened to the waste products?), listen to the radio (the guy-wired broadcast towers would remain standing?), and come out unscathed into a brave new world when it had all "blown over."

The immediacy of the threat of nuclear war triggered the sense of mortality in a generation. It may even have catalyzed the beginning of widespread concern about the environment—as though since we and our world could no longer be taken for granted, maybe we had better begin to take better care of it.

A more sensitive and rational application of earth-shelter design started in the 1960s and '70s, with Malcolm Wells and others, including Philip Johnson, whose design for a sod-roof house in Cincinnati is upstaged by his Glass House and more monumental buildings.

But even with the interest and effort and innovations in earth-shelter building, there were less than 50 fully sheltered homes in the United States by 1970. Everything that has happened since—the tax credits, code approvals, new waterproofing and damp-proofing materials, the run of how-to books and articles—has all happened since 1970. So it is not surprising that many 7- and 8-year-old earth-sheltered homes have developed problems with condensation, leaks, and structural faults. There hasn't been much time for trial and error. But there has been enough time to work out the major problems because the interest is so intense.

There are still a lot of people exploring alternatives, innovative ideas and designs not yet tested by time. So in this book, you'll see the major distinctions, how different designs work on different sites, and how different plans work for the people inside and the environment outside.

Earth-sheltered construction now includes large, elaborate homes like this one by Terra Domes, which has built schools and firehouses. Aside from energy efficiency, the reinforced-concrete domes offer 40-foot clear spans.

DESIGN DECISIONS

Nomenclature may vary from source to source, but this diagram lays out the basic choices. The different gradations of exposure range from all-underground to an envelope design that may have two fully exposed facades and wrap-around clerestory light-vent access. Every plan can be expressed two ways, in chamber or berm shelter, which doubles your options.

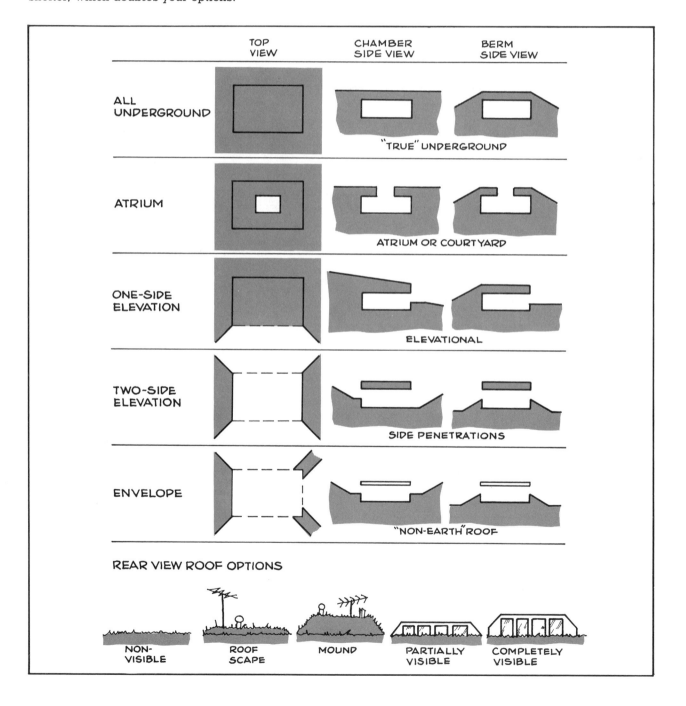

Coverage may be limited to a sod-covered roof and very modest earth-berming, where most of the house is built with traditional wood timbers.

Below: Traditional earth-shelter materials used in this Minnesota house built by Joe Topic are: precast concrete walls and roof planking, enclosing 2,400 square feet with 20 inches of soil on the roof over 4½ inches of rigid insulation. (Courtesy Topic Earth Homes, Shakopee, MN 55379.)

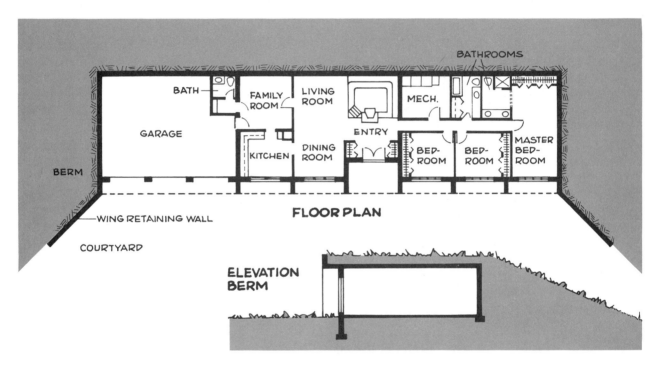

Site. As you can see in the sample drawings, there are two very basic types of earth-shelter plans: berm and chamber. Berms (just mounds of earth) may exist on site already, or you can build them by pushing the dirt around. Bermed houses would sit largely above grade just like conventional houses, were it not for the earth sloped up from the natural grade elevation to cover part of the walls, or all of the walls, or all of the walls and the roof.

Chambers are more like natural caves—what some authorities call "true" earth sheltering. A variety of chamber designs, such as atriums, elevations (cut into slopes), and envelopes (like roofed atriums providing clerestory lighting), are cut into the landscape. You can nestle down into the site or nestle the site up around the house, depending on your concern for view, water table, storm protection—you name it.

Coverage. A split-level development house with a full cellar backfilled on one side and fully exposed on the other is an earth-sheltered house. The owners may not consider it earth-sheltered, but it is. Homes can be sheltered on one side, with three full exposures, or on three sides, with one exposure. There are no absolutes.

But if you feel the need for some criteria on what exactly makes a house earth-sheltered (the backfilled split-level doesn't qualify), it should be this: Earth sheltering is a primary design consideration, planned for, and integrated with, every component of the project from construction materials to aesthetics, and making a significant impact on the finished site, with overall energy efficiency being the guiding factor. If one little earth berm can do all that, it's bermed up against an earth-sheltered house.

For conventional builders and homeowners, it may be difficult to think of the soil as a major building component requiring close attention. Here there's need for large gravel trenches and one or two elevations of drainpipe, and roof terracing, particularly until the sod takes hold to prevent roof erosion.

Earth-shelter architect Don Metz employs a distinctive wing wall (in foreground) extending from the main elevational design. The angled wing creates a wider, more open courtyard just outside the glazing wall enclosing the living space at "ground" level.

Soil. Slope, groundwater flow, surface-water flow, water-table level, and soil composition are all parts of a unique puzzle. On nearly flat sites, the water table is likely to be a primary concern. On a steep-slope site near the ridge, surface water is crucial. On a site near the foot of the slope, there may be accumulations of surface and groundwater. For bearing capacity, from best to worst, you look for sedimentary rock, dry clay, compacted gravel, compacted sand, soft clay, and organic soils (the swamp). That's clear enough, except that when you analyze the site for drainage characteristics, the order of the list changes. Clean gravel and clean sand are ideal, silts and clays are variable but generally mediocre, and peat is unsuitable. Generally, well-graded sands and gravels are excellent for building loads and drainage, silts and clays are variable enough to require close analysis, while soft clays, peat, and highly organic soils should be avoided.

Light and air. Taken for granted on most homes, well-planned natural illumination and ventilation are crucial in earth-shelter designs. They are as important as the waterproofing. The more sheltered you are, the more important they get. You know about deciduous trees to the south for winter-season solar gain through glazing, about evergreens to the north to break prevailing winter winds, about the five-to-one windbreak ratio—trees shelter a ground area extending from the tree away from the wind for a distance five times the height of the tree. Light and ventilation are code-con-

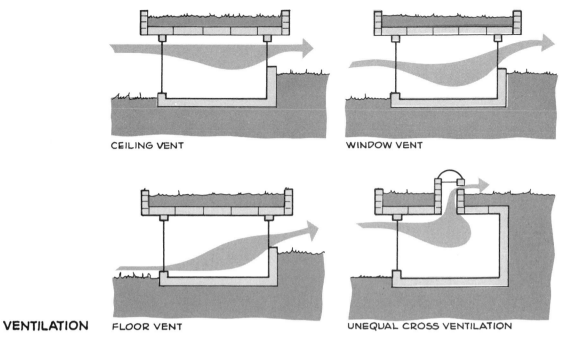

CEILING VENT

WINDOW VENT

VENTILATION FLOOR VENT

UNEQUAL CROSS VENTILATION

Roughly equal inlet and outlet vent areas is the basic rule, fine-tuned by breaking large glazing facades with floor vents and clerestory runs with ceiling vents.

Before building, you can gain so much practical information by investing in a scale model—a fairly large, very basic cardboard version, cut away so that you can see how much light will filter in through different wells.

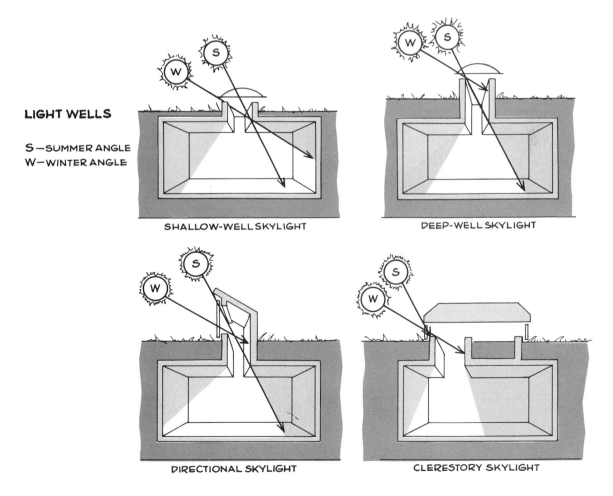

LIGHT WELLS

S—SUMMER ANGLE
W—WINTER ANGLE

SHALLOW-WELL SKYLIGHT

DEEP-WELL SKYLIGHT

DIRECTIONAL SKYLIGHT

CLERESTORY SKYLIGHT

LIGHT AND VENT WELLS

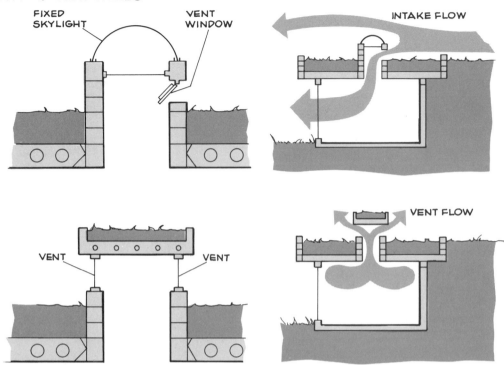

Light and vent wells offer the same design options as earth-sheltered buildings. The wells can be of bermed, elevation, and envelope construction.

Shallow light wells can be built through poured-concrete roofs, concrete-plank roofs, or more conventional heavy-timber frames. Try for at least a few inches of clearance between frame and ground on the outside.

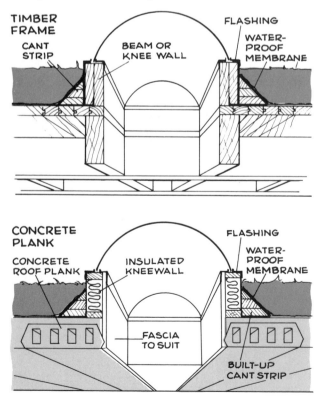

trolled, but the minimums are more for health reasons than for aesthetics or interior environment.

You can get light and air through wells and skylights in addition to at least one wall of exposed glazing. If the finished ceiling is within a few feet of grade, you'll have no problems. A 4-foot-square or 4-foot-diameter skylight on a 2-foot-deep box will flood a large room with light. On a 20-foot-deep box (that would be unusual), it will shed a small, bright square of light on the floor at midday and a small square of reflected light the rest of the time. Deep wells never seem to perform up to expectations. Working like a telephoto lens, they narrow the field; in the case of operator skylights, they can create a minor jet stream.

Cross-ventilation is most effective and efficient when inlets and outlets are roughly the same size. If your main exposure is filled with windows or sliding glass doors, it is difficult to equalize cross-flow with individual vent wells. That would be like reducing pipe diameter from ¾ to ⅜ inch and expecting the same rate of water flow out the small end as goes in the big end. It just doesn't work that way. More continuous runs of skylights do a better job, but there is another option that will appeal to people who do not fancy their sod-covered roof dotted with plexiglass bubbles.

Consider a site with one exposure, where the house is cut into a slope, where the natural grade line runs down the hill and right across the roof like a continuous ski slope. Secondary light and ventilation has to come through roof wells. But if the roof is raised just a few feet, interrupting the continuous slope, you get the plastic off the grass and a continuous clerestory protected by the roof overhang all along the high side of the slope. Tucked out of the weather, the entire run can be awning or hopper vents, up near the ceiling where they should be to evacuate warm summer air.

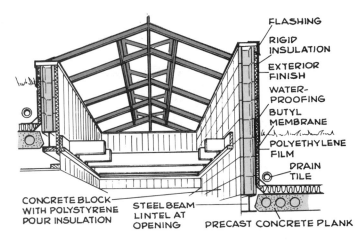

Once you put a hole in the roof, it doesn't matter too much how big the hole is. Instead of scrimping on light and air, continuous-light wells, shown below, with pull shades or controllable louvers, shown above, offer full or limited exposure, depending on the weather outside and the amount of light you want inside.

It is important to design a versatile system. For instance, instead of large, fixed-glass frames that feature the stunning view, break the panel into sections, with a vent window toward the floor on the sun side of the house and toward the ceiling on the shade side. Instead of fixed glass on light wells, use operator skylights with louvers or shades for sun screening. To get light and vent control in the same well, you can move the skylight up on a deeper box that is framed for vent windows on the sides of the well projecting above grade. Again, there are a lot of options.

AWWF WOOD FRAME CLERESTORY

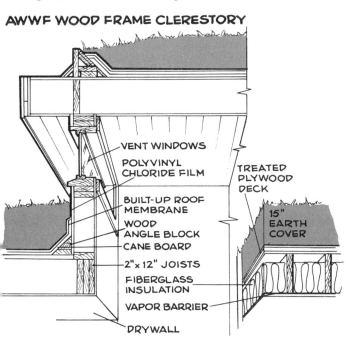

VENT WINDOWS
POLYVINYL CHLORIDE FILM
BUILT-UP ROOF MEMBRANE
WOOD ANGLE BLOCK
CANE BOARD
2" x 12" JOISTS
FIBERGLASS INSULATION
VAPOR BARRIER
DRYWALL
TREATED PLYWOOD DECK
15" EARTH COVER

There are as many types of light and vent openings as there are earth-sheltered structures. Above is an all-wood frame clerestory in an AWWF earth-shelter system. Below is an insulated well cut through concrete planking on a 1,387-square-foot condominium unit designed by Close Associates (3101 East Franklin Ave., Minneapolis, MN 55406).

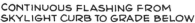

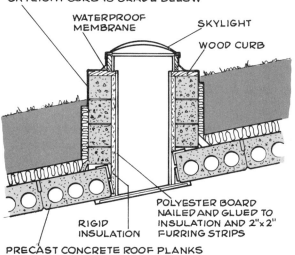

CONTINUOUS FLASHING FROM SKYLIGHT CURB TO GRADE BELOW
WATERPROOF MEMBRANE
SKYLIGHT
WOOD CURB
POLYESTER BOARD NAILED AND GLUED TO INSULATION AND 2" x 2" FURRING STRIPS
RIGID INSULATION
PRECAST CONCRETE ROOF PLANKS

Heat. If only you could be as flexible with heating! Too bad there is not a modular heating plant where you could plug in BTU units as needed, as you got used to the house and discovered, after all the discussions of R-values and cubic feet per minute of air, how much heat you really need. With an earth-sheltered home you are likely to need less than you think.

This is the place to set right the most common misconception about energy efficiency in earth-shelter design—namely, that fuel costs are so low because all that earth provides so much insulating value. No. Heat, like water, seeks equilibrium. If you put a block of ice in hot water, it melts rapidly. But if you put it in nearly freezing water, the ice will stay solid for quite a while. This cannot be explained by saying that ice cubes or buckets of hot water are good insulators. They're not.

The soil outside an earth-sheltered house is temperate. On a cold day, the temperature inside the house is closer to the earth temperature than the outside air temperature. And on a bitterly cold day, it is even closer to the earth temperature and further from the outside air temperature. The closer your inside air is to equilibrium with the temperature on the other side of the wall (in the soil), the less heat in winter and less cooling in summer you will lose through the walls as the two sides work endlessly towards equalization.

Stable, moderate soil temperature is about the best reason for building an earth-sheltered home. The Underground Space Center at the University of Minnesota conducted a study of fluctuations in air temperatures and in soil temperatures at various depths. They documented conditions in the Minneapolis-St. Paul area, where changes are likely to be about as severe as you'll encounter.

They measured an outside temperature range from $-30°$ F. to $100°$ F., and annual range of $130°$ F. But at only 8 inches below grade, that swing was limited to $40°$ F. ($30°$ F. to $70°$ F.). At 2½ feet below grade, the swing was cut further to $33°$ F. ($32°$ F. to $65°$ F.). Don't those animals who burrow down into the soil for winter hibernation begin to look pretty smart? At 10½ feet below grade, the swing was only $18°$ F. ($40°$ F. to $58°$ F.)—and remember, this is an annual swing. Daily fluctuations were minuscule. Finally, at 26 feet below grade, the swing for the entire year, with $130°$ F. swings on the surface, was only about $5°$ F.

In January, when it's $-30°$ F. and the city has frozen solid and stopped cold, 26 feet down under, just outside the living room walls, it's about $45°$ F. And in August, when it's $100°$ F. and the city has melted and stopped dead in a pool of sweat, outside the living room walls it's still about $45°$ F.

To ice the cake for you, the Underground Space Center published outside-to-inside air temperature relationships over a 4-day period on an earth-sheltered

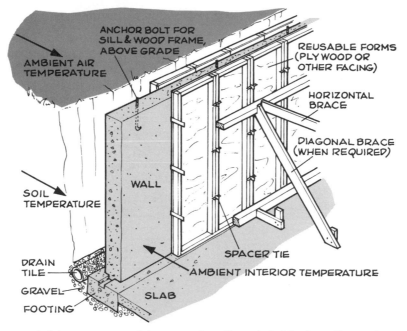

ANCHOR BOLT FOR
SILL & WOOD FRAME,
ABOVE GRADE

AMBIENT AIR
TEMPERATURE

REUSABLE FORMS
(PLYWOOD OR
OTHER FACING)

HORIZONTAL
BRACE

DIAGONAL BRACE
(WHEN REQUIRED)

WALL

SOIL
TEMPERATURE

SPACER TIE

AMBIENT INTERIOR TEMPERATURE

DRAIN
TILE

GRAVEL

FOOTING

SLAB

Homemade or commercial forms are used for poured walls, while block walls are formed by hand. Both have so much mass and make up so much of the interior surface that they can store internally produced heat and coolness, and can also store and reradiate external, moderating temperatures in the soil.

home in Rolla, Missouri. The test was conducted without any internal heating to raise or maintain inside air temperature. On day 1, the house was 58° F.; outside, it was 32° F. On day 2, the house was still 58° F.; outside, the temperature had dropped to 24° F. On day 3, the house was down a bit to 55° F.; outside, it was down to 5° F. On day 4, the house was 53° F.; outside, it was just below 0° F.

There are reports from homeowners whose furnaces broke down for a few days—creating only a marginal and gradual heat loss inside. There are reports of vacation homeowners leaving for a week, and without any ongoing, maintenance-level fuel costs, returning to find that the inside air temperature had dropped only about 5°.

In simple terms, the soil has a moderate and relatively steady temperature day to day and season to

season. No freak storms before you get the wood in; you can count on it. The trick is to get down in there, with all the tremendous energy-saving numbers, and to take enough light and fresh air with you to make space that is not only attractive but also physically and psychologically comfortable.

ENERGY MAINTENANCE AND ENERGY GAIN

I want to re-emphasize the unique double benefit offered by energy-efficient earth-shelter design. First, earth sheltering makes it relatively easy to maintain inside temperature. Inside the protected walls and roof, temperatures drop slowly and then only marginally during the winter because the soil stays warm. In summer, when inside temperatures rise, even with good shading and ventilation a lot of heat is given up through the walls and roof because the earth on the other side is still relatively cool.

The other side of the coin is the energy benefit based on the starting point, the ambient temperature, of the inside air. Let's take a sample Case A, where you are saving money in a conventional home by turning the thermostat down to 40° F. on cold nights. You do save a bundle. The furnace will cycle on and off a few times during the night to maintain the 40° F. First thing in the morning, of course, you'll freeze. When you bump the thermostat back up, the furnace will make a long burn for 10 to 15 minutes solid to raise the temperature back up to 68° F.

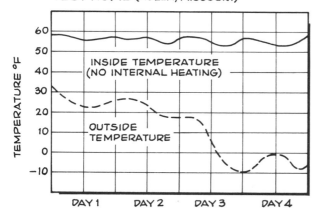

UNDERGROUND SPACE CENTER
TEST HOME (ROLLA, MISSOURI)

INSIDE TEMPERATURE
(NO INTERNAL HEATING)

OUTSIDE
TEMPERATURE

TEMPERATURE °F

DAY 1   DAY 2   DAY 3   DAY 4

Just the tip of the iceberg (right photo), this butterfly-roofed light and vent well is about all that's visible above the roof scape on one of architect Don Metz's designs. Below is another interesting detail by Metz. A combination of bermed earth and low retaining walls frames a below-grade entrance made inviting and recognizable by the arched opening.

In the earth-sheltered version of Case A, first of all the furnace won't cycle on and off during the night because the soil temperature is probably about 50°—well above your minimum, money-saving, freeze-your-toes-off furnace setting in the conventional house. You save that nighttime maintenance fuel, and you're more comfortable, too.

When you get up in the morning, the conventional house is starting at 40° F., maintained by the furnace. The furnace burns to reach 68° F., raising the temperature 28° F. In the sheltered version, the house is starting at 50° F., maintained only by the wall temperature, which in turn is maintained by the soil temperature. Getting from 50° F. to 68° F. requires a burn for only 18° F. That uses less fuel because it's a shorter trip.

Now let's take Case B, where you're saving money in your summer home by turning off the central air conditioning, which has been working to maintain 72° F. when it's 95° F. outside, while you're back in the city apartment tending to business for a week.

In the earth-sheltered version of Case B, you really don't need air conditioning, but let's say you didn't know that, and you're finicky about temperature, and so you have an air conditioner. When you return to the house from the sweltering city, outside temperatures are still running about 95° F. The air conditioner in the conventional house needs to pump from 95° F. down to 72° F., a 23° F. drop. But in the sheltered version, the earth mass in summer is up only some 10° F.—for example, from winter temperatures to 60° F.

Now you tell me how much cooling you'll need in a house with one exposure of glass shaded from direct midday sun, when it's 95° F. outside and all the walls of the house are 60° F. How much? Did you say a few cents for electricity to push the fan that circulates the slightly stale air? Did you say a few dollars on the service contract for the unused air conditioner.

With earth-shelter design, you use less energy to maintain comfort, and you need less energy to get

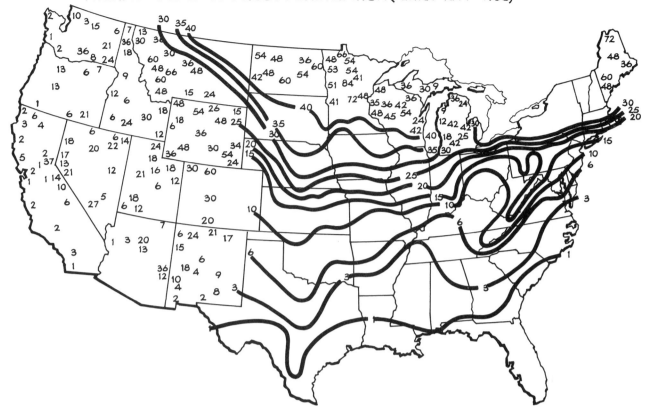

comfortable in the first place, no matter what's happening outside. Aboveground houses are subject to aboveground ambient air temperatures. The time and money spent on weatherstripping and caulking and insulation is devoted to isolating the house environment from the outside environment. When the ambient air outside is 0° F., you don't want any part of it. The same goes for 110° F. in the summer.

But no matter what you do with insulation or superinsulation or superduperinsulation, the process of seeking temperature equilibrium inside to outside is inexorable. You may install excellent barriers that provide isolation as a practical matter. But the battle is still one of extremes, getting from 0° F. to 68° F. in winter, from 100° F. to 72° F. or so in summer, a battle with 95-100° F. between the two sides.

With earth-shelter design, the "battle" is more a question of adjustment, probably no greater than from 45° F. to 68° F. in winter (23° F.), and nothing or next to nothing in summer. Closing the gap costs money, sends oil tankers around the world and pipelines through virgin tundra. It just makes sense to close a 25° F. gap instead of a 90 or 100° F. gap.

FOUR FINAL GOODIES

Rob Roy reports that his Log-End Cave house in New York State near the Canadian border requires about three cords of medium-grade firewood per year, figured at only 50-percent burn efficiency. You can do a lot better than that using one of the new airtights with an integral catalytic combustor to reduce creosote levels and increase combustion efficiency. That's north, and cold, with 40° F. subterranean temperatures in dead winter. Roy converts his wood consumption for conventional fuel users as follows: 360 gallons of fuel oil, figured at 65-percent efficiency; 9,500 kw of electricity, figured at 100-percent efficiency; 375 therms of natural gas, figured at 75-percent efficiency.

Every degree of heat you do not have to produce in winter saves between 2.5 and 3 percent of your heating costs. Aboveground, you play this game with the thermostat, a wool bathrobe, and a pair of furry slippers. Underground, you don't have to play.

Every degree of cooling you do not have to produce in summer saves roughly 5 percent of your cooling costs. With efficient ventilation in the earth-shelter design, you should not even feel the need for air conditioning.

Using national averages, which I try not to do since they are not likely to apply to your specific site, at 4-inch depths, soil temperature runs between 18 to 25° F. ± ambient air temperatures. Get down 10 feet, and soil temperatures are relatively constant when taken on a national average, with minimal seasonal variation and a maximum range of 10° F. ± ambient air temperatures.

PREPARATIONS FOR CONSTRUCTION

Since most or all of an earth-sheltered house is underground, soil analysis and soil handling are important, because of compressive strength and loading and the need for good drainage. It is always more profitable to adjust a basic plan to the site than to impose a detailed, inflexible plan that may work against the site. You can fine-tune most plans to fit the site, which beats looking endlessly for the near-perfect match.

CONTOURS

Even if your plan is simple, it will be difficult to look at a natural berm or slope and get an accurate idea of how many yards of earth you need to excavate (including at least 4 feet of working room outside the walls), where you will put it, how much of it you will have to move a second time and over what distance, and what the contours across your site will look like when you're finished.

You can take the adventurous approach, then discover that you have soil for a giant earth berm left

Berms can be held with stone, poured concrete, reinforced-block walls, railroad ties, and similar materials. Berming can create entrances, open courts, access to underground carports, and more, in addition to sheltering the building proper.

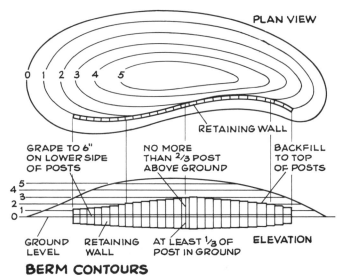

GRADE TO 6" ON LOWER SIDE OF POSTS

NO MORE THAN 2/3 POST ABOVE GROUND

BACKFILL TO TOP OF POSTS

RETAINING WALL

PLAN VIEW

GROUND LEVEL

RETAINING WALL

AT LEAST 1/3 OF POST IN GROUND

ELEVATION

BERM CONTOURS

over after you have graded the site just the way you want it (CLEAN FILL FOR SALE is what you can print on the sign). It is difficult to make a big pile of dirt disappear, particularly when you know better than to bury the bases of nearby trees in a foot or so of excess backfill.

You could pay an engineer or surveyor to give you close contours from which you can get reasonably close cubic-volume figures for earth moved, house enclosed, and earth repositioned. On thick, scrubby sites, you may have to do quite a bit of clearing just to see how the land lies.

A cursory survey, even if you haven't done one before, can be a tremendous benefit. Allowing for error (you're not surveying with precise vectors and initial points for a recorded deed), you can make a rough grid, then a survey map, working with a helper, a story pole, and a transit (which you can borrow or rent). A rough approximation can be made as follows:

1. By eye, pick a point on the site roughly halfway between the highest and lowest elevation, from which the transit can be rotated to view the entire site.
2. Tape a ruler onto a long 2 × 4 to make a story pole. (You read the ruler increments through the transit eyepiece.)
3. Lay out a 4-foot grid (or, for greater accuracy, a 2-foot grid), across the site with string lines or with stones or stakes marking the grid intersections.
4. Duplicate this grid on graph paper, marking prominent features such as large trees and boulders for reference.
5. With the transit elevation as 0 registered on the story pole, take a reading of the pole through the eyepiece (up a foot from 0, down 6 inches from 0, etc.) at each grid intersection. When you play connect the dots, tracing all the similar elevations, contour lines will emerge.
6. Using, say, the 4 × 4-foot grid (16 square feet, 64 cubic yards), you can estimate the average elevation of each 4-foot quadrant, how much of the soil in it will be removed, and where to establish window elevations and other details from the plan.

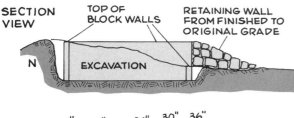

SECTION VIEW

TOP OF BLOCK WALLS

RETAINING WALL FROM FINISHED TO ORIGINAL GRADE

EXCAVATION

N

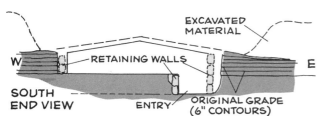

6" 12" 18" 24" 30" 36"

1"

N

HIGH RETAINING WALL

LOW RETAINING WALL FOR ENTRY

LOW SOUTH WALL (42") FOR GLAZING

RETAINING WALL

EXCAVATION PERIMETER

TOP CROSS SECTION

ORIGINAL CONTOURS

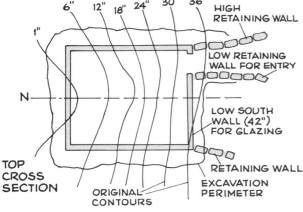

EXCAVATED MATERIAL

W RETAINING WALLS E

SOUTH END VIEW

ENTRY

ORIGINAL GRADE (6" CONTOURS)

EXCAVATION

Since excavation is one of the first expenses in the job sequence—and one of the largest—it is natural to think of conserving resources for the long haul and to scrimp. Don't. Pay for professional help and hydraulic equipment. Digging is not the most thrilling part of the project. If you take too long with it, everything else gets behind schedule, so that you risk bumping into bad weather at the end of the project. Since you almost always pay for this work by time instead of by contract, thorough preparation saves money. You can do things such as setting flagged stakes just outside the excavation area to mark the contours, so that you don't lose track of what dirt is going where in the excitement of bulldozers and bucketloaders; lay out areas for boulder storage (where you will build retaining walls or terraces); decide on a resting place for excess fill (where you can berm it for a woodshed or shop or studio or the back wall of a future greenhouse).

Above is an example of cut and fill (berming with excavated material), showing how Rob Roy built his earth-sheltered house. He established 6-foot contours to map the site. Below, the large rocks hold back the berm at the southwest corner of the house. Close contours can even tell you where the biggest boulders will fit, minimizing later experimental placements.

Before you get too far along, take the time to look at the soil being excavated to determine if it has suitable drainage characteristics and if it will suffice as backfill. Refer back to the heading "Soil," in the section "Design Decisions," or consult one of many more detailed references listed at the end of this section, such as *The Earth Shelter Handbook* or the updated and revised edition of *Earth Sheltered Housing Design*. (Look for an expanded version of the Underground Space Center's report, being compiled at this writing, *Preliminary Design Guidelines for Earth Contact Structures*, including the latest information on insulation configurations and thermal performance evaluated through complex computer simulations. It will appear in the revised edition of *Earth Sheltered Housing Design*.)

A suggestion: If your site is on a slope, and you have a shovel load of dirt, and you are not exactly sure where to put it, dump it uphill. It is fairly easy, even working alone with a spade-point shovel, to move a lot of loose dirt a few yards downhill. It takes forever to move the same amount in the other direction.

Another suggestion: Assuming you are in a code-controlled area, be sure the building authorities have

The on-site soil absorption test, below, was devised by Tri/Arch Associates, a design firm. Test early, and have your results confirmed to determine what type of drain system will work, if it is code-approved, and where it can be built economically.

SOIL ABSORPTION TEST

 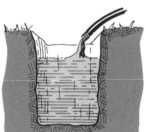 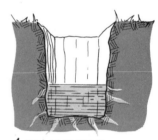

1 DIG 12" SQUARE HOLE 18" DEEP.

2 FILL HOLE WITH WATER.

3 LET WATER DRAIN AWAY.

4 FILL THE HOLE AGAIN 6" DEEP.

Below: There is no reasonable do-it-yourself, owner-builder substitute for earth-moving machinery. Even if you can drive a backhoe, as shown below left, you may not have the experience needed to finesse the bucket and bucket teeth to move large rock, even if you carry them around the site with a chain hitch, shown below right. Some contractors use minidozers to get in and around footings.

approved all phases of your plans and building. Even with an architect's or engineer's stamp on the plans, go into specifics: second egress for fire, underground entrances, septic and leaching field plans, perc tests, front-yard minimums (sunken courts in many earth-shelter designs) for "basement" rooms where ceiling height is code-regulated at, for example, 4 feet above "grade." Interpretations of some code language that was never intended to cover earth-shelter work can cause problems. Better to pick it all apart on paper rather than on site.

I urge caution with code issues, because you can expect that your house will be the first earth-sheltered house plan for your inspector. As this book goes to press, the Underground Space Center estimates that there are only about 3,000 true earth-sheltered homes in the United States. This does not include several thousand semisheltered designs, but only plans where earth sheltering is a major part of the planning and construction. Earth sheltering is growing but is still likely to be new territory, particularly to suburban and rural planning, code, and building-department professionals.

Worries about light and air, and about views through windows and doors, present stumbling blocks for potential earth-shelter owners. The courtyard below and the atrium at left by architect Don Metz should help dispell those fears. Sunken courts and atriums can be protected from the weather with overhangs or glazing, or simply left exposed and built with central drains.

CONSTRUCTION

All the attractive energy-saving numbers are down there waiting for you. Good earth-shelter design relies on making a pleasant environment with ingenious provision for light and air and mechanical systems and for such special circumstances as soil pressure and the increased potential for leaks.

There are countless varieties of wall and roof systems. Sometimes, it seems that each home has its own

Basic types of retaining walls for berms and terraces can support simple changes in grade elevation or a change in elevation and slope. Thorough site planning lets you pour house and area walls at the same time.

RETAINING WALLS

GRAVITY RETAINING WALLS

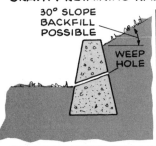

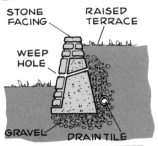

T RETAINING WALLS

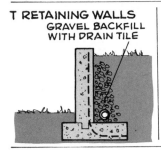

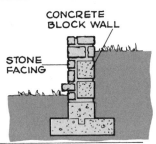

L RETAINING WALLS

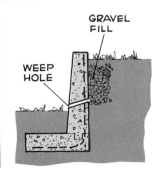

special structural system. Most designs, though, use poured concrete or reinforced concrete block to separate the living space from the surrounding soil.

There is nothing exceptional about concrete work in earth-shelter construction. You can use *Architectural Graphic Standards, Building Construction Handbook, Concrete Masonry Handbook,* and specific how-to books (see "Information Sources") for the nitty-gritty particulars of mix ratios, pouring-in lifts, mix segregation, rebar placement, tying, formwork, curing, and more. Certainly there is a large book's worth of detailed information on getting reinforced concrete into its forms. Concrete in earth-shelter building does get special treatment where it meets the soil.

But first, even though you'll probably wind up using some form of concrete masonry, take a look at a few of the unique possibilities that have been used successfully and economically.

AWWF (all-weather wood foundation).

This works on the standard principles of wood-frame construction with a twist—all the wood components are pressure-treated with chemical preservatives. A typical AWWF wall is made of 2×6 dimensional timbers 24 inches on center. As with conventional footings, the AWWF can be bolted through the sill to a concrete footing. Unlike conventional footings, AWWF has been used during dead winter, set complete with first-floor framing in 12 hours on a pressure-treated mudsill laid over a compacted gravel trench footing.

On the inside, an AWWF wall is insulated, channeled with plumbing and wiring, and finished to suit. On the outside, pressure-treated plywood is waterproofed as thoroughly as a concrete wall. It's almost as good as reinforced concrete, but not quite, and therefore is recommended for designs with shallow roofs (not the 24-inch-thick jobs) and for sites near the crest of a slope but not at the foot of a slope, where earth pressures are the greatest.

The footing is protected with conventional drain tile in gravel. The advantages of AWWF wall include: increased flexibility under pressure that can crack concrete; quicker, less-expensive installation; conventional space for plumbing and wiring systems and insulation; and conventional nailing for interior finishing. AWWF is still experimental in earth-shelter construction but has undergone extensive monitoring, 30-year burial tests, and more for standard wood-frame housing, where its use permits low-cost, fast, year-round construction. No concrete freeze-thaw and admixture worries here. Also, AWWF has the appropriate blessings from HUD (Department of Housing and Urban Development) and the FHA (Federal Housing Administration).

Steel culvert. In its earth-shelter application, this is the basic quonset hut one flight down. The corrugated steel shell used on large storm drains and small overpasses serves as earth-shelter walls and roof literally rolled into one. Very few earth-sheltered homes,

much less any other kinds of homes, have been built with culvert. Culvert offers a series of unastonishing advantages and disadvantages, with one really unique advantage. With steel culvert you get both form and structure in one package. But there is more to it than that. Large-span culvert is fabricated in sections. If you set up the job carefully, you can pour trench footings, then drive the truck loaded with culvert sections between the footings, off-loading as you go. The individual sections are bolted together, and the sections are bolted or anchored to the footings. From that point on, culvert works like many other earth-shelter systems, with drain tile at the footings and backfilling.

The steel is covered with foam insulation—not UFFI (urea formaldehyde foam insulation) which has produced significant outgas problems and, at least for now, has been banned by the U.S. Consumer Product Safety Commission. Then come a fibrous asphalt, multiple layers of plastic film, a sand backfilling of several inches

If you don't know about AWWF (all-weather wood foundation), it would be reasonable for you to think that burying wood underground is a crazy idea. But AWWF is all pressure-treated with wood-preserving chemicals. It is low-cost and fast to install year-round hot or cold. It is definitely a foundation wall most do-it-yourselfers could install, and it enjoys widespread code approval.

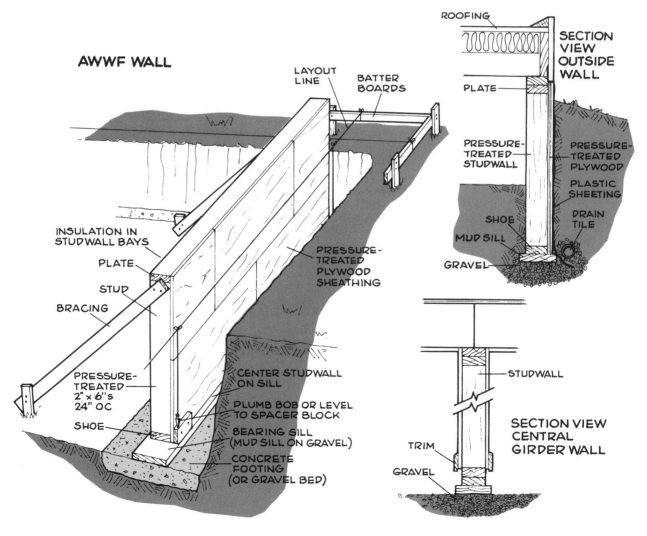

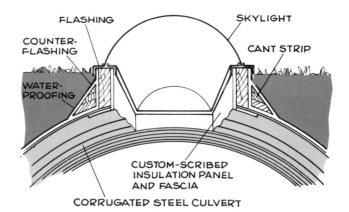

FLASHING SKYLIGHT
COUNTER-FLASHING CANT STRIP
WATER-PROOFING

CUSTOM-SCRIBED
INSULATION PANEL
AND FASCIA

CORRUGATED STEEL CULVERT

Culvert assembly (below) is practically automated, which is why Quonset huts make sense for temporary shelter. Using stock industrial components saves time on assembly, but this method comes back to haunt you with complex details like cutting skylight openings (at right) with a torch and fabricating custom insulating panels, headers, and trim.

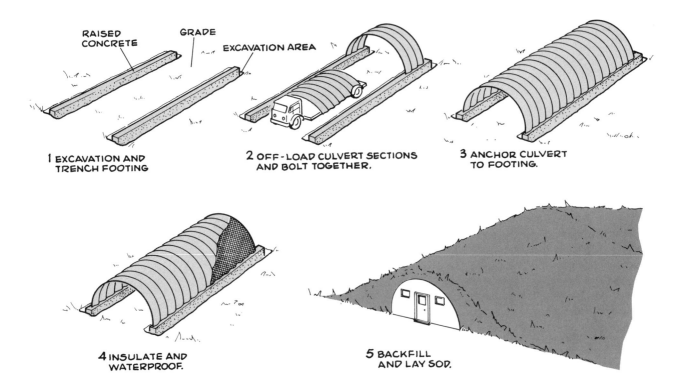

RAISED CONCRETE GRADE EXCAVATION AREA

1 EXCAVATION AND TRENCH FOOTING

2 OFF-LOAD CULVERT SECTIONS AND BOLT TOGETHER.

3 ANCHOR CULVERT TO FOOTING.

4 INSULATE AND WATERPROOF.

5 BACKFILL AND LAY SOD.

to guard against puncture of the membrane, and finally soil backfill to a finished grade. (Water getting through the membrane would cause rust and corrosion in the steel.)

The speedy, one-shot system is difficult to work with in some respects. For instance, you need a cutting torch to make skylight openings. And the corrugated pattern produces unfamiliar, irregular wall and ceiling surfaces. But with hangers for interior framework bolted to the steel, you can create channels for plumbing and wiring and points of attachment for conventional surfaces.

Concrete. Once you're bringing in concrete to pour footings, why not use it for the walls, retaining walls, the slab, and the roof as well? Concrete involves a lot of time and money for formwork, but you recoup by getting it all done at once instead of setting one block

at a time. In an earth-sheltered house, you can use every variety of concrete masonry: rebar-reinforced concrete footings and foundation; welded-wire-reinforced floor slabs; rebar and web-tie-reinforced concrete block bearing walls (or surface-bonded block); and prestressed, precast concrete roof planks.

Great compressive strength in concrete permits multistory underground design, allows forming or block-building for all sorts of air and light access, and with creative formwork can offer many combinations of interior finishes and patterns (repeated patterns like striating or beading from molding or plank seams in the forms).

Concrete is the ultimate for resistance to many home problems: water damage, dry rot, wet rot, rust, warping, splitting, delaminating, thermal expansion and contraction, fire, insects, vermin—the full list of win-

Unusual designs such as Don Metz's wing retaining wall (left) require a lot of formwork and steel reinforcement. Owner-builders may prefer to build block walls, or subcontract the concrete work, before setting wood sills, beams and decking themselves.

Concrete is the most commonly used structural material in earth sheltering. And this may take many forms, including reinforced, poured walls and domed roofs, and as a binder in soil cement to make rammed-earth wall panels, as in the earth-sheltered home, below, built by Rammed Earth Works.

ners. And because of its mass and density, concrete efficiently soaks up and recycles hot air in winter and cool air in summer.

Even though reinforcement is time-consuming and adds to the cost of concrete, it is a job and an investment that can only be made once—before pouring. Reinforcement can give concrete the extra strength needed to resist cracking under soil pressure, which can lead to leaking. Even though you can probably do the job less expensively in concrete block, including rebars, I prefer pouring, unless getting concrete to the site requires some extraordinary effort such as elaborate chute construction. To me, every block has four potential cracks, which means potential leaks, built in— one on each side of the rectangle.

To prevent cracking under soil pressures, you still must use rebars, set in solid grout voids every third

block or so, set large steel lintels (or concrete) at openings, and clean up the mortar as you go. I think concrete block is at least as much trouble as poured concrete in the end, and given that, I'd rather wind up with a monolithic structure—a kind of all-or-nothing proposition. Take the trouble up front, and you don't have to worry about all the potential weak joints in a block wall.

I don't want to overstate my preference. I should let you know that at least some small part of that choice is based on my predilection for carpentry and, therefore, for formwork as opposed to masonry and block work. I take a look at some of the possible underlying reasons for opinions but, in the end, pay more attention to my experiences on the job, building block walls and pouring concrete, and tearing down masonry walls and frame walls.

TIPS ON BUILDING
WITH CONCRETE BLOCK

Block work, shown in the three photos at right, isn't as easy as it looks. Even with constant checking against corner-to-corner string lines and against levels, it takes experience to keep a long wall straight. To stabilize the first course and help resist leaks, the first row is set in full-bed mortar. Butt edges are buttered. As shown at far right, corners rise first and walls between are filled in. Lots of joints. (The mason is checking layout by finding block centers along the diagonal.)

As shown in the three photos at right, one of the ways to improve block walls is to employ reinforcement. In addition to vertical rebar in grouted voids, specialty ties like this angle bar (far right) will join partition walls to outside walls. Wire mesh allows you to fill only one course of voids.

On conventional 8 × 8 × 16-inch block or double-wide block to get more mass and more strength in an earth sheltered wall, headers for openings are set in mortar bedding on a slip plate, shown below left. This permits expansion and contraction and load carrying without stress cracking and fracturing joints in the wall.

Can't you just get out the radial-arm saw, as shown below left? Well, not quite, but masonry contractors will save time and improperly broken block by carbide cutting. The owner-builder alternative, below right, is scoring the block on both sides with a brickset (through one of the voids), then hitting a few hard blows to sever the connection, which is substantial in double-width blocks.

On block, a smooth pipe or other jointing tool is used to finish exterior mortar.

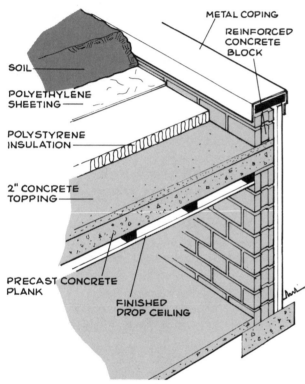

A typical earth-shelter concrete composite structure combines wide, reinforced block walls, high enough to create a parapet for soil retention, with concrete planking set into the wall. Planking is tied with rebar and a 2-inch concrete topping. With a stiff mix, a 4-to-2-inch differential can be used to create just the hint of a slope.

CONCRETE PLANKING

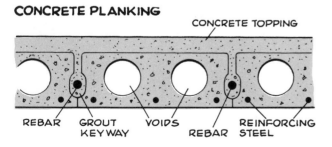

Prestressed, precast concrete. No, you won't find this at your local lumberyard or home center. Yes, it is a little remote for owner-builders, both physically and emotionally. That means you may have difficulty finding this material. Or at best it may be a struggle to get a commercial supplier to sell you a small load for only a single roof. And precast concrete may be a bit intimidating to work with once you get it, because it is an unfamiliar, industrial-grade product.

Precast concrete planks are the masonry version of tongue-and-groove roof decking, except the concrete comes in bigger, heavier, harder pieces. You can't balance four or five 16-footers on your shoulder, or toenail or facenail the planks in place. As different as they feel, concrete and wood planking still do the same job. You shouldn't automatically shy away from unfamiliar materials. Check the pros and cons, and if it seems like a reasonable choice but a little beyond your capabilities (and there is nothing wrong with that kind of honesty, which produces a better, more durable home in the end), you can subcontract that part of the job.

Concrete planking, normally 2 to 8 feet wide and available in various lengths (including major spans, up to 50 feet long), is factory-made, uniform, and prefinished (although you may not like raw concrete, even painted).

If you stood concrete planks on edge, the high compressive strength of concrete (10:1 compressive to tensile, roughly) would support massive loads. Laid flat and tension-stressed, the planks would bow, crack, and probably break under earth loading. That's why the planking is cast with pretensioned steel cables set in pairs top and bottom on each side of every longitudinal void.

You can think of it as a flexed muscle implanted in the concrete, literally building in not just the extra strength of rebars, but also the particular type of muscle that works on tension forces in the flat concrete plank. Concrete generally looks and acts like a rock. Pretensioned steel in concrete provides some of the elastic quality of wood grain in structural beams. The planks may crown slightly when they are installed but straighten out under loading.

Don't worry about the hollow voids cast along the length of the planks. Structural performance would not be improved by the extra bulk if these voids were filled in, the same way a steel I-beam would not be improved (in fact, it would be a disaster) if the I-shape outline were filled in to form a solid steel rectangle. And just when you started thinking that precast concrete planking was going to create all kinds of problems you discover that the voids can be used for wiring, ducting, and venting.

That's the pretensioned part of concrete planking. It is also prestressed, which imparts added compressive strength and compacts the cement binder and mix aggregate, leaving the finished plank impermeable to water. Sealing seams between planks is accomplished with cement grout beneath one of the many waterproofing and rigid-insulation combinations to make a high-quality, really industrial-quality roof.

The disadvantages? If your site is far from a plank-fabricating plant, you'll pay dearly for trucking. Also, because the planks are heavy, they must be set by machine, such as a mobile crane. This may take the material out of the ballpark of some rough, rural, low-cost, owner-builder jobs.

There are varieties of concrete plank, trade names

to look for like Flexicore and Twin Tees. David Carter has used 15-foot lengths of Flexicore set by mobile crane, completing a 2,000-square-foot roof deck in two hours. More time is required to turn the individual planks into a single roof unit—grouting the seams, laying 1-inch rebar in grout along the keyways provided at right angles to the plank seams to tie the planks together, then pouring a concrete cap sloped 2-4 inches to prevent puddling and to better distribute concentrated loads. Twin Tees is cast like its name, in the section view like two tees joined together. It is more industrial looking than Flexicore and other planks finished flat on both surfaces, but the distinctive design has about 3 feet between the tee stems—enough room to cut through for light and vent wells without destroying the structural integrity of the beam. Obviously, trying to cut a 2-foot light well through a 2-foot-wide Flexicore plank would cut the plank in half.

Poured roof slabs. If concrete planking won't work for you, consider pouring your own before opting for the familiar and accessible wood decking. Poured roofs should be formed with a slight slope to prevent puddling (standing water that hangs on until it evaporates), using plywood and either 2 × 4s or proprietary steel staging to support the forms and the plastic loads of concrete moving into the forms—and the concrete itself, of course. On a typical 16-foot span, the reinforcement, which is the key to a sound poured-concrete ceiling, would consist of something on this order: #6

Don Metz of Earthtech presents three roof options and these reinforcement details. Carpenters may prefer to stick with wood if they can find the heavy timbers. The only reason not to use wood would be fear of leaks and resulting water damage, which could be more structurally severe on a wood deck.

ROOF OPTIONS

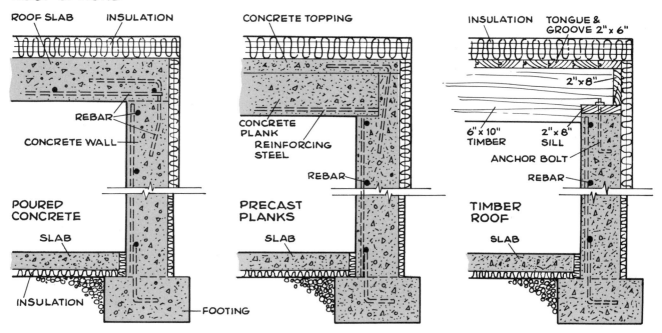

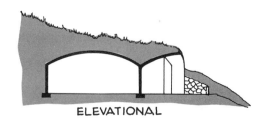

ELEVATIONAL

The Terra Domes pour creates a swale where dome modules meet. Collected water is drained through PVC pipe laid above the waterproofing membrane.

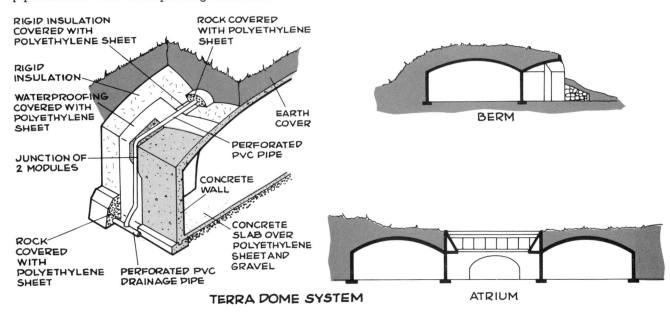

RIGID INSULATION COVERED WITH POLYETHYLENE SHEET

ROCK COVERED WITH POLYETHYLENE SHEET

RIGID INSULATION

WATERPROOFING COVERED WITH POLYETHYLENE SHEET

JUNCTION OF 2 MODULES

EARTH COVER

PERFORATED PVC PIPE

CONCRETE WALL

ROCK COVERED WITH POLYETHYLENE SHEET

PERFORATED PVC DRAINAGE PIPE

CONCRETE SLAB OVER POLYETHYLENE SHEET AND GRAVEL

TERRA DOME SYSTEM

BERM

ATRIUM

rebar (¾-inch diameter) 4 inches on center, suspended on commercial holders, called chairs, 1½ inches above the floor of the forms, with #3 rebar (⅜-inch diameter) laid and wire tied at right angles 4 feet on center. Using a deformed rebar with a tensile strength of 40,000 pounds per square inch, in a 4,000-psi mix, provides the best possible bond between steel and concrete because of the rebar's textured surface.

Pouring your own is like making one giant roof plank. I have seen do-it-yourself instructions for attempting to pretension with cable on site, but I wouldn't try it. It's up to you to decide how involved you will be. You can do all the work on site. But you can also subcontract excavation for instance, or excavation and masonry, then go ahead with insulation and waterproofing and finishing. You have to keep the idea of being an owner-builder in perspective. There will be enough to do even if you don't do it all. If the point of the project, the reason you're building, is to go through the owner-builder process, fine—stay away from materials and methods you can't handle confidently. But if the point of the project is to wind up with the best possible house, it would be unwise to dismiss materials and methods just because you can't handle them yourself.

WATERPROOFING

The most intelligently and ingeniously planned earth-sheltered home, full of light and open space, with breathtaking energy efficiency, would be a failure if it leaks. A little tar, a few new shingles, reroofing, pitch pockets—none of these fixes common to conventional houses would work. You can go up top to mow the grass and plant petunias, but you can't take 15 minutes during half-time of the game on TV to throw a little asphalt on a spot leak. Repairs are quite a production under a foot or two of sod. And on flat or nearly flat roofs, water that makes it through the waterproof (or formerly waterproof) membrane can migrate to the other side of the roof. Locating the leak can be even more of a production than fixing it. The answer is to get it right during construction, to do more than you have to, more than is recommended, so that you will positively, absolutely, never have to uncover the roof.

It is not terribly difficult to build a house that stands up. It may be ugly and an energy glutton, but you really have to try hard to build a floor or a wall or a roof that just falls down. Really. But it is more challenging to build a watertight roof, particularly a roof

At left, Terra Domes shows up with form sections, on a truck, ready to be interlocked. The monolithic, reinforced pour may be accomplished through pumping (below left) or building chutes (below) fed from a truck uphill of the site. The mix is hand-tamped and formed, then wet-cured. Dome sections are used in multiples. Bottom photo: The curvilinear Terra Domes system complements earth berms and sculptured sites.

with skylight and vent openings, which are, after all, holes in the roof. And it is more challenging to build a watertight wall when it is a subterranean wall.

There is a lucrative and growing industry in this country that exists only because so many builders do such a poor job on waterproofing and drainage. There are firms selling clay injection and pressure-pumping services where sodium bentonite is shot into the soil around a leaky foundation. (Bentonite is a clay that can be used as a waterproofer when it is applied in specific, controlled ways.) It is supposed to flow through the soil and rocks and root systems to settle against the foundation, bridge the cracks, and stop the leaks. Sure. And the moon is made of cheese.

There are firms that will rip up your landscaping, dig full-depth perimeter trenches, lay hot-mop tar and plastic sheeting on the walls, and then backfill. The bill may run only $4,000 or so, and your yard will be left looking like a baseball infield. And best of all, there are firms that offer no help with waterproofing (even though they are called waterproofing firms) but have devised expensive systems to cope with the symptoms of waterproofing problems—cutting miniature moats

around the basement slab to carry water to a sump pump. It's a $2,000 to $3,000 Band-Aid that doesn't even stop the bleeding, much less close the cut.

If that strikes you as sorry commentary on the quality of construction, this one will really get you. Consumer's Checkbook, a consumer-advocacy organization in Washington, D.C., got together with a group of construction experts to set up a test house with a leaky basement. They purposely created leaking that could be completely stopped by extending the leaders and regrading around the foundation, for a job price of around $300. Nine of 10 waterproofing firms contacted to bid on the job recommended work costing from $1,200 to $2,000. A real circus.

I don't want to panic you. It doesn't take a magician to keep out the water. But in most residential construction, waterproofing is an afterthought. Not on your earth-sheltered house.

Here are some of the prominent dos and don'ts before we go to specific applications.

GENERAL DRAINAGE. Before you build, and as you excavate, watch where the water is, where it

Water will always take the easiest route, seek its own level, flow downhill. This house uses a low-slope sod roof to begin controlling rainwater as soon as it hits, creating gradual percolation. Free-drain areas (large gravel and rock beds against earth-sheltered walls) provide another control, encouraging the water to collect in places where you can make provision, through drain tile, to move it away from the house.

gathers, and where it travels in a rain. If your site has one trouble spot, a concentration of ground or surface water coming from one direction, include in your excavation a plan for trenching, drain tile, and gravel for an area drain. The idea is simply to collect the water before it gets into the site, then channel it to a point on the side or past the house where it cannot flow back against the walls.

SPECIAL DRAINS. Provide drains for open atriums and sunken courtyards below grade. They will collect their own share of rainwater and groundwater from surrounding terrain at higher elevations. Treat these areas like a slab that cannot absorb water.

TWO-TIERED DRAINS. When the walls and the roof are sheltered, place two tiers of drain tile in gravel, one at the footing and one at the roof line.

The more water you remove from the building, the more effective your waterproofing will be. Taken to its ultimate conclusion, this means that if your area drain and drain tile and swales and courtyard drains work so incredibly well that the house walls never get wet, no water will ever get through the waterproofing

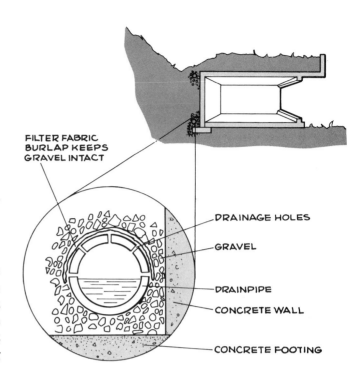

FILTER FABRIC BURLAP KEEPS GRAVEL INTACT

DRAINAGE HOLES

GRAVEL

DRAINPIPE

CONCRETE WALL

CONCRETE FOOTING

There are different types of drain tiles and different ways to install them. But the idea is always the same—to gather water through a filtering system that keeps out most of the soil and silt that could choke the system, and that feeds into a pipe that will carry the water away. Two drain systems are recommended when there is soil between roof and slab. One drain line is generally sufficient when a large free-drain area is created from roof to footing.

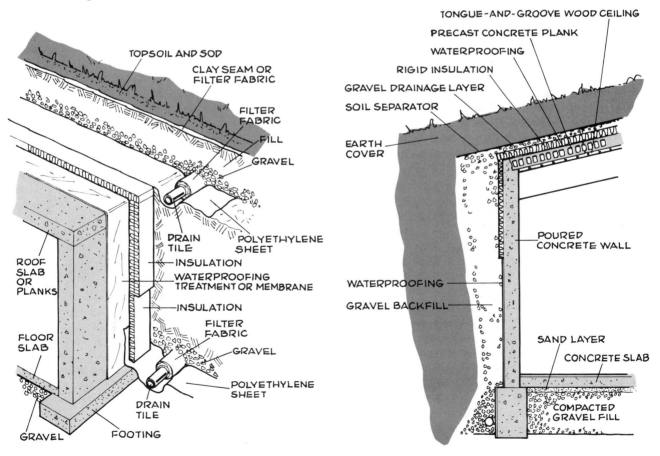

TOPSOIL AND SOD
CLAY SEAM OR FILTER FABRIC
FILTER FABRIC
FILL
GRAVEL
DRAIN TILE
POLYETHYLENE SHEET
INSULATION
WATERPROOFING TREATMENT OR MEMBRANE
INSULATION
FILTER FABRIC
GRAVEL
POLYETHYLENE SHEET
DRAIN TILE
ROOF SLAB OR PLANKS
FLOOR SLAB
GRAVEL
FOOTING

TONGUE-AND-GROOVE WOOD CEILING
PRECAST CONCRETE PLANK
WATERPROOFING
RIGID INSULATION
GRAVEL DRAINAGE LAYER
SOIL SEPARATOR
EARTH COVER
POURED CONCRETE WALL
WATERPROOFING
GRAVEL BACKFILL
SAND LAYER
CONCRETE SLAB
COMPACTED GRAVEL FILL

barrier. It's like the elephant joke. A: What are you doing ripping up little pieces of paper and scattering them all over the floor? B: Keeping the elephants away. A: You're crazy. Scattering paper can't keep elephants away. B: You don't see any elephants, do you?

The foolproof waterproofing system is one that never gets wet.

Since this subject is central to successful earth-shelter construction, it gets a lot of attention. There are several varieties of waterproofing materials and many different ways to combine them and apply them. They have not been used in enough houses, on enough distinctively different sites, or for a long-enough time for one or two clear winners to emerge. (If you're dry, your system is one of the winners.)

ROOF CEMENT AND POLYETHYLENE SHEET-ING. This is the thin version of a built-up roof, with roof cement (not tar or liquid roof coating) troweled on over double-cover felt paper or roll roofing to prevent nailhead or splinter punctures on a wood deck, then 6-mil black poly sheeting, a second layer of cement, a second layer of black poly, then a few inches of sand to prevent punctures from surface traffic and root systems, then sod. Single layers are used to waterproof the walls. It is a manageable job. But it is also a dirty, tiring, and incredibly hot job for owner-builders.

BENTONITE. This clay is a natural waterproofer that expands when wet to become impermeable. Success with bentonite depends on an even and complete application. Manufactured bentonite panels appear to be an improvement over the raw clay, partly because the application is controlled by the uniformity of the product. Clay is embedded in cardboard panels, applied to the walls, and soaked. Bentonize, a trademark of Effective Building Products, combines the clay with a kind of paste that is sprayed onto walls and roof in a ⅜-inch-thick layer. The one-step Bentonize system mentioned in many earth-shelter books was removed from the market by Effective Building Products in early 1984 (it was officially removed as of October '83), to be replaced by a two-step system. One of the main differences is that the company no longer sells the two-stager as an owner-builder product, but only through licensed applicators (roofers). The names to look for now are Bentonize, as before, and Poly-G, the tag for their proprietary combination of some 90-percent bentonite clay and polymer mix. It is installed with a spray gun simultaneously with an adhesive, which keeps the material on the wall.

MULTIPLE MEMBRANES. Commonly put together with a minimum of four plies, this built-up system can consist of sandwiched layers of asphalt or pitch, and glass fiber mat or roll roofing.

One system well within the reach of owner-builders is roof cement used with built-up layers of roofing felt or black poly sheeting. This is a laborious job, working with incredibly hot and sticky materials. But as long as you are careful with foot traffic and soil and sod application so as not to start with a puncture, the system works well and lasts.

The idea of free-drain areas also works on the roof. Assuming built-up cement and poly or multiple membranes that keep out the water, and assuming that sod will create controlled percolation, water will still flow downhill. A gravel trench backed by flashing along the structural fascia on this house guides and controls runoff.

Roofing and waterproofing are the most experimental aspects of earth-shelter construction. There are so many options. At left, plastic sheeting over insulation and waterproofing should send groundwater directly down to the foundation drain. Below, roll roofing will be only part of a built-up roof. Layers of hot tar, flashing, and counterflashing will follow.

After patching voids, clearing any sharp protrusions, and curing for at least 7 days, Bituthene primer is applied directly over concrete with a lamb's-wool roller.

ELASTOMERIC MEMBRANES. One of the prominent trade names, Bituthene (a W.R. Grace product), is a rubberized asphalt with a polyethylene coating. It has good bridging characteristics and has been used successfully with a small roof slope to discourage puddling. Butyl rubber membranes, the generic term, have the flexible characteristics of butyl caulk. And as butyl and silicone caulks are the most elastic, bridging gaps subject to thermal expansion and contraction, they are the most expensive. So it is with butyl membranes.

It is common to apply rigid insulating board with spot cement over the waterproofing prior to backfilling. The thickness of the board depends on ambient soil and air temperatures, wall mass, energy source, and the thermal values required to keep you reasonably comfortable.

The possibilities of styrene-type rigid insulating board sank in with me when I saw it used in place of plywood sheathing on a 24-inch-center, wood-frame demonstration house outside Chicago in the early '70s. The house was a code-cruncher, set up by a developer who had plans for hundreds more. The 2-foot-wide panels were tongue and groove laid over diagonal wind bracing, the kind of let-in braces you see in old building books when 1 × 6 T&G was used for sheathing instead of plywood. The energy-crisis hoopla about combining insulation and sheathing got a lot of press. I sure wrote it up. There were lots of good facts and figures about reduced air infiltration and such, but I remember being struck by the fact that the board was laid over the wood frame, over the seams between the wood frame, the sill, and the foundation, and on down the foundation below grade. No gaps. No muss. No fuss. And no rot.

Turns out that rigid insulating board does not stay intact and at design capacity forever. But the two principal materials do pretty well. Styrofoam, the trade name for Dow Chemical's closed-cell, extruded polystyrene, performs at roughly 90 percent of its thermal capacity (R-5.5 per inch) over 20-year subterranean tests. Beadboard, the common name for expanded polystyrene, stays within roughly 80 percent of its design rating of about 4.5 per inch. Both are extremely lightweight, easy to handle and install, and rigid enough to support foot traffic.

In the beginning of this section, I associated several professionals with their views of earth-shelter benefits. For some of the options in waterproofing materials and applications, here are a few of the people and specific applications they have used successfully.

ROB ROY. On Roy's house, which was the subject of an article in *Underground Houses*, Roy used a wooden-plank roof deck, 15-pound felt paper, a layer of hand-troweled plastic roof cement, a layer of 6-mil black plastic sheeting, a second course of cement and plastic sheeting, then sod.

MALCOLM WELLS. Wells has waterproofed with 1/16-inch butyl rubber sheeting embedded in roof cement—certainly not the least-expensive system.

DAVID CARTER. Over 15-inch-thick block, Carter has used 2-inch-thick, sprayed-on polyurethane, two courses of polyethylene sheeting, then 4 inches of coarse gravel to create a free-draining area next to the block, with fiberglass cloth matting between the gravel and the backfill to prevent the free-drain area from choking with sand and silt.

JOHN BARNARD. Barnard has used three hot-mopped courses of 60-pound asbestos felt roofing.

On sloped roofs from low to high so laps shed water (left), self-adhesive Bituthene membrane is rolled in place with 2½-inch overlaps. Over walls (right) the membrane (factory-coated with a thick, even layer of rubberized asphalt) should cover wall-to-slab and wall-to-footing joints. Hand rolling seals the seams.

It is appropriate to close this crucial subject by leaving you just a little up in the air. If that makes you uneasy, think twice about earth-shelter construction, because it is new and changing, incorporating new materials and applications rapidly, and will continue to do so for the next decade at least. Roy, Carter, and just about all the other designers and builders who have put their experiences into print have added that, with hindsight, they would have done at least some part of the waterproofing job differently.

It's not that they discovered leaks, structural faults, drainage problems, or chronic condensation. It's just that all the pieces of the earth-shelter puzzle are not yet in place. Some of them are not yet available. Some are not yet imagined. That's the adventure of designing and building with an alternative still in development. You are in the small but select company of earth-shelter owners and builders.

The proprietary Bituthene system is used over concrete and under protective insulating board. A ⅛- or ¼-inch asphalt hardboard is used over decks with heavy traffic during construction. Application details are available from the manufacturer (Bulletin 8205 from W.R. Grace & Co., 62 Whittemore Ave., Cambridge, MA 02140).

INFORMATION SOURCES

A Golden Thread, Ken Butti and John Perlin (Cheshire Books). The subtitle tells it all: 2,500 years of solar architecture and technology. Not a how-to book on earth sheltering, it does offer a fascinating tour through the evolution of passive solar design and housebuilding, an important component of many earth-shelter designs.

Build It Underground, David Carter (Sterling Publishing). A very thorough introduction to the subject, covering many projects, materials, and methods, from reinforced concrete down to a few slightly questionable details, such as laying electrical conduit on top of the roof deck under the waterproofing. Wide-ranging coverage. A lot of specifics.

Constructing Earth Sheltered Housing With Concrete (Portland Cement Association, 5420 Old Orchard Rd., Skokie, IL 60076). A technical look at earth-sheltered housing using cast-in-place and precast and poured concrete; drainage and waterproofing systems; formwork; joint detailing, and more. Written for a professional audience of architects and engineers.

The Earth Shelter Handbook, Tri/Arch Associates (Tech/Data Publications, 6324 W. Fond du Lac, Milwaukee, WI). An extremely thorough and sensible tour of earth-shelter design that moves logically through the design process, presenting options without a blitz of fine-print statistical tables. A solid combination of design and theory, this book is short on how-to, long on design options.

Earth Shelter Fact Sheets, Department of Energy, and Underground Space Center, University of Minnesota, Minneapolis, MN 55455. A series of consumer-type bulletins on earth-shelter waterproofing and other subjects.

Earth Sheltered Homes, Underground Space Center (Van Nostrand Reinhold). A thorough plan and design book filled with pictures and drawings, section views, details, but with a modest text. An eyeful.

Earth Sheltered Housing Design, Underground Space Center (Van Nostrand Reinhold). A thorough design tour and companion piece to *Earth Sheltered Homes* and *Earth Shelter Residential Design Manual*, though more specific and construction detailed than

either of these two. (All USC books are also available directly from the source: Underground Space Center Publications, 790 Civil and Mineral Engineering Bldg., 500 Pillsbury Drive SE, University of Minnesota, Minneapolis, MN 55455.)

"Earth Sheltered Structures," Lester L. Boyer (School of Architecture, Division of Engineering, Oklahoma State University, Stillwater, OK 74078). An interesting article on earth-shelter design, owner preference, energy performance, and (just in case you want to spend the next six months reading about earth-sheltering) a 155-item bibliography. Have fun.

"Heavy Loads," Don Metz *(New Shelter, Jan. '82)*. One of the very few magazine articles that spares some of the gee-whiz stuff to concentrate on practical details: poured walls and roofs, concrete roof planks, 6×10 roof timbers—as the title says, heavy-duty earth-shelter construction. Written by Don Metz, who runs Earthtech. Write for reprints or go through your library. Worth the effort.

Home in the Earth, Larry S. Chalmers, Jeremy A. Jones (Chronicle Books). A selection of 40 houses. Each is presented with a perspective and plan drawing and a minuscule text; without how-to details; but with interesting and varied plans.

Let's Reach for the Sun, George Reynoldson (Space/Time Designs, Inc., PO Box 1989, Sedona, AZ 86336). An interesting plan book of solar and earth-sheltered homes; 30 plans with perspective, plan, and section views.

Underground Houses, Rob Roy (Sterling Publishing). Roy's book is pinned to the design and development of his Log-End Cave residence. It's a clear, step-by-step case history that includes the problems, options, second thoughts, and afterthoughts generated by a thoughtful project. The details are for a specific house, so you have to do some extrapolating. But I think it is important to read at least one case-history book to get a smell for the job—things that have little to do with construction details. Try this one.

The Underground House Book, Stu Campbell (Garden Way Inc.). About the most thorough book available (with a nice shot of one of Don Metz's houses on the back cover), and a bargain in paperback at about $10 the last time I saw the price. Subjects include finding and buying land, heating, interior treatments, and, of course, the heart of the book, which is all about design. Most topics are covered with artwork and not enough pictures of the real thing—about the only drawback.

ILLUSTRATION ACKNOWLEDGMENT

The following contributed illustrations to this part of the book:

RAMMED EARTH WORKS, Blue Mountain Rd., Wilseyville, CA 95257. ERT specializes in rammed-earth residential construction. Many of their projects are at least partially earth-sheltered and sod-roofed.

EARTHTECH, PO Box 52, Lyme, CT 03768. Don Metz's earth-shelter design and construction firm.

EARTHWOOD BUILDING SCHOOL, RR1, Box 105, West Chazy, NY 12992. Rob Roy's cordwood facility, where he conducts seminars and hands-on classes in earth-shelter design and construction. Write for the brochure.

EFFECTIVE BUILDING PRODUCTS INC., 28001 Chagrin Blvd., Suite 207, Cleveland, OH 44122. Manufacturer of Bentonize waterproofing products and application machinery.

ELLISON DESIGN & CONSTRUCTION CO., 2001 University Ave., Minneapolis, MN 55414. Tom Ellison's firm, which designs and custom-builds earth-sheltered homes.

MALCOLM WELLS, Architect, PO Box 1149, Brewster, MA 02631. This name will pop up again and again as you get involved with earth-shelter and solar-efficient building.

TERRA DOME, 14 Oak Hill Cluster, Independence, MO 64050. A surprisingly varied group of projects produced from the same mold by Terra Dome, which uses a proprietary process of formed dome shapes of different diameters and poured concrete.

UNDERGROUND SPACE CENTER, University of Minnesota, 500 Pillsbury Drive SE, Minneapolis, MN 55455. The Center does not design and build homes but serves as a kind of clearing-house of information, in addition to doing a great amount of research, monitoring, and analysis.

W.R. GRACE CO., Construction Products Div., 62 Whittemore Ave., Cambridge, MA 02140. Manufactures Bituthene waterproofing membrane.

WOOD PRESERVATIVE

This book contains frequent warnings on the potential health hazards of wood preservatives during and after application and advises the reader to consult the latest guidelines of the EPA (U.S. Environmental Protection Agency) before using preservatives or commercially treated wood. As this book went to press, the EPA restricted preservatives containing pentachlorophenol, creosote, and inorganic arsenicals to use by certified people only. Book production scheduling did not allow this change to be incorporated into the text itself. Affected pages are 71, 72, 165, and 184–188. For more information, consult a library copy of the *Federal Register* of July 13, 1984, Vol. 49, No. 136, pages 28666–28689, or contact the EPA Registration Division (TS 767C), 401 M. St., SW, Washington, DC 20460.

INDEX